W9-CNM-076

Basic Human Anatomy and Physiology

Clinical Implications for the Health Professions

Paul D. Anderson, Ph.D.

Massachusetts Bay Community College

Wadsworth Health Sciences Division

Monterey, California

Wadsworth Health Sciences Division
A Division of Wadsworth, Inc.

© 1984 by Wadsworth, Inc., Belmont, California 94002. All rights reserved. No part of this book may be reproduced, stored in a retrieval system, or transcribed, in any form or by any means—electronic, mechanical, photocopying, recording, or otherwise—without the prior written permission of the publisher, Wadsworth Health Sciences Division, Monterey, California 93940, a division of Wadsworth, Inc.

Printed in the United States of America

10   9   8   7   6   5   4   3   2   1

Library of Congress Cataloging in Publication Data

Anderson, Paul D.
  Basic human anatomy and physiology.
  Bibliography: p.
  Includes index.
  1. Human physiology.   2. Anatomy, Human.
I. Title.   [DNLM: 1. Anatomy.   2. Physiology.   QS
4 A548b]
QP34.5.A5   1984   612   83-23511

ISBN 0-534-03089-0

Subject Editor: *Adrian Perenon*
Manuscript Editor: *Deborah Gale*
Designer: *John Edeen*
Cover and Chapter Art: *Vincent Perez*
Illustrations: *Gayanne DeVry*
Composition: *Computer Typesetting Services,
Burbank, California*
Production Service Coordinator: *Marlene Thom*

Book Produced by Ex Libris □ Julie Kranhold / Sara Hunsaker

Preface

Many college students take a course in anatomy and physiology as part of their education for one of the health professions, such as nursing, respiratory therapy, or radiologic technology. The material in their anatomy and physiology course should provide the groundwork on which their professional courses can be based. A knowledge of how the body functions in normal situations is later applied to understanding what happens to the body during disease and what kinds of procedures can be used in treating the disease. Unfortunately, many texts contain a great deal of anatomy that the student will never use again and often do not adequately explain the physiological processes that will serve as the basis for the student's professional studies.

The aim of this textbook is to present the core material of anatomy and physiology in such a way that students can understand it and put it to use. The depth of coverage is based on recommendations of representatives of the various health professions. The names of important anatomical structures are given, but the emphasis is on physiological processes. Microscopic anatomy is included when it is relevant to understanding physiological processes. Every attempt has been made to emphasize material that students will use in professional courses, professional licensing examinations, and careers.

Two major themes dominate this text: homeostasis and pathology. Throughout the book, the student is shown how normal structure and function are maintained by dynamic counterbalancing forces. Pathology is viewed as a disruption in homeostasis. Accordingly, I have presented a large number of clinical topics and have contrasted them with specific normal processes. In addition, concepts of radiology and the biologic effects of radiation are introduced (Chapter 20) to emphasize the important role that radiology plays in modern health care and in the differentiation of normal from abnormal structures.

The subject matter of human anatomy and physiology is an exceedingly large and complex body of knowledge to present in an introductory course, so I have concentrated on providing an intelligible, lucid, readable, and up-to-date account of the important principles. I have emphasized unifying concepts and data considered critical to a basic understanding and working knowledge of the human body.

Pertinent questions follow each chapter. These are intended to test knowledge and comprehension and to provoke discussion. Detailed outlines are also included to assist the student in reviewing important material.

Although the instructor and the books used are significant in the phenomenon of education, it is the learner who must ultimately assimilate, integrate, and synthesize the information. It is my hope that this book will make that process an enjoyable one.

Paul D. Anderson

Brief Contents

Detailed Contents

13   The Heart

14   Circulation   385

18 Excretion   511

19 Reproduction, Growth, and Development   541

20 Radiologic Health   581

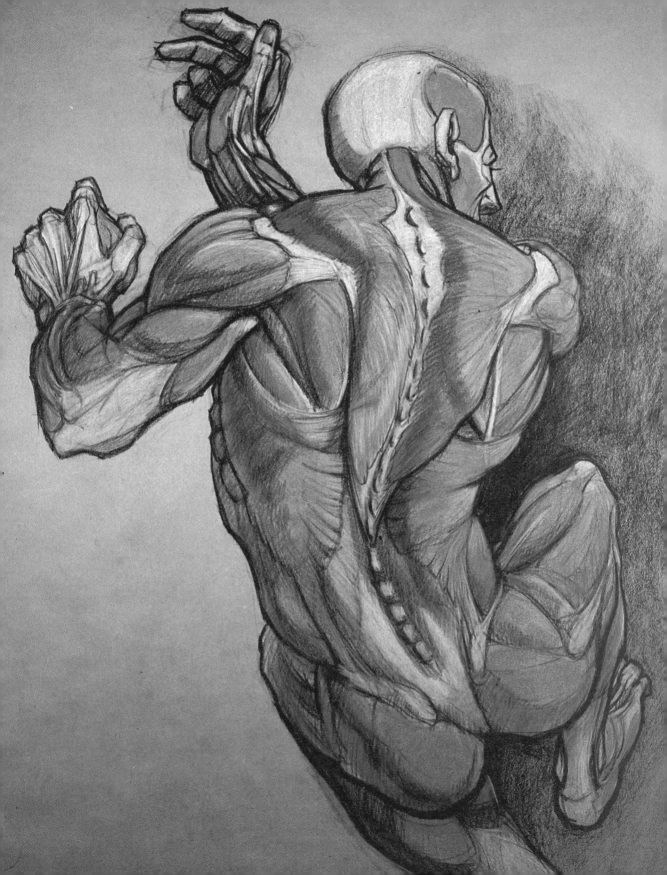

1

Basic Anatomy and Physiology

LEARNING OBJECTIVES

- Differentiate between an ionic and a covalent bond

- Define the terms acid, base, and buffer

- Distinguish between electrolytes and non-electrolytes

- Describe the pH scale and relate its significance

- Differentiate between the English and metric systems of measurement by comparing the different units of weight, length, and volume

- Identify the different subdivisions of anatomy and physiology

- Use certain important anatomical terms

- Describe the structure of an atom

- Explain the relationship between atoms and elements and between molecules and compounds

- Explain how the octet theory predicts the combining properties or valence of elements

- Name the planes of the body using proper anatomical terms

- Name the important regions of the body and body cavities

IMPORTANT TERMS

acid	bond	electron	metric	valence
acidosis	buffer	elements	mixture	ventral
alkalosis	compound	homeostasis	molecule	
atom	covalent	ion	neutron	
base	dorsal	ionization	proton	
bilateral	electrolyte	isotope	radical	

1

To study and understand our body and its activities, we must first know something of the structure and functions of the body's basic unit, the living cell, and its composition and normal working patterns. The human body is made up of trillions of such minute cells, and the study of these structures is called *cytology*. Groups of cells that are alike form tissues; *histology* concerns itself with the study of these. Tissues that serve a common function form organs, most of which are large enough to be visible to the naked eye. Their structure is described and analyzed in *gross anatomy*, whereas consideration of their functions is the major concern of *physiology*. There is enormous overlap among these fields of science, and research into them grows more complex every year. The latter two, gross anatomy and physiology, are the chief topics of this textbook.

CHEMISTRY OF LIFE

The cell is quite complicated but also fascinating. Cellular mechanisms are 95 percent chemical, and the introduction of chemical terminology in this chapter is essential to your later understanding. A working definition of living things is that they are chemical machines that grow and reproduce themselves. Chemistry is the study of matter—the analysis of its structure and of changes occurring in that structure. *Matter* is anything that has weight and occupies space. Chemistry also considers the amounts and kinds of energy needed to effect changes in matter, and the physical laws that govern such processes. Through a knowledge of chemistry, we are able to understand some of the normal and abnormal functions of the body. All body processes—for example, the digestion of food, the production of urine, and the regularity of breathing—are based on chemical principles. Drugs, antiseptics, and other agents used in treating disease are chemicals.

To emphasize the importance of chemistry in our daily lives, this section will briefly describe some fundamental chemical concepts—atoms, elements, compounds, and mixtures—and how they are interrelated.

The Atomic Theory

It is difficult for a beginning science student to realize that such things as atoms actually exist, especially since no one has ever seen an atom, even with the most powerful microscope. Scientists such as Rutherford (1910) and Bohr (1915) developed theories of matter in general and atoms in particular without having seen them. Their scientific investigations led them to conclude that an atom has a structure. Recognition of this fact is fundamental to even an elementary understanding of chemistry.

Atoms can be thought of as the "building blocks" of matter—the smallest complete units of which all matter is made. An atom has a nucleus that contains positively charged electric particles called **protons** and noncharged (neutral) particles called **neutrons** (Figure 1-1). The positively charged proton (p) has weight and is commonly called the "fingerprint" of the atom because the number of protons in an atom's nucleus provides a means of identifying it, referred to as the *atomic number*. No two atoms can have the same number of protons.

The neutron (n) has weight but is not charged; therefore, it is said to be neutral. The number of neutrons in the nucleus is usually the same or nearly the same as the number of protons. In heavier atoms, there are more neutrons than protons. Protons and neutrons together determine the weight of an atom (atomic weight).

The nucleus of an atom is surrounded by one or more **electrons** (e), negatively charged particles circling in well-defined orbits, or energy shells (Figures 1-1 and 1-3). There are as many electrons present in an atom as protons. However, because the weight of the electrons is so minute, they are not considered in the total weight of the atom. Each energy shell can contain only a certain number of electrons: The first energy shell can contain no more than two electrons; the second, no more than eight electrons;

the third, no more than 18; the fourth, no more than 32, and so on. Each energy shell is identified beginning with the shell closest to the nucleus and proceeding outward:

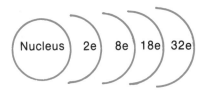

An **element** is composed of atoms with the same particle arrangement. No two elements have the same atomic composition, nor can an element be transformed or broken down by chemical means to form another element. In addition, the number of electrons in each energy shell differs for each element. For example, an atom of hydrogen has one electron whirling around one proton in the nucleus, whereas an atom of oxygen has eight electrons whirling around eight protons in the nucleus. The first energy shell contains 2 electrons, and the second contains 6 electrons (Figure 1-1).

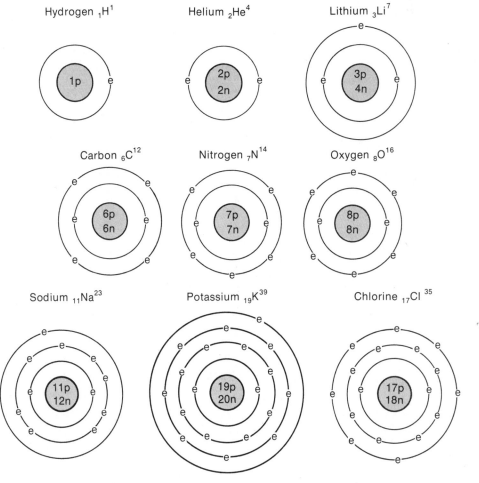

Figure 1-1 Diagrammatic representation of elements selected from the periodic table. Symbols: p, proton; n, neutron; e, electron.

The periodic table lists the elements in order of the number of protons each carries in a single atom. A hydrogen atom has one proton and one electron, and thus its atomic number is 1. The helium atom has two protons; carbon six; nitrogen seven; and chlorine seventeen (Figure 1-1). The atomic structures illustrated also show that the electrons are located in certain prescribed energy shells, indicated diagrammatically by circles.

The number of neutrons present in an atom can be calculated by subtracting the atomic number from the atomic weight. Helium has an atomic weight of 4 and an atomic number of 2; it therefore has two protons in the nucleus and two electrons outside the nucleus. Thus two neutrons are present (4 − 2). The sodium atom has an atomic weight of 23 and an atomic number of 11; therefore twelve (23 − 11) neutrons are present in its nucleus.

Every phase of professional life requires a specific set of terms for everyday use. In chemistry, this special language is short, precise, and to the point. The chemical symbol is an abbreviation of the name of the element—the large "X" shown in the diagram below. The atomic number (Z) has the subscript position immediately preceding the symbol; the atomic weight (A) is placed in the superscript position following the symbol:

$$_ZX^A$$

In Figure 1-1, you can see the terminology in use:

$$_6C^{12}$$

This indicates that the element is carbon, that the atomic number is 6, and that the atomic weight is 12. Therefore, the nucleus contains six protons, six electrons are in the outer shells, and six (12 − 6) neutrons are in the nucleus.

Table 1-1   Common Chemical Elements

Element	Symbol	Atomic Number	Atomic Weight	Valence
Hydrogen	H	1	1	±1
Carbon	C	6	12	±4
Nitrogen	N	7	14	−3
Oxygen	O	8	16	−2
Sodium	Na	11	23	+1
Magnesium	Mg	12	24	+2
Phosphorus	P	15	31	−3
Chlorine	Cl	17	35	−1
Potassium	K	19	39	+1
Calcium	Ca	20	40	+2
Iron	Fe	26	56	+3
Cobalt	Co	27	59	+2
Copper	Cu	29	64	+1
Iodine	I	53	127	1

Some of the common chemical elements that will be referred to throughout this text are listed in Table 1-1.

Valence

An element's **valence** is its combining power in the formation of compounds. This capability is entirely due to the number of electrons present in the atom's outermost shell. When all the shells of an atom contain their maximum number of electrons, the atom is stable, or nonreactive. For example, helium (He) has two electrons in its K shell (its only shell) (Figure 1-1); therefore, it will neither accept nor release electrons—it will not react at all. Nitrogen (N), on the other hand, has two electrons in its K shell and five electrons in its outermost shell. To be stable, the outermost shell should contain eight electrons. Nitrogen can thus do one of two things: it can either release its five outermost electrons and revert back to a stable K shell; or it can accept three electrons from a nearby atom to complete the outer shell. To simplify these concepts, an *octet theory* was formulated. It states that:

1. All outermost shells (except the K shell) will contain at most eight electrons; a full shell is stable and nonreactive.

2. If an atom loses electrons from its outermost shell, it will acquire a positive (+) charge.

3. If an atom obtains electrons to complete its outermost shell, it will acquire a negative (−) charge.

4. To conserve as much energy as possible, the atom will release or accept the least number of electrons that it can.

For example, nitrogen has five electrons in its outermost shell. It can either release the five electrons and acquire a +5 charge or valence, or it can gain three electrons and acquire a −3 charge or valence. As the octet theory states, the least energy is lost when the three electrons are taken on; thus, the common valence of N is −3.

There are always exceptions to rules, and hydrogen (±1) and carbon (±4) are prime examples. When the number of electrons in the outermost shells of reacting atoms is equal, an atom will either lose or gain electrons, depending on the atom with which it is reacting. For example, hydrogen, with only one electron in a K shell that needs two electrons, can either release the electron or acquire an electron:

1. $H^+ Cl^-$—In this case, hydrogen lost its electron to chlorine and became +1.

2. C^+H^-—In this case, hydrogen gained electrons from carbon and became −1.

The Chemical Bond

As we have seen, an atom is chemically stable when its outermost shell is completely filled. If this shell is not full, the atom will tend to react with any other element that will provide the shell with its full complement of electrons. An atom in this situation is considered to be a charged particle, or an **ion.**

An *ionic (or electrovalent) bond* occurs when a chemical reaction involves a loss or gain of electrons in which electrons are transferred from one atom (+ charge) to another (− charge). The formation of sodium chloride (NaCl) typifies such an ionic bond. Sodium is a highly reactive metal with an atomic number of 11. Its outermost shell has one free electron that is easily removed in ionic bond formation (Figure 1-2). Chlorine (Cl) is a gas of atomic number 17, with seven electrons in its outermost shell. When these two ions react with each other, the single electron in the outermost shell of the sodium atom is transferred to the outer shell of the chlorine atom, thus constituting a chemical reaction.

A **covalent bond** is a sharing of electrons between two atoms. Oxygen (O) has six electrons in its outermost shell—two electrons short of a completed shell. Therefore, oxygen has a combining power of −2 and tends to gain two electrons in a reaction. Figure 1-3 illustrates the formation of water (H_2O), which is an example

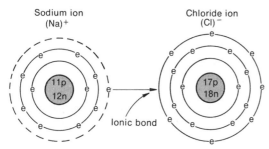

Figure 1-2 Union of oppositely charged ions to form a compound. Sodium loses an electron in its outer orbit to chlorine, making sodium positively charged. Chlorine accepts the electron in its previously incomplete outer orbit and becomes negatively charged. The combination of the two oppositely charged ions forms an ionic bond and a molecule of sodium chloride (table salt).

of covalent bonding. One oxygen atom combines with two hydrogen atoms, each of which has one electron in its outer shell. In forming the compound H_2O, all three atoms are actually sharing electrons so that each atom has completed its outer electron shell to the maximum number—eight for oxygen and two for hydrogen. Most gases composed of only one element, such as hydrogen (H_2), are also covalently bonded.

Molecules, Compounds, and Mixtures

What are the substances formed by chemical reactions? A **molecule** is formed whenever two or more atoms join, whether these atoms are dissimilar or whether they are individual atoms of the same element. A **compound** is a union of atoms of two or more different elements. Compounds can either be organic or inorganic. Organic compounds contain carbon; inorganic compounds do not. Nitric oxide (NO) is made up of one atom of nitrogen and one atom of oxygen. It is thus both a molecule and a compound. Hydrogen (H_2) and oxygen (O_2), on the other hand, both consist of two identical atoms. They are molecules but not compounds.

Combinations of two or more dissimilar substances that do not unite chemically produces a **mixture.** Blood plasma, for example, is a mixture of many different substances that are essential in their separated form and not in one large chemical compound. The major difference between a mixture and a compound is that the components of a mixture can be separated by mechanical processes. For example, boiling a salt water mixture will separate the salt from the water, but will not break down or separate

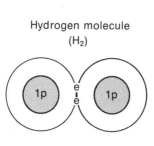

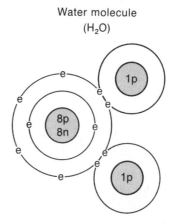

Figure 1-3 Formation of a molecule of water. The covalent bond between oxygen and hydrogen represents a set of electrons shared between the two. Also illustrated is covalent bonding forming a hydrogen molecule.

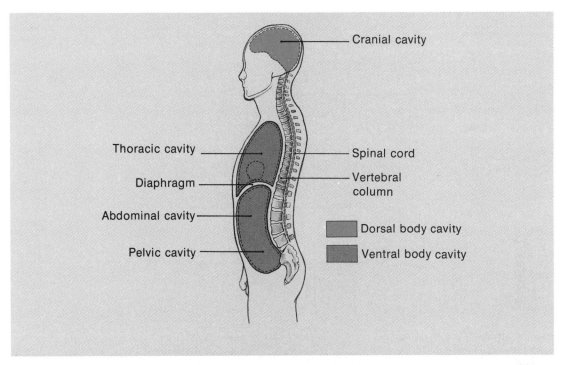

Figure 1-6 Sagittal section of the body, showing the dorsal and ventral body cavities.

Because these two subcavities are continuous and interconnected, and because both are encased entirely within a series of membranes (meninges), they are collectively regarded as one cavity.

Ventral Body Cavity

The ventral cavity is subdivided into the thoracic cavity and the abdominal-pelvic cavity. Each of these cavities is much larger than the entire dorsal cavity.

The thoracic cavity is separated from the abdominal cavity by the broad, muscular diaphragm. It is lined with a membrane called the **pleura.** The lungs and large vessels are located here, as is the heart, which actually is found in its own pericardial sac or cavity between the lungs (Figure 1-6).

The upper region of the abdominal-pelvic (or abdominopelvic) cavity contains the major organs of metabolism and the kidneys. The lower pelvic region contains the urinary bladder, the rectum, and the internal organs of the reproductive system (Figures 1-6 and 1-7). The abdominal-pelvic cavity and all organs within it are covered with a membrane, the **peritoneum,** which secretes a fluid that protects and lubricates the organs and spaces contained within.

Abdominal Regions

The abdomen is so large that it is divided, for convenience of study, into nine regions (Figure 1-7). A horizontal line at the level of the ninth rib creates the right hypochondriac, epigastric, and left hypochondriac regions. A horizontal line at the level of the iliac crest creates the right and left lumbar, umbilical, right and left inguinal, and hypogastric regions.

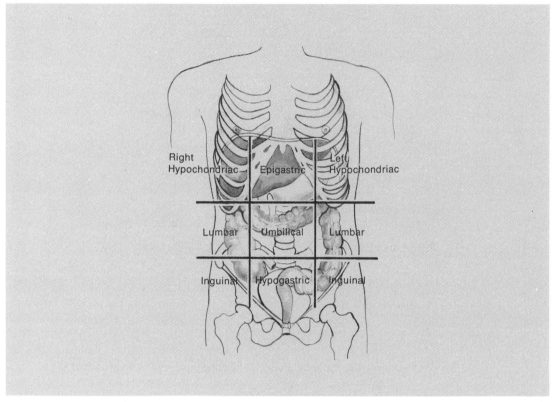

Figure 1-7 Abdominal regions.

THE HUMAN BODY—AN ORGANISM

According to Webster, "... an organism is any living plant or animal made up of many complicated parts ..." The human organism fits this description to a "T". You have now finished the first chapter of this book and have been presented with an assortment of facts that may seem completely unrelated. What does chemistry have to do with the planes and cavities of the body? These facts do appear to be—and, in a sense, are—rather isolated. Still, we must start somewhere. They are the foundation for anatomical and physiological processes and concepts.

The chapters that follow build on the premise that a knowledge of the structure and function of a cell is of paramount importance in developing an understanding of the human organism, a mass of interrelated cells.

This text will be clinically oriented, which means that you will learn a great deal about the abnormalities of the body (pathology) and about the many different methods of detecting disease. One must first learn the normal before understanding the abnormal. Therefore, we shall begin with the cell and develop the "human organism" from there. The human body is an organized assemblage of highly specialized tissues, which together perform the necessary functions that contribute toward the state of *homeostasis*—a constant internal environment.

OUTLINE

I. Introduction
  A. Related sciences
    1. Cytology
    2. Histology
    3. Anatomy
    4. Physiology
II. Chemistry of life
  A. Matter
  B. Atomic theory
    1. Rutherford and Bohr
    2. Nuclear components
      a. Proton
      b. Neutron
      c. Atomic weight
    3. Energy shells
      a. Electrons
      b. Energy shells
    4. Periodic table
    5. Atomic structures
    6. Labeling configuration
  C. Valence
    1. Combining power of an atom
    2. Octet theory
      a. All outer rings can have eight electrons at most, except the first, which has two.
      b. An atom that loses electrons acquires a positive charge
      c. An atom that gains electrons acquires a negative charge
      d. An atom will gain or lose electrons using the least energy possible
  D. Chemical bond
    1. Full outer shells imply stability
    2. An ion is a charged atom
    3. Ionic bonding—electrical attraction of oppositely charged ions
    4. Covalent bonding—sharing of outer electrons
  E. Molecules, compounds, and mixtures
    1. Compound
    2. Molecule
    3. Mixture
    4. Radical—(examples: hydroxyl, OH^-; amine, NH_2^-; carboxyl, $COOH^-$)
  F. Acids, bases, and salts
    1. Acid
    2. Base
    3. Salt
  G. pH—the hydrogen ion concentration of a solution
    1. More H^+ than OH^- results in an acid solution (pH from 1.0 to 7.0)
    2. More OH^- results in an alkaline solution (pH from 7.0 to 14.0)
    3. Blood plasma pH is 7.3 to 7.44
      a. Any decrease below 6.8 results in acidosis (and possibly death)
      b. Any increase above 7.8 leads to alkalosis (and possibly death)
  H. Buffers—chemical compounds that prevent a major shift in pH by acting as weak acids or weak bases and their respective salts
  I. Ionization
    1. Anion—negative ion
    2. Cation—positive ion
    3. Electrolytes—ion-containing solutions
  J. Chemical reactions
    1. Oxidation-reduction reaction
      a. One element loses an electron—is oxidized
      b. Another element accepts an electron—is reduced
      c. Both independent particles must be available to complete reaction
    2. Recombination reaction
      a. Exchange of atoms between molecules—compounds split and recombine without a gain or loss of electrons

K. Isotopes
  1. Forms of a chemical element with the same number of protons but different numbers of neutrons
  2. Atomic weights different from element

III. Measuring devices
  A. English system of measurement
  B. Metric system of measurement
    1. Based on factor of 10
    2. Length
    3. Weight
    4. Volume
  C. Temperature measurement
    1. Fahrenheit scale—32°F is freezing point and 212°F is boiling point of water
    2. Celsius (centigrade) scale—0°C is freezing point and 100°C is boiling point of water
    3. Conversion formulas
       Fahrenheit = (9/5 × °C) + 32;
       Celsius 5/9 (°F − 32)
       Fahrenheit = (40 + °C) 9/5 − 40;
       Celsius = (40 + °F)5/9 − 40

IV. Directions, planes, and cavities
  A. Directions in the body
    1. Superior–inferior
    2. Anterior–posterior
    3. Medial–lateral
    4. Cranial–caudal
    5. Proximal–distal
  B. Planes of the body
    1. Sagittal
    2. Frontal
    3. Transverse
  C. Body cavities
    1. Dorsal cavity
       a. Combined cranial and spinal cavity covered with meninges
    2. Ventral cavity
       a. Thoracic cavity lined with pleura (includes heart in its pericardial sac)
       b. Abdominal-pelvic cavity lined with peritoneum
  D. Abdominal regions
    1. Abdominal area divided into nine basic anatomic regions

V. The human body—an organism
  A. An organism is made up of many complicated parts
    1. Necessary items for a beginner
    2. Cellular mechanisms are primarily chemical
    3. Many cells make up a tissue, an organ, or organism
    4. To learn the abnormal you must know the normal

STUDY QUESTIONS AND PROBLEMS

1. Of what is living matter composed?
2. What is the internal environment of the body? Explain the term homeostasis.
3. Differentiate between parietal and visceral.
4. What is an ion? What are electrolytic solutions?
5. Are the electrons haphazardly grouped on the energy shells? If not, how are they arranged?
6. Define molecule, compound, and mixture, and show their interrelationship.
7. Define pH. Why is it important?
8. What is a buffer? How does it attempt to prevent an acidotic or alkalotic state?
9. What is an acid? If acid A has a pH of 3.0, and acid B has a pH of 5.0, which is the stronger acid? Why?
10. Most body reactions are of the oxidation-reduction type. What does this mean?
11. If normal body temperature is 98.6°F, what is it in °C?
12. Describe the different directions of the body and identify their opposites.
13. What are the three main body planes? Why is it important that we know them?
14. Correlate the lining of each body cavity with the cavity name.
15. Why are precise terms for body structures and their positions essential for health science workers?
16. List the body structures that correspond to bilateral symmetry of the human body and those that deviate from bilateral symmetry.
17. If a doctor asks you if you had any epigastric pains, to what region is he or she referring?
18. Define tissue, organ, and system in reference to the entire organism.

2

The Living Cell

LEARNING OBJECTIVES

- List the properties of life and describe the action of one or more organelles as they relate to those living properties

- Describe the basic cellular organization

- Give one or more important functions of the organelles

- Describe mitochondria, ribosome, and lysosome and show their relation to cellular metabolism

- Explain the possible roles of microtubules and microfilaments in the shape of a cell

- Explain the complementary pairing of nucleic acids in the structure of DNA

- Cite the main function of DNA contained in the nucleus of a cell

- Distinguish between DNA and RNA

- Relate the function of the nucleolus with chromosomal duplication

- Describe the chemical and mechanical phases of mitosis

- Differentiate between m-RNA, r-RNA, and t-RNA

- Define necrosis

- Describe what happens to a cell when it dies

- Differentiate between monosomy and trisomy, with examples

- Describe the anatomical and behavioral effect of Klinefelter's syndrome

IMPORTANT TERMS

adenosine tri-phosphate (ATP)	catabolism	familial	matrix	pinocytosis
aerobic	centromere	gene	mitosis	polymerize
aging	chromatin	glycoprotein	monosomy	protoplasm
anabolism	chromosome	helical	necrosis	replication
anaerobic	cisternae	hermaphroditism	nucleic acid	somatic
assimilation	codon	heterolysis	nucleoplasm	transcription
autolysis	cytoplasm	hypoxia	organelle	translation
autosome	deoxyribonucleic acid (DNA)	interstitial	phagocytosis	trisomy
		intoxication	pigmentation	

Living things or organisms display a remarkable and fundamental similarity in both structure and function. All living forms are essentially made up of one or more basic units or structural compartments called the *cell.*

The living substance of the cell is called **protoplasm.** It refers to the living matter within the cell or plasma membrane. Within the protoplasm are substances or structures called **organelles,** each playing a major role in the total physiology of the cell.

PROPERTIES OF LIFE

Chemically, a cell is composed of such familiar materials as oxygen, hydrogen, carbon, nitrogen, and many other substances that may be found in nonliving structures. Within the cell, however, these materials combine in a special way, establishing the unique distinction of being called alive. To be alive, the cell must have all the following functions or properties:

RESPIRATION We usually think of respiration as breathing, but in our present discussion, the term refers to the cell's ability to extract energy from food substances supplied to it. When a cell obtains energy without the aid of oxygen, its respiration is called **anaerobic;** when oxygen is required, as is the case with most cells, **aerobic** respiration occurs.

METABOLISM This function of the living cell is rather nonspecific, since it refers to the life processes in general—all the chemical processes that occur within the cell. **Anabolism** is the building of new cellular materials; **catabolism** is the breakdown of these materials into simpler substances. Metabolic functions include such processes as nutrition, digestion, and assimilation.

GROWTH Normal cells continually increase in volume or mass until they divide, and this necessary growth is closely associated with processes of metabolism. If the rate of anabolism is greater than that of catabolism, growth takes place.

EXCRETION Like growth, this process is closely related to metabolism. The products of metabolism can be unnecessary—even poisonous—and therefore must be disposed of, or excreted, from the cell.

REPRODUCTION Life cannot continue without reproduction, a physiologic function closely related to growth. Reproduction results in growth, since the embryo develops through the multiplication of cells—a form of growth. Growth, in turn, generally requires reproduction, since cells multiply by dividing, thus reproducing themselves. This ability of cells to multiply asexually by means of mitotic division, as occurs in the growth of tissues, should not be confused with the ability of an organism to reproduce itself sexually, thus perpetuating a species. The physiology of sexual reproduction—the development of sperm and egg and the fertilization process that follows—will be discussed in Chapter 19.

MOTILITY Movement is common in unicellular (one-celled) forms of life. For example, the ameba and paramecium move by means of a unique protoplasmic streaming and the beating of hairlike appendages called *cilia,* respectively. In humans only two types of cells are capable of movement: the male sperm cell, which is propelled by its whiplike tail, and certain white blood cells, which pass through capillary walls by a process called **ameboid movement.**

IRRITABILITY This physiologic property involves the ability of the cell to respond to an external stimulus or to changes in the organism's internal environment. All living cells will respond to certain stimuli. Their ability to do so is regulated by the nucleus—the "control center" of every cell.

SIZES AND TYPES OF CELLS

As we have found great variation in the size of different kinds of cells, so we will also find a greater variation in the shapes of cells. Since cells are relatively soft jellylike bodies that are closely packed together to form sheets of cells, called tissues, they are subject to the law of mechanics and will form many-sided figures of various types. However, cells that usually exist alone, uninfluenced by pressures from surrounding cells, will demonstrate a spherical shape. Egg cells are examples of this typically spherical shape.

Most cells are far too small to be seen without a microscope. Some, however, are large enough to be studied by the unaided eye. One example of a very large (macroscopic) cell is the unfertilized ostrich egg. Its yolk is three times the size of a hen's egg.

Some cells, especially nerve cells, are very long. The nerve cells that govern the movements of our big toes, for example, are nearly three feet long. The nerve fibers in the neck of the giraffe are more than eight feet long.

The shape of a cell is usually determined by the function that cell must perform within the organism. That is why nerve cells, which must carry electrical impulses over long distances, usually have such long extensions. The cells of muscles must also be long so that muscular contractions can be carried out efficiently in a single direction. The tissues that provide the body's covering (the skin) and line its cavities are composed of cells that tend to be relatively box shaped and are arranged close together. The cells of connective tissues, on the other hand, are grouped loosely among the strong elastic fibers that support the body.

Microtubules and Microfilaments

Microtubules are found in all cells. They are especially prominent in nerve axons. They are similar to cilia in structure and provide the cell with its shape. They are also essential for movement of organelles in the cytoplasm *(cytoplasmic streaming)* and may participate in the production of synaptic vesicles. Microtubules consist of thirteen rows of protein tubulin in a helical (spiral) arrangement, forming a cylinder.

Microfilaments are much smaller than the microtubules and are found related to supporting tissue such as glial cells. Examples include neurofilaments, epithelium, and astral rays from the centrosome during cell division. There are numerous filamentous diseases affecting collagen, reticular, and elastic tissues. Microfilaments' primary role is in the movement and contraction of a cell (actin and myosin in muscle).

The cytoplasm, long assumed to be a homogeneous, protein-rich solution, is now believed to be a highly intricate structure. Its latticelike

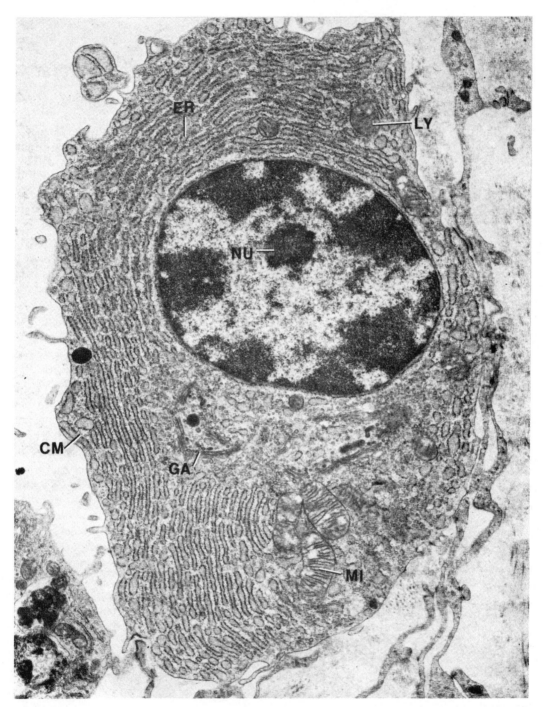

Figure 2-1 An electron micrograph of a protein-synthesizing cell in section showing abundant rough endoplasmic reticulum (ER), Golgi apparatus (GA), mitochondria (MI), lysosomes (LY), a cell membrane (CM), and a nucleus with a nucleolus (NU). (From *Gray's Anatomy* by R. Warwick and P. Williams. Copyright © 1973 by Churchill Livingstone, Edinburgh. Reprinted by permission.)

organization of microtubules and microfilaments affords the cell a functional unity and directed transport within the cell. It is becoming increasingly apparent that the control of cell shape and cell movement depends on the integrated functioning of the microtubules, microfilaments, and the cytoplasmic organelles suspended within them.

CELLULAR COMPONENTS

Cell Membrane

At the outermost surface of the cell is a very thin, elastic membrane, called the *cell membrane* or *plasma membrane*. This membrane is so thin that, although its existence had long been hypothesized, its detail was never actually seen until the invention of the electron microscope. The functions of the cell membrane will be discussed in more detail in Chapter 3.

Protoplasm

The cell membrane encloses a clear, jellylike substance called **protoplasm.** Protoplasm is usually thought of as the living matter of the cell. It is 60 to 70 percent water, 16 percent protein, and 14 percent fat, or lipid. The remaining protoplasmic components are dissolved gases, inorganic salts, metals, and a small amount of carbohydrate (sugar).

There are two principal kinds of protoplasm: the **nucleoplasm,** the fluid substance of the cell's nucleus, and the **cytoplasm,** the denser material outside the nucleus (Figures 2-1 and 2-2). The cytoplasm contains the organelles (or cytoplasmic inclusions) mentioned earlier (Figure 2-2).

Cytoplasmic Organelles

Organelles are minute organs, each enveloped in its own membrane, suspended or surrounded by cytoplasm. The biochemical functions of a cell for the most part are performed by the organelles, each one specialized for a particular function.

Endoplasmic Reticulum and Ribosomes

Within the cytoplasm is a system consisting of pairs of parallel membranes enclosing narrow cavities of varying shapes. This system is known as the *endoplasmic reticulum* **(ER)** (Figures 2-3 and 2-4). In other words, the ER is a network of canals running through the entire cytoplasm. These canals are continuous with both the plasma membrane and nuclear membrane. It is believed that the ER provides a surface area for chemical reactions, a pathway for transporting molecules within the cell, and a storage area for synthesized molecules and, together with the Golgi apparatus, secretes certain chemicals. (This last function will be discussed in the next section.) Attached to the outer surfaces of the ER are exceedingly small, dense, spherical bodies called *ribosomes.* In these areas, the ER is referred to as granular, or rough, reticulum. Portions of the ER that have no ribosomes are called agranular, or smooth, reticulum.

Ribosomes (Figures 2-3 and 2-4) are the site of protein synthesis. They contain ribonucleic acid (RNA) and a variety of proteins. Some ribosomes are free in the cytoplasm; most are attached to interior membranes (endoplasmic reticulum). Protein that is destined to be used outside the cell is prepared at the ribosome and temporarily stored in the reticulum. Protein that is produced by the freely floating ribosomes is probably used within the cell as structural protein or as enzymes used in metabolic reactions.

Golgi Apparatus

Still another membranous structure, or series of small spaces enclosed by membranes, is the Golgi apparatus. It is believed that the Golgi apparatus and the ER are continuous; the Golgi apparatus appears to be a specialization of the ER (Figure 2-2). It possesses a number of flat-

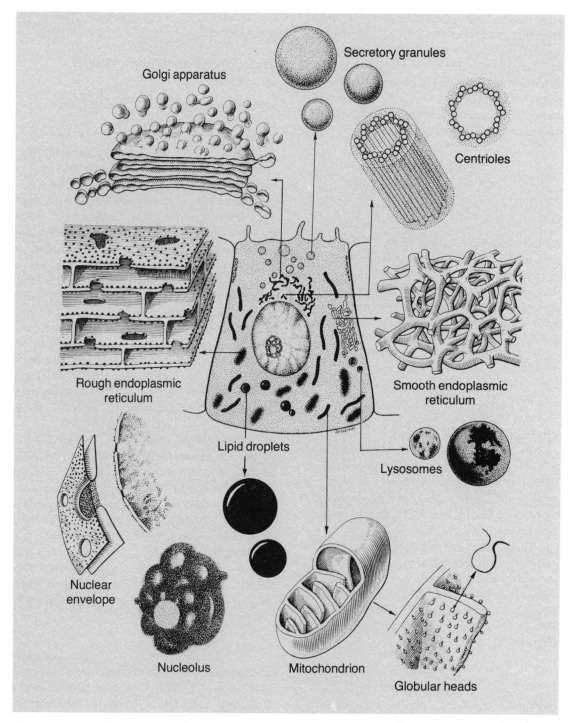

Figure 2-2 Diagram of a cell (center). Enlargements of the various organelles are shown around the central figure, as seen with the electron microscope. (From *A Textbook of Histology*, 10th Ed., by W. Bloom and D. W. Fawcett. Copyright © 1975 by W. B. Saunders Co. Reprinted by permission.)

6. The lysosome sacs fuse with the endocytic vacuole where initial digestion occurs.

7. Products of digestion diffuse out of the vacuole, into the cytoplasm, and eventually into the mitochondria where final digestion and extraction of ATP occur.

8. Waste products, not wanted by the cell, remain in the vacuole; travel to the cell membrane, and are excreted outside the cell through the cell membrane.

9. The release of wastes (excretion) by "rupture" of the food vacuole is reported to be an exact opposite mechanism as the formation of the food vacuole. In each case, the vacuole wall begins and ends as a part of the cell membrane.

Lysosomal enzymes will digest and destroy the cell (**autolysis**) if released from the lysosomes. After the death of an animal, decomposition (decay) of tissue takes place partly because of lysosomal hydrolytic action. These enzymes, released into the surrounding tissues, as well as enzymes in alimentary tract secretions and invading bacteria are jointly responsible for eventual tissue breakdown. The aging of meat in cold storage is probably the consequence of the action of these enzymes.

Aging is related to changes in the intercellular substances and the response by lysosomal enzymes. During normal metabolism or cell division, when certain proteins, carbohydrates, nucleic acids, and other materials are not needed by the cell, the digestive enzymes of the lysosomes start disposing of the material (autophagy). They do this by *catalyzing* the breakdown of the materials into their component building blocks. As the body ages, certain of its cellular components take on a foreign character, triggering an immunological response through which the foreign tissues are attacked, destroyed, and excreted. The role of the lysosomal enzymes is to begin a partial destruction of foreign material.

One of the more prominent foreign materials discovered is a pigment called *lipofuscin*. It is an indigestible fat waste product that attaches firmly to the inside of the cell membrane. The build-up of lipofuscin is said to be a cause of death of the cell. The higher the metabolic rate of the cell, the quicker the build-up of lipofuscin, and the shorter the life span of the cell. As cells die, acids are released into the extracellular fluid, which in turn becomes acid. The sequence of cell death promotes production of the hormone bradykinin, which is responsible for the inflammatory reaction—redness, pain and swelling. For this reason, an aching joint can be described as getting "old."

Lysosomes may also play a role in drug action. For example, vitamin A destroys lysosomal membranes. When animals receive excess vitamin A, a pathological condition known as vitamin A intoxication results, which is associated with spontaneous fractures and other lesions in bones and cartilage. Cortisone has the opposite effect; it stabilizes the lysosome, and this property may account for the anti-inflammatory effect of this drug. The other roles of these newly discovered organelles await further research.

Peroxisomes

Peroxisomes are organelles found primarily in liver and kidney tissue that are often mistaken for lysosomes. They are the same size and shape as lysosomes and appear to originate from terminal dilations of the ER, but they contain different enzymes. Peroxisomes contain oxidative enzymes (catalase, peroxidase, and uricase) that convert hydrogen peroxide (H_2O_2), formed by hydrolysis of water and toxic if excessive, to water and other compounds harmless to the cell. Overexposure to external energy sources (radiation) can cause excess hydrolysis of H_2O_2 in the cell. Peroxisome enzymatic action may prevent cellular death from this cause.

Centrioles

The high power of the microscope reveals a dense area near the center of the cell where the nucleus is found. This dense area, called the *centrosome* (central body), contains two centrioles.

Each centriole is about 3 to 5 nm long and 1 to 2 nm wide. More than two centrioles may be in a cell, but there must be at least two in any dividing cell. They always occur in pairs, are cylindrical in shape, and are positioned at right angles to each other (Figure 2-2). Electron microscopy of a cross section of one centriole reveals nine groups of cylindrical bodies arranged in a circle, each group commonly containing two or three cylindrical fibers or microtubules.

Centrioles are self-replicating bodies, with no limiting membrane. Interestingly the basal bodies of cilia, flagella, and sperm tails have a similar structure. Centrioles are known to migrate from the area around the nucleus to the cell membrane and can form cilia and flagella.

The main function of centrioles is in cell division. They control the formation of the spindle during cell division in animal and some plant cells. Their role will be discussed in more detail in the section on mitosis and meiosis.

Nucleoplasmic Organelles

The nucleus is bounded by a membrane consisting of an outer portion that is continuous with the endoplasmic reticulum and an inner portion (Figures 2-2 and 2-6). Electron micrographs indicate that these layers are porous. The pores are related in some way to nuclear permeability, since it is known that large molecules can enter and leave the nucleus. Inside the nucleus is the nucleoplasm, which contains darkly stained, threadlike material called **chromatin** and di-

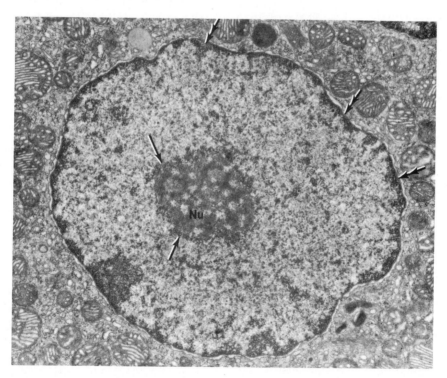

Figure 2-6 Section of a nucleus showing the nucleolus (Nu) and chromatin. The nucleolus has a thin rim of chromatin (arrows)—the nucleolar-associated chromatin—and a fibrillar portion in the center surrounded by a granular portion. The layer of chromatin lining the internal surface of the nuclear envelope and seen in the optical microscope as nuclear "membrane" is shown by the double arrows. 12,500×. (Courtesy of J. Long. Reprinted by permission.)

verse bodies of granular material called **nucleoli.**

Nucleolus

The nucleolus has no membrane surrounding it (Figures 2-2 and 2-6). Sometimes several nucleoli are found in a nucleus. Internally it is composed of dense granules and fibers of ribonucleic acid (RNA). The nucleolus participates in the production of ribosomal RNA (r-RNA) and is therefore important in protein synthesis. Some investigators consider the nucleolus as reserve chromosome material that is utilized in nuclear division, since it disappears during the mechanical phases of cell division.

Chromosomes

The human cell has forty-six chromosomes, made of chromatin material that contain the hereditary genes of the cell and the organism. Genes are responsible for development and regulation of the metabolic processes performed by the cell. Enzymes are composed of protein, and, therefore, genes control the formation of hundreds of enzymes.

The principal chemical substances in chromosomes are proteins (histones) and deoxyribonucleic acid (DNA). Of the forty-six chromosomes, twenty-two pairs are exactly alike in the two sexes and are called **autosomes.** The remaining pair, the sex chromosomes, are alike in the female (two X chromosomes), but in the male, one is an X chromosome and the other is a smaller, atypical form (Y chromosome) (Figure 2-7).

DNA

Each chromosome consists primarily of a long spiral ladder, or double helix, of deoxyribonucleic acid (DNA). The sides of the ladder are formed by alternating units of deoxyribose (a five-carbon sugar) and phosphate; the rungs are pairs of nitrogenous bases. In most cases there are four kinds of bases: two purines (adenine and guanine) and two pyrimidines (thymine and cytosine). The purines and pyrimidines have double-ring and single-ring structures, respectively (Figure 2-8). The four base substances can conveniently be referred to as A, G, T, and C.

The pairing of the bases is highly specific: A is constructed so that it will fit only with T; G will fit only with C. This specificity of pairing is important because the sequence of bases along one side of the ladder necessarily determines the sequence along the other side. This fact helps us understand how DNA duplicates itself during cell division.

The ingenious model that established the physical structure of the complex DNA molecule was developed by J. D. Watson and F. H. Crick. In their first paper, these Nobel prize winners theorized the double helix structure of DNA; they also postulated that it was wide. Genes, they suggested, were made up of 500 to 2000 bases. The total number of bases per chain was estimated to be approximately 200,000 in DNA of viruses.

Watson and Crick's discovery of the structure of DNA was particularly significant because of the implication that the coiled molecule could uncoil itself. The two chains of the molecule, their bases still intact, can, by uncoiling, make complementary copies of themselves. The eventual result is the formation of two new DNA molecules, each exactly like the "parent"—a process known as **replication.** This concept of the "unzipping" of corresponding nitrogenous bases and the matching of pairs provided the key to our understanding of how chromosomes are duplicated (Figure 2-9).

As discussed in the next section, the replication of DNA occurs at interphase, just before mitotic division. Thus the genetic code of each cell is duplicated every time the cell divides, the original DNA molecule acting as a template for the new one. The hereditary characteristics of the genes are therefore carried forward to each new generation of cells.

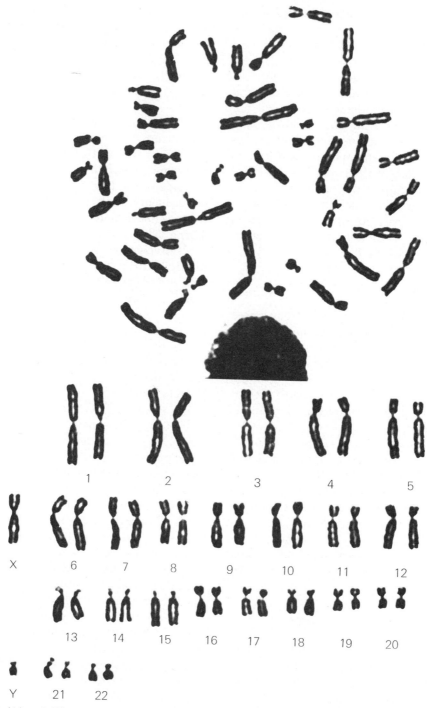

Figure 2-7 (Above): Photomicrograph of chromosomes of a human cell obtained during metaphase. (Below): The chromosomes grouped according to their morphologic characteristics. (Courtesy of G. Gimenez Martin. Reprinted by permission.)

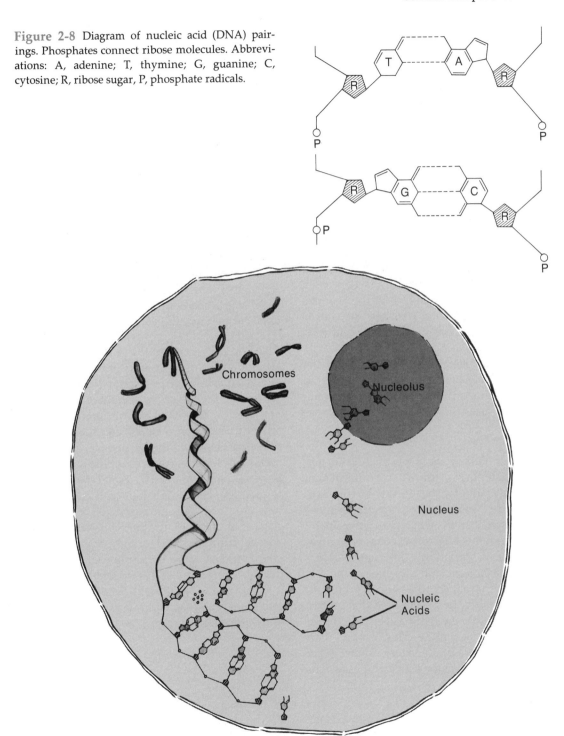

Figure 2-8 Diagram of nucleic acid (DNA) pairings. Phosphates connect ribose molecules. Abbreviations: A, adenine; T, thymine; G, guanine; C, cytosine; R, ribose sugar, P, phosphate radicals.

Figure 2-9 Diagram of DNA molecule, showing replication in the nucleus.

CELL DIVISION

Mitosis

The most common form of cell division is **mitosis.** In this process, each chromosome in the cell divides into identical halves, separates, and forms "daughter" nuclei. Thus, each of the two daughter cells receives exactly the same number and kind of chromosomes that characterized the parent. The various stages of mitosis are illustrated in Figure 2-10; following is a brief account of what happens:

INTERPHASE   When a resting nucleus is examined under a light microscope at its highest magnification, the nucleoplasm is seen as a dense, shapeless mass with little or no organization. The first indication that cell division is about to take place occurs when the centrioles divide.

At the same time, the chromatin material of the chromosomes begins to form a long, spiral thread containing histone and DNA. These substances organize themselves to form dense threads.

PROPHASE   During early prophase, the centrioles migrate further toward opposite ends of the nucleus. Simultaneously, the dense **chromatids**—the two halves of each chromosome—become distinct entities but are still held together by a connecting constriction, or **centromere.** Later in prophase, the centrioles are located at the extreme opposite ends of the cell, where they send out spindle-shaped protein fibers that surround the nucleus and that replace the nuclear membrane that has begun to disintegrate. At this stage, if the centrioles and their fibers are stained carefully, a star-shaped apparatus called an *aster* can be seen at opposite ends of the cell.

METAPHASE   By now, the nuclear membrane is gone, and the spindle fibers maintain the integrity of the nucleus. The chromosomes form a line across the middle of the cell and attach themselves to the spindle fibers. Microscopically, this gives the appearance of a solid dark band across the cell called the *equatorial plate* (Figures 2-10 and 2-11).

ANAPHASE   By a means not fully understood, the chromatids pull apart entirely, and the now separated chromosomes begin their migration toward either end of the cell. According to one theory, the centrioles retract their fibers and the chromatids respond with them. Another theory proposes that the two chromatids of each chromosome are identical chemically and therefore identical electrically. The law of electrical signs would then apply (opposite signs attract; like signs repel), and the identical chromatids would repel each other. Both theories have their adherents, but neither has been proved.

TELOPHASE   Early in this final stage of mitosis, there is a "dimpling-in" of the sides of the elongated cell, with an aggregation of chromosomes at each pole. As telophase progresses, new nuclear membranes form around the two sets of nuclear components, and the cell membrane carries forward the division until cross membranes are laid down within the cell and between the two nuclei. Two cells are fully formed, and mitosis is complete. Each of the two new daughter cells now reenters the resting stage of interphase—the period between cell divisions during which normal physiologic processes proceed. Since the chromosomes doubled during interphase and subsequently divided into separate halves, the nuclei of the two new cells contain the same number of chromosomes as did the parent nucleus.

Protein Synthesis: DNA and RNA

All the chemical reactions of living systems are controlled by enzymes, which are protein substances, and much of the structure of organisms

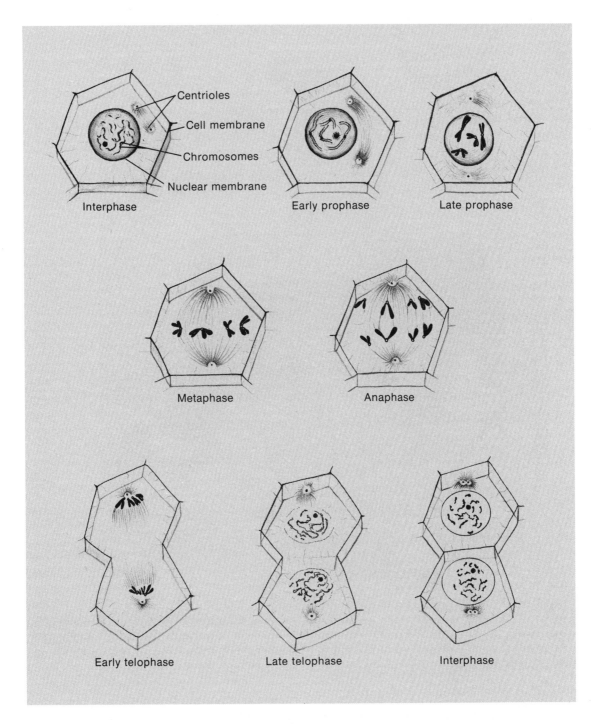

Figure 2-10 Schematic representation of mitosis of a cell containing four chromosomes. See text for an account of what happens during each of the various stages.

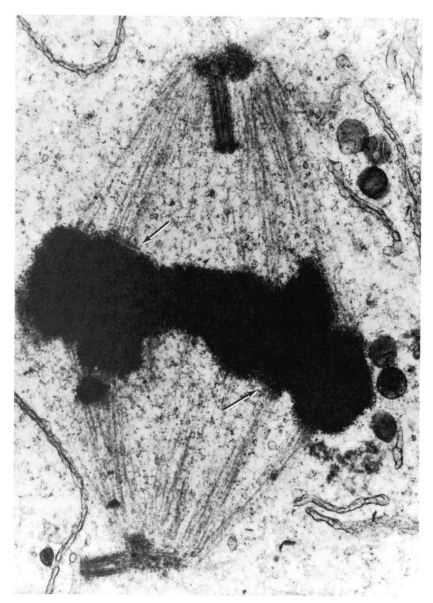

Figure 2-11 Electron micrograph of a section of a rooster sperm atocyte in metaphase. Observe the presence of two centiroles in each pole, the mitotic spindle formed by microtubules, and the chromosomes in the equatorial plate. The arrows show the insertion of microtubules in the centromeres. Reduced from 30,000×. (Courtesy of R. McIntosh. Reprinted by permission.)

consists of proteins. Proteins, therefore, determine the form and function of an organism. Proteins are made of combinations (polymers) of amino acids. Twenty different amino acids commonly occur in nature. The sequence of these largely determine the specific properties of protein. What then determines the amino acid sequence of a protein?

According to the *one gene–one polypeptide hypothesis* (a revision of the one gene–one enzyme hypothesis), a sequence of DNA bases (about 1500) contains the code for the primary structure (amino acid sequence) of all or part of a protein. This theory suggests that genes control the production of enzymes. A sequence of three nitrogenous bases in a DNA strand produces the code for a specific amino acid. The genetic code thus is a triplet code; each triplet is called a **codon** (m-RNA).

Since there are four different bases that function in groups of 3, there are sixty-four (4^3) different codons. Since only twenty amino acids exist, there is some redundancy in the code; several codons may code for the same amino acid.

A few codons function as punctuation; designating the beginning or end of a polypeptide sequence of amino acids.

Only one of the two strands of a DNA molecule serves as a code for protein synthesis. How the cell knows which strand to read is not yet known.

The nucleic acid RNA differs from DNA in three ways: it has a different sugar (ribose instead of deoxyribose), uracil *(U)* replaces thymine, and it is single-stranded rather than double helical (Figure 2-12). RNA exists in several distinct forms in the cell: messenger RNA (m-RNA), ribosomal RNA (r-RNA), and transfer RNA (t-RNA). All these perform complementary functions in protein synthesis.

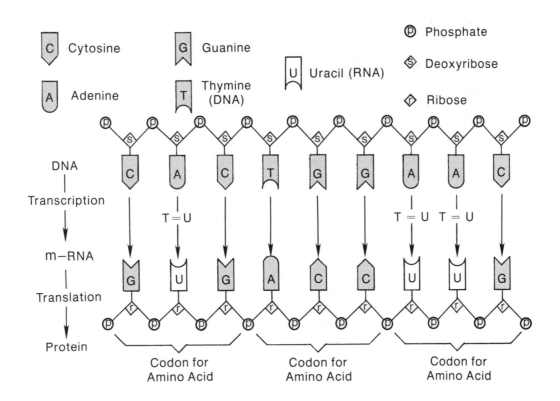

Figure 2-12 Transcription from one DNA strand to m-RNA showing codons that specify amino acids.

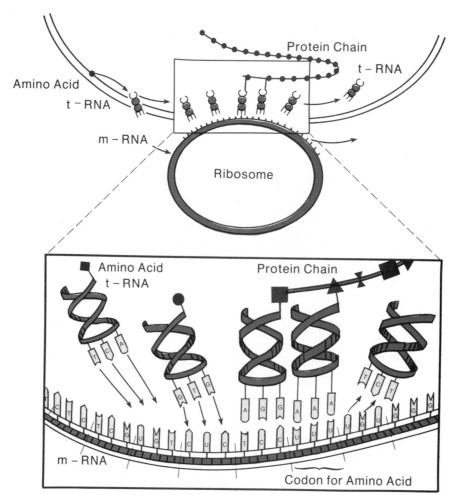

Figure 2-13 Scheme of translation and protein synthesis.

A genetic message is first passed from DNA to m-RNA by the process of **transcription** (Figure 2-13). First a segment of the DNA helix unravels. Then, ribonucleotide bases pair with the complementary DNA bases of one of the DNA strands. The ribonucleotides then polymerize to form a m-RNA strand with a base-pairing sequence complementary to the DNA template strand. A transcription enzyme controls the process.

The m-RNA strand then passes from the cell nucleus into the cytoplasm carrying the in-

structions for one polypeptide sequence of a protein. The m-RNA codons have been standardized as representing the genetic code. These codons are complementary to the DNA codons.

Located in the cytoplasm are small molecules of cloverleaf-shaped t-RNA. At one end, each t-RNA bears a specific anticodon that matches a m-RNA codon. The other end bears a specific attachment for one of the twenty amino acids.

The granules (ribosomes) on the rough

OUTLINE

I. Introduction
  A. Properties of life
    1. Respiration
    2. Metabolism
    3. Growth
    4. Excretion
    5. Reproduction
    6. Motility
    7. Irritability
  B. Sizes and types of cells
    1. Macroscopic
      a. Ostrich egg
    2. Cell shape correlates with function
      a. Nerve cells—long appendages
      b. Muscle cells—elongated
      c. Skin cells—box-shaped
      d. Connective tissue cells—fibrous
    3. Microtubules and microfilaments
      a. Cell size
      b. Cell shape
  C. Cellular components
    1. Cell membrane
    2. Protoplasm—living matter of cell
      a. 60 to 70 percent water
      b. 16 percent protein
      c. 14 percent fat
      d. Trace elements
II. Cytoplasmic organelles
  A. Endoplasmic reticulum
    1. Irregular canals that may serve for intracellular communication and transport, and may connect nucleus with the outside
    2. Rough reticulum stores and transports protein
    3. Smooth reticulum stores carbohydrates
  B. Ribosomes
    1. Protein synthesis
  C. Golgi apparatus
    1. Excretion
    2. Carbohydrate storage
    3. Produce glycoproteins
    4. Lipid secretion
  D. Mitochondria
    1. Matrix—center of mitochondria
    2. Cristae—folds of mitochondria
    3. Enzymes in matrix responsible for cellular respiration
    4. "Powerhouse" of the cell
  E. Lysosomes
    1. Contains enzymes
    2. Cellular digestion of complex compounds
    3. Aging
  F. Peroxisomes
  G. Centrioles
    1. Restricted to animal cells
    2. Nine microtubules
    3. Two units positioned above the nucleus at right angles to each other
    4. Cell division
III. Nucleoplasmic organelles
  A. Nuclear membrane
    1. Double-layered, porous
  B. Nucleolus
    1. Prominent in period preceding cell division
    2. Nucleic acid and protein synthesis
  C. Chromosomes
    1. Histone and DNA
    2. Genes
    3. 46 in each human cell
IV. DNA
  A. Double helix
    1. Deoxyribose (five-carbon sugar)
    2. Two purines—adenine and guanine
    3. Two pyrimidines—thymine and cytosine
  B. Coding—respective pairings
    1. Adenine with thymine
    2. Guanine with cytosine

V. Mitosis
  A. Interphase
     1. Visibly an inactive phase
     2. Biochemically the most active phase
  B. Prophase
     1. Centrioles migrate to positions at opposite ends of nucleus
     2. Chromosomes duplicated
     3. Nuclear membrane disintegrates
  C. Metaphase
     1. Chromosomes line up in midline of cell—"equatorial plate" phase
  D. Anaphase
     1. Chromatids pull apart
     2. Theories for chromatid separation
        a. Fibers pull chromatids to each pole
        b. Chromatids repel each other
  E. Telophase
     1. Cell wall "dimples in" to form two new cells
     2. Two new nuclear membranes form
     3. Daughter cells have same number of chromosomes as original parent cell

VI. Protein synthesis
  A. Nucleic acid code
     1. Arrangements of twenty amino acids
  B. Types of RNA
     1. Messenger (m-RNA)—codon
     2. Ribosomal (r-RNA)—alignment
     3. Transfer (t-RNA)—anticodon
  C. Transcription—codon activity in nucleus
  D. Translation—anticodon activity at the ribosome
VII. Clinical implications
  A. Chemical poisons and toxins
     1. Cytoplasm altered
     2. Cell degenerates
  B. Necrosis
  C. Morphological changes in cell death
  D. Genetic factors
     1. Monosomy
     2. Trisomy
     3. Abnormalities of sex

STUDY QUESTIONS AND PROBLEMS

1. Describe the properties of life and indicate which cellular organelles perform the cellular functions.
2. What is protoplasm?
3. What are some of the physiological activities performed by living cells?
4. What are the functions of the endoplasmic reticulum? What do they mean by "rough" ER?
5. What do the ribosomes do? Where are they located?
6. What do the mitochondria have to do with metabolism of energy?
7. What kind of enzymes are found in lysosomes? How has the function of the lysosomes been linked with aging?
8. What is the importance of the nucleus? What does the nucleolus do?
9. Centrioles have what function in the cell? Discuss their role in cellular reproduction.
10. What is DNA? How is the DNA molecule constructed?
11. Describe the DNA coding mechanism.
12. Trace the chromosomes through the phases of mitosis.
13. Discuss the roles of the ribosomes and Golgi apparatus in protein synthesis and liposomal development.
14. Relate the roles of m-RNA, r-RNA, and t-RNA in protein synthesis.
15. List examples of genetic abnormalities related to autosomes.
16. Differentiate between monosomy and trisomy.
17. Define nondisjunction and relate it to Klinefelter's syndrome.

3

The Cell Membrane and Permeability

LEARNING OBJECTIVES

- Describe the molecular structure of a cell membrane

- Define semipermeable membrane in terms of its role in governing the exchange of chemical compounds between cell and compartment

- Define diffusion and osmosis in terms of movement of particles or liquid in response to a concentration gradient

- Differentiate between facilitated diffusion and active transport

- Define phagocytosis and pinocytosis

- Define isotonic, hypotonic, and hypertonic and their effects on cell volume

- Define dialysis and differentiate between peritoneal and hemolytic methods

- Define edema and give at least three reasons why it can occur

- Relate water content and dehydration with electrolyte imbalance

IMPORTANT TERMS

active transport	diffusion	homeostasis	mosaic	semipermeable
autolysis	equilibrium	hydrophilic	osmosis	solute
concentration	filtrate	hydrophobic	permeability	solvent
gradient	filtration	hypertonic	permeable	thermodynamics
crenation	glycocalyx	hypotonic	phagocytosis	
dehydration	glycoprotein	isotonic	pinocytosis	
dialysis	hemolysis	lipid	protein	

For the process of cell metabolism to occur, food must enter the cell, and waste products must be discharged from it. The maintenance of **homeostasis,** the stable internal environment necessary for an organism's well-being, depends on the movement of substances through the plasma membranes that separate the cells. The exact nature of the physical processes governing the movement of materials across a cell membrane is not completely understood. The basic modes of transport are known, however, and will be discussed now.

THE PLASMA MEMBRANE

The plasma membrane is a delicately balanced, functional organelle that separates the cell from its environment and allows materials to pass across it in both directions. By controlling ionic composition and water content, the membrane prevents the cell from swelling or shrinking and, therefore, is directly responsible for maintaining homeostasis.

The plasma membrane is composed of inner and outer protein layers separated by two lipid layers (bilayer lipid leaflet) (Figures 3-1 and 3-2). The thickness of the membrane is approximately 8 nm (80Å).

In protein-carbohydrate complexes, **glycoproteins** not only lie on the surface of the lipid bilayer, but may enter the **hydrophobic** (water-repelling) interior or even penetrate the membrane from one side to the other (Figure 3-3).

The outer covering of the glycoprotein is layered with **glycocalyx,** a carbohydrate-rich substance that gives the cell its antigenic (immune) property, allowing other cells to adhere to it. This property is lost in cancer cells, and, as

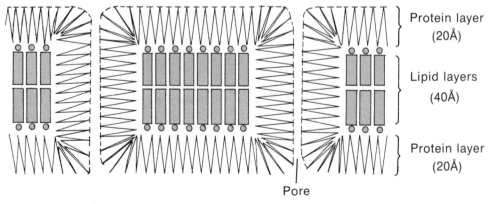

Protein layer (20Å)

Lipid layers (40Å)

Protein layer (20Å)

Pore

Figure 3-1 Diagram illustrating the layers of a cell membrane.

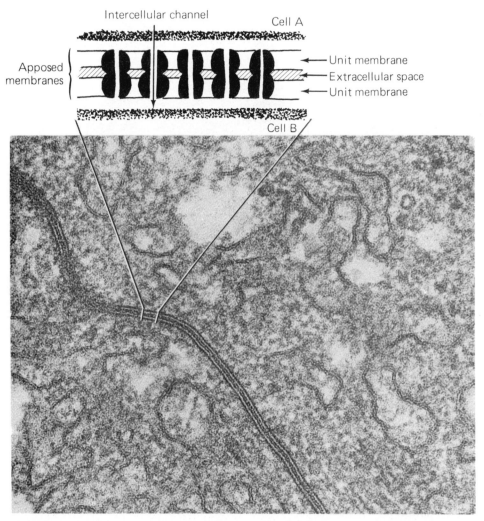

Figure 3-2 Gap junction from rat liver. Two apposed membranes are separated by an electron-dense space or gap 2mm wide. In the upper portion is a drawing representing the structure of a gap junction that basically consists of short tubular components that cross the membranes of adjoining cells. 193,000✕. (Courtesy of M. C. Williams. Reprinted by permission.)

a result, the cancer is able to spread, or *metastasize*, to other tissues.

In 1972, Singer and Nicolson suggested a fluid mosaic model in which glycoproteins or lipoproteins are oriented within the lipid bilayer. The protein molecules in the membrane are primarily round and are of two types: *Intrin-* *sic proteins* (also known as *endoproteins*) are deeply imbedded in the lipid bilayer and often are large enough to stretch across the membrane. These proteins are an integral part of the membrane and are difficult to remove without destruction of the membrane itself.

Because of the fluid nature of the lipids, in-

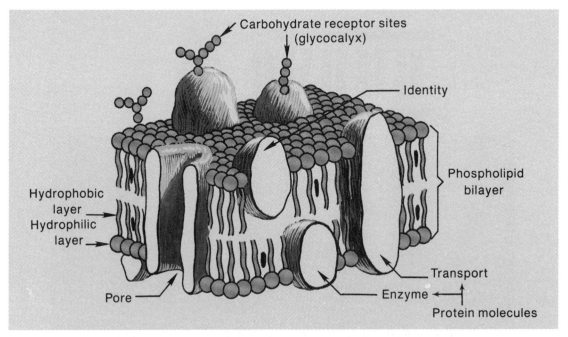

Figure 3-3 Fluid-mosaic model of the membrane (Singer-Nicolson). A double layer of lipids forms the main continuous part of the membrane, the lipids are mostly phospholipids (the brown balls represent their hydrophilic heads and the wavy lines their hydrophobic tails. In plasma membranes of higher organisms, cholesterol (solid bars) is also present. Proteins (shown here as blocks, but really coiled polypeptide chains) occur in various arrangements. Some are entirely on the surface of the membrane. Others are partly embedded in the lipid layers, some of these may penetrate all the way through the membrane. There is also evidence that at least some of the pores in the membrane are entirely protein-lined (lower left).

trinsic protein molecules are thought to "float" sideways in the membrane, a fact that may prove to be significant for proper functioning of the protein. Among their functions, these proteins serve as structural supports for the membrane, receptors for various hormones, and "carrier" molecules.

Extrinsic proteins (also known as *ectoproteins*) are toward the outer or inner surface of the membrane and can be easily removed without damaging the membrane structure. These proteins serve as surface receptors for certain substances and as a cytochrome (enzyme) in the electron transport system.

The mosaiclike distribution of the proteins in the membrane and the possibility for lateral movement of these proteins are the basic properties of the fluid mosaic model of membrane structure. Currently, the fluid mosaic model is the most widely accepted model for membrane structure.

The main barrier to substance exchange across the membrane is undoubtedly the lipid layer. A molecule must be able to pass through the small pores of the cell member or dissolve in the lipid layer and then diffuse through the membrane. Water passes readily through all membranes. Small, positively charged ions, however, move through the membrane very slowly due to the postulated positive charge of the lipid pores. Like charges repel each other. Negatively charged ions pass more easily.

PASSIVE TRANSPORT

Diffusion

A solid, a liquid, or a gas may be dissolved in a liquid. The material dissolved is called the **solute;** the liquid in which it is dissolved is the **solvent.** Molecules of the solute are dispersed throughout the solvent much as if they were molecules of gas filling an enclosed space.

The term **diffusion** refers to this spreading or scattering of molecules of gases or liquids. If two gases are mixed, their molecules will move continuously and soon become uniformly dispersed. The molecules always move toward an area where fewer molecules exist—in other words, from an area of greater concentration of molecules to an area of lesser concentration of molecules. When a cube of sugar is dissolved in a beaker of water, the concentration of sugar in the immediate area of the dissolving cube remains high at first. Within a few hours, however, the sugar will have scattered throughout the water until an even distribution of sugar per volume of water exists (Figure 3-4A). Thus, diffusion can be defined as the passage of solute particles from an area of higher concentration to an area of lesser concentration. The number of particles per volume is called the **concentration gradient.**

Diffusion occurs in all the body's life-maintaining processes. Oxygen in the blood moves from the plasma into the extracellular fluid and then into the cell as a result of diffusion. The exchange in the lungs of the gases, oxygen and carbon dioxide, during inspiration and expiration is another example of diffusion.

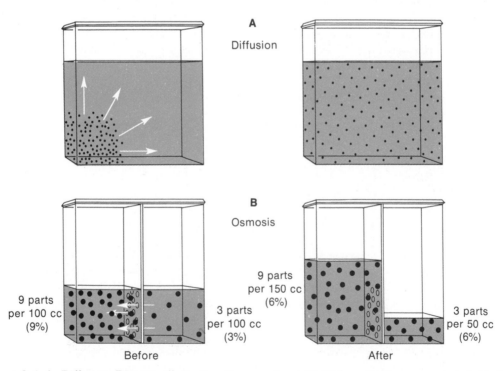

A
Diffusion

B
Osmosis

9 parts per 100 cc (9%)

3 parts per 100 cc (3%)

Before

9 parts per 150 cc (6%)

3 parts per 50 cc (6%)

After

Figure 3-4 **A**, Diffusion. Diagram illustrating the diffusion or scattering of particles from an area of greater particle concentration to an area of lesser particle concentration, until equal parts per volume are achieved. **B**, Osmosis. Arrows indicate the passage of solvent (H_2O) from the less concentrated side to the more concentrated side to achieve equal parts per volume. In this case the particles could not penetrate the semipermeable membrane.

Osmosis

The cell membrane permits some molecules to enter the cell but does not admit others. Therefore, it is called selectively permeable or **semipermeable.** This means that it allows the passage of molecules up to a certain size or diameter but restricts others. Sugar and water molecules are small enough to pass through the membrane; larger molecules are partially or completely prevented from passing through the membrane pores. In diffusion the molecules or particles move across the membrane according to a concentration gradient established by the difference in quantity of particles on either side of the membrane.

In another process, **osmosis,** the same kind of concentration gradient exists, but the particles in the intracellular fluid are too large to pass through the membrane. Since homeostasis must be maintained, water (solvent) penetrates the cellular membrane to effect an equal concentration on either side (Figure 3-4*B*).

The force that causes this movement of water is called *osmotic pressure* and varies directly with the number of solute particles in the solution. For example, a thistle tube containing 20 percent sugar (twenty particles per 100 mL of water) can be covered by a semipermeable membrane, inverted into a beaker of 10 percent sugar (ten particles per 100 mL of water), and allowed to stand for a few hours. The membrane is impermeable to sugar in this instance and will not allow diffusion to occur. However, the concentration gradient is clearly greater in the thistle tube than in the beaker. Therefore, water will pass from the beaker into the tube, causing the level of fluid in the tube to rise.

Osmosis can be thought of as the passage of solvent from an area of lesser concentration to one of greater concentration of solute. In a sense, osmosis represents an attempt to dilute the more concentrated solution to achieve chemical balance, or **equilibrium.**

The principle of osmosis can be demonstrated clearly in the laboratory by using red blood cells suspended in salt solutions of various concentration. That homeostasis is maintained in the bloodstream becomes evident when we observe that the concentration of dissolved substances (such as salts) in the plasma is the same as the concentration of those substances in the red blood cells. When two solutions demonstrate the same concentrations of dissolved substances, they are **isotonic** (*iso* means same or equal). To show the effect of osmosis, red blood cells may be placed in two separate solutions, one having a lower concentration of salts than red blood cells and the other having a greater concentration. The solution having the greater concentration of salts is called **hypertonic** to the blood cell; osmotic pressure forces water to leave the cell and to pass into the solution, causing a shrinking or **crenation** of the cell. In this situation, water enters the cell until the cell finally bursts—a condition called **hemolysis** (Figure 3-5).

The exchange of water and dissolved nutrients in the capillary network of the circulation is a prime example of osmotic pressure in action.

Dialysis

Dialysis involves the movement of solutes across a selectively permeable membrane from a region of high solute concentration to a region of low solute concentration.

Several abnormal physiological conditions can be corrected, at least temporarily, by the appropriate use of methods involving osmosis and dialysis. Patients with kidney failure are faced with the problem of accumulating large quantities of metabolic wastes in their blood. These wastes contain toxic substances, which, if allowed to remain in the blood for an extended period of time, will cause serious damage and even death.

Hemodialysis and peritoneal dialysis are used to remove toxic substances and body wastes that are normally excreted by the kidney. The main indication for dialysis is a high and rising level of serum potassium. Hemodialysis, also known as the artificial kidney, uses a

Isotonic

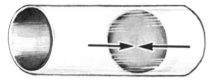

Hypotonic

Hypertonic

Figure 3-5 Effect of salt solutions of different concentrations on red blood cells. Isotonic: the cell is surrounded by a liquid of concentration equal to that of the cell. Hypotonic: the cell is surrounded by a liquid having a lower salt concentration than itself; liquid enters the cell and causes it to swell and eventually burst (hemolysis). Hypertonic: the cell is surrounded by a liquid having a greater salt concentration than itself: liquid from the cell passes out into the surrounding medium, causing the cell to shrink (crenation).

cellophane membrane to separate the patient's blood from an electrolyte fluid bath, or *dialyzing fluid.* The membrane has pores similar to those of the glomerular capillaries. Blood is pumped from the patient's radial artery through the cellophane membrane, which is immersed in the electrolyte solution. The nonprotein nitrogen retention products pass through the semipermeable membrane of the cellophane tube into the dialyzing fluid and are flushed away. The blood continuously circulates through the patient and through the dialyzing tube for a period of 4 to 7 hours. The composition of the dialysis fluid is determined before dialysis to

establish a correct concentration gradient of metabolites and waste materials.

Peritoneal dialysis is also based on the principle of diffusion of substances across a semipermeable membrane. In this technique, a sterile dialyzing fluid is introduced into the peritoneal cavity at intervals. The surface area of the peritoneum acts as the semipermeable membrane. A catheter is inserted into the peritoneal cavity, and dialysis fluid is allowed to run into the peritoneal cavity. After equilibrium between the dialysis fluid and the highly vascular peritoneal membrane takes place, the fluid is allowed to drain by gravity into a closed collecting bottle. The advantage of this technique is simplicity. The disadvantage is that it takes five to six times longer and is often painful.

Filtration

Until now, we have considered only those transport processes involving the passage of particles or water through a membrane according to a concentration gradient. Another passive mode of transport, **filtration,** occurs when water and particles are literally forced through a permeable membrane from an area of higher pressure to an area of lower pressure of fluid. The force is *hydrostatic pressure.*

In the laboratory, the principle of hydrostatic pressure can be used to advantage when a fluid containing dissolved particles is poured through a filter paper in a funnel. The weight of the fluid in the funnel forces the fluid from an area of lower hydrostatic pressure in the beaker below. The filter paper is not permeable to some of the particles in the fluid, and these remain in the funnel.

In the human body, this principle can be demonstrated by the action of the kidney capillaries. The hydrostatic pressure of blood is greater than that of the surrounding tissue fluids; therefore water and dissolved substances filter through the capillary membrane into the tissue fluid. Some particles, such as blood cells and proteins, are too large to be permeable and remain in the bloodstream. A small quantity of

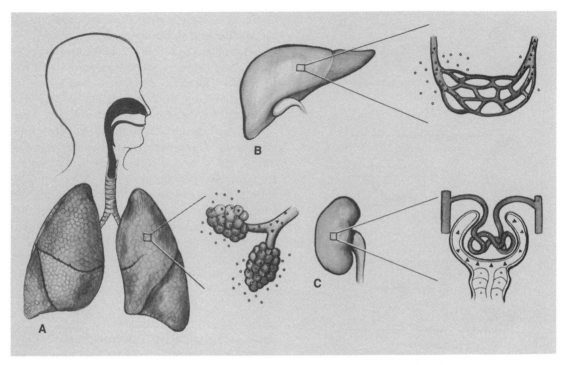

Figure 3-6 Types of permeability. **A**, Diffusion of gases in the lung. **B**, Osmosis passage of nutrients and waste materials in and out of the capillaries by osmotic pressure. **C**, Filtration: passage of materials from the bloodstream into the kidney tubules by a filtration force.

albumin may be forced through capillary membranes, but only when large quantities of protein (proteinuria) or blood (hematuria) are evident in the urine can kidney disease be diagnosed. (Examples of diffusion, osmosis, and filtration are shown in Figure 3-6.)

Facilitated Diffusion

Glucose and amino acid molecules are insoluble in lipid and are too large to pass through the pores of a membrane. Since glucose is the main energy source for most cells, it must be able to enter the cell in some way. There is evidence that, in such cells, the glucose molecule combines with a carrier molecule in the membrane. The carrier molecule, thought to be protein, is free to move across the membrane, taking the glucose molecule along with it and releasing it inside the cell (Figure 3-7). This kind of diffusion is called *facilitated diffusion.* It does not consume metabolic energy. It only takes place down a diffusion gradient, where there is a higher number of particles outside the membrane than inside, and there is high molecular specificity between the carrier and the molecule carried.

According to the membrane carrier theory, the cellular membrane houses certain molecules that move from side to side in the membrane in response to certain chemical stimuli. The most likely molecular candidates for this carrier role are the protein molecules housed in or on the cell membrane. Proteins are relatively large molecules that can undergo structural changes under certain circumstances. A protein mole-

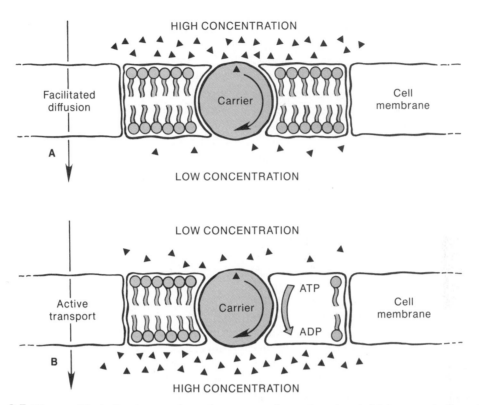

Figure 3-7 Diagram illustrating two-carrier molecular system. **A**, Movement of molecule (Δ) across the membrane from higher to lower concentration of molecule. **B**, Movement of a molecule (Δ) across a membrane from low to high concentration of molecule with the expenditure of energy to afford the transport.

cule situated along the exterior surface of the cell membrane is in a position to react or bind with extracellular materials. When this occurs, the protein molecule is thought to undergo a structural and positional change. Instead of facing the exterior surface of the cell, the protein, plus the extracellular material bound to it, is repositioned toward the interior surface of the cell. In this new position, the bound material can be released to the interior of the cell to complete the membrane transport process. Once the bound material is removed from the carrier protein, the protein undergoes a reverse structural change to return to its original position in the membrane. This return process probably is accompanied by the transport of some intracellular material to the extracellular environment.

ACTIVE TRANSPORT

In many cases, the cell expends energy to carry substances across membranes. This is known as **active transport,** and it continues only as long as the cell is metabolizing and producing energy (ATP) for the process. Active transport permits the cell to move substances against the concentration gradient from lower to higher concentration of substance. A specific carrier is involved, but just where and how the energy is used is not known.

The cell is constantly expending energy to maintain the distribution of electrolytes and other solutes found on the two sides of the cell membrane in the face of chemical, ionic, temperature, and osmotic gradients. Several mechanisms have been proposed to explain the active transport processes that account for this intracellular and extracellular distribution of materials.

The basic mechanism of active transport is thought to be a carrier system (Figure 3-7B) that involves a number of enzymes called *permeases.* Sodium and potassium are exchanged across the cell wall by active transport in all cells of the body. Other substances, such as glucose and other sugars, are actively transported in only a few areas (such as the intestine and the kidney) to prevent loss of energy.

The mechanism that transports sodium ions (Na^+) and potassium ions (K^+) across cell membranes is essential in maintaining a physiologic balance between the concentrations of each of these ions within the cell and in the surrounding tissue fluids. It is important to remember that physiologic balance does not necessarily mean "equal concentration." On the contrary, in this situation, proper cell function exists only if the concentration of Na^+ outside the cell is greater than the concentration within the cell, and if there is a greater concentration of K^+ inside the cell than outside.

Under normal conditions, Na^+ tends to pass into the cell, where it is less concentrated, and K^+ tends to leak out into the extracellular fluid where similarly, it is less concentrated. Proper cell function, however, requires that these ions be returned to their original loca-

tions. Homeostasis is maintained, according to a widely accepted theory, by the action of a sodium-potassium "pump" that shuttles these ions back and forth across the cell membrane by special "carriers." These carriers transport Na^+ out and return unloaded for another sodium ion. According to this theory, each carrier is specific for a particular ion, and energy is expended as the carrier releases the ion at its appropriate destination. This mechanism of ion exchange is particularly important in the transmission of nerve impulses and will be discussed again when we consider the physiology of the nervous system.

Phagocytosis and Pinocytosis

Phagocytosis, the process by which a cell engulfs and digests a substance, is also relevant to the present discussion of cell physiology. The transport of materials from the extracellular fluid into the cell's interior can be accomplished by engulfing them as a portion of the cell membrane folds inward (Figure 3-8). The pouch that results from this process then breaks loose from the cell membrane and forms a vacuole within the cytoplasm. By this process, solid particles such as bacteria and foreign bodies are kept separate from the cell's normal constituents. The principal cells of the body capable of phagocy-

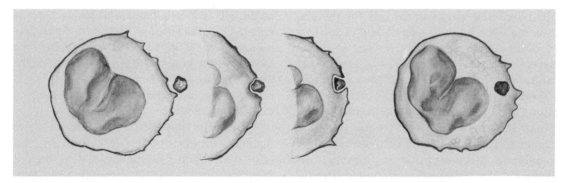

Figure 3-8 Phagocytosis. Portions of cytoplasm stream out, surround the particulate matter, and draw this material into the interior of the cell, where it is digested by cellular enzymes. Here the cell is a white blood cell (neutrophil) that is ingesting a bacterium.

tosis are the white blood cells and macrophages (histiocytes).

Pinocytosis involves essentially the same mechanism as phagocytosis. The only difference is that a liquid rather than a solid particle is ingested by the cell. A pinocytic vacuole contains small molecules dissolved in water. These vacuoles play a major role in the life-sustaining processes of metabolism and excretion.

CLINICAL IMPLICATIONS

The supreme importance of water, which makes up about 70 percent of the body weight, has already been mentioned. Water is present in the blood vessels (in the form of plasma), in the extracellular fluids, and in the cells. An exchange of fluid among these three major compartments goes on continually. Water lost from one compartment can be restored from another compartment.

For the most part, water balance is regulated by water loss, which averages about 2500 mL (2½ quarts) per day in an adult. This loss occurs through the intestines, the lungs (in expired air), the skin (as perspiration), and the kidneys. In a healthy person, the main responsibility for removing excess fluid belongs to the kidneys. However, large amounts of water may be lost in severe and continued diarrhea or in profuse perspiration.

Electrolyte Balance

The electrolytes of the body exert an important influence on the amount of water contained in the various compartments because of the osmotic pressure they produce. The laws of physics dictate that whenever water is shifted within the body, electrolytes are also shifted. Thus an abnormal water balance leads to a depletion of electrolytes in one area and to an excessive accumulation in another area. Both changes are potentially hazardous to health.

To illustrate this principle, let us consider potassium, which is the major intracellular cation. The concentration of potassium within the cells is roughly thirty-five times greater than the concentration in the tissue fluids. Potassium deficiency is more common than potassium excess. Such deficiencies can result from vomiting, diarrhea, and excessive glandular activity, leading to severe acidosis that results in progressive mental disorientation, muscular weakness, and eventual paralysis. Excessive potassium (intoxication) is a different matter. Malfunction of the kidneys can elevate potassium levels; an increase in potassium blood levels from the normal average of 4.5 mEq/L to just 8 mEq/L is sufficient to cause heart stoppage (cardiac arrest). (Milliequivalent, or mEq, is a measure of the chemical reactivity or combining power of ions.)

A slight increase in the concentration of sodium increases muscle excitability, resulting in convulsive contractions called *spasms*. Too low a concentration of calcium in the blood can lead to convulsions of the entire body (tetany); too high a calcium level can cause the formation of stones (calculi) in the gallbladder and kidney.

Obviously, potassium, sodium, and calcium are cations of major clinical importance. Many anions can also cause serious illness if they are present in abnormal concentrations. These facts clearly illustrate the significance of electrolyte imbalance.

Dehydration

Normally the body takes in more water than is needed for nutrition and cellular metabolism, and the excess is excreted in the urine by the kidneys. However, if water intake is insufficient, the loss due to evaporation through the skin and other causes results in a negative water balance and the state known as **dehydration.** Generally, when the body's water loss reaches 15 percent of body weight, death follows.

A decrease in body water level from any cause is seen first in the circulatory system, where there is a reduction in blood volume, carrying with it the possibility of shock (systemic collapse). Then the regulating mechanism of the hormonal system comes into play, withdrawing fluid from the extracellular tissues and finally from within the cell **(crenation).** It is this final depletion of cellular fluid that really defines cellular dehydration, creating a thirst that can be fatal. These disturbances of water and salt balance may be caused by prolonged exposure, kidney failure, severe burns, or acute intestinal obstruction.

Edema

Edema may be defined as an abnormal collection of fluids in the extracellular spaces. Such fluids can move from one place to another, so that when the swollen (edematous) part is pressed, a "pit" is left. Accumulations of fluid are most marked and readily recognized in the subcutaneous tissues. Edematous tissues have a pale, watery appearance and, if the cause is not corrected, may come to resemble a jelly.

Principal causes of edema are:

1. An increased permeability of the capillary walls, usually due to inflammation

2. A decrease in the osmotic pressure of the plasma caused by a loss or deficiency of the protein albumin, which acts to establish or maintain osmotic pressure

3. Increased pressure inside the capillary, as occurs in chronic heart failure, in which blood accumulates in the vessels leading back to the heart

4. Lymphatic obstruction, which may be caused by a tumor, an inflammation, or a parasitic infestation

Although we usually think only of the unfavorable effects of edema, it can produce benefits as well. The shift of large amounts of fluid into the extracellular spaces may dilute bacterial poisons (as in infection); such a shift may also reduce a state of hypertonicity, thus protecting the cells from electrolyte imbalance.

OUTLINE

I. Introduction
   Metabolism—the sum of all the physical and chemical processes by which living substance is produced and maintained and also the transformation by which energy is made available for the uses of the organism
II. Structure of the plasma membrane
   A. Molecules up to a certain size pass through pores; larger ones are not permitted to pass
   B. Membrane consists of inner and outer protein layers separated by two lipid layers
   C. Glycoproteins oriented within the lipid bilayer
III. Passive modes of transport
   A. Diffusion
      1. Solute
      2. Solvent
      3. Concentration gradient
      a. Movement of solute from area of greater concentration of solute to area of lesser concentration of solute (example: oxygen and carbon dioxide exchange in blood cells and lungs)
   B. Osmosis
      1. Passage of solvent through membrane from area of lesser concentration of solute to area of greater concentration of solute
         a. Isotonic
         b. Hypotonic and hypertonic
            (1) Hypotonic—hemolysis
            (2) Hypertonic—crenation
   C. Dialysis
      1. Hemodialysis—use of cellophane membrane
      2. Peritoneal dialysis—use of vascular peritoneum

D. Filtration
  1. Forced diffusion through a permeable membrane
E. Facilitated diffusion
  1. Use of a carrier molecule down a diffusion gradient
  2. Carriers are globular proteins in the plasma membrane
IV. Active transport
  A. Carrier system
    1. Sodium normally higher in concentration outside the cell
    2. Potassium normally higher in concentration inside the cell
    3. Carriers called permeases
    4. Sodium and potassium exchange against a concentration gradient occurs in the presence of energy (ATP)
V. Phagocytosis and pinocytosis
  A. Phagocytosis
    1. Engulfing solid particles through a folded portion of cell membrane
    2. Formation of vacuoles inside the cell
    3. Performed by white blood cells and macrophages
  B. Pinocytosis
    1. Engulfing liquid
    2. Lysosome activity
VI. Clinical implications
  A. Water content
    1. 70 percent of body weight
    2. Water loss about 2500 mL daily
    3. Water loss balanced by fluid intake
  B. Electrolyte balance
    1. If water is shifted, electrolytes in tissue fluids are shifted
      a. Excessive potassium can be lethal
      b. Increase in sodium increases muscle excitability
  C. Dehydration
    1. Decrease in body water level may promote systemic shock
  D. Edema—abnormal collection of fluids in extracellular spaces caused by following:
      a. Increased permeability of capillary walls
      b. Decreased osmotic pressure
      c. Increased capillary pressure
      d. Lymphatic obstruction

STUDY QUESTIONS AND PROBLEMS

1. What does a plasma membrane look like?
2. List at least two functions of cell membranes.
3. Contrast the fluid mosaic model of the cell membrane with the "sandwich" model initially proposed on the basis of electron micrographs. Are the two models mutually exclusive?
4. What is the pore theory? How was it discovered?
5. What does permeable mean? Semipermeable?
6. Define the three basic means of passive transport through a membrane and give an example of where each occurs.
7. Why may crenation occur in the presence of a hypertonic solution?
8. Are blood cells crenated in the urine in the case of kidney retentions?
9. What is meant by osmotic pressure?
10. How does filtration differ from osmosis? Where does it occur?
11. What does active transport mean? Describe the permease carrier system.
12. How is energy used in the active transport system? In what form?
13. How does a cell phagocytize?
14. What happens to the food particles after phagocytosis?
15. Explain the importance of maintaining the body's electrolyte balance.
16. How is the body's water content normally maintained?
17. What is edema? How does it occur?

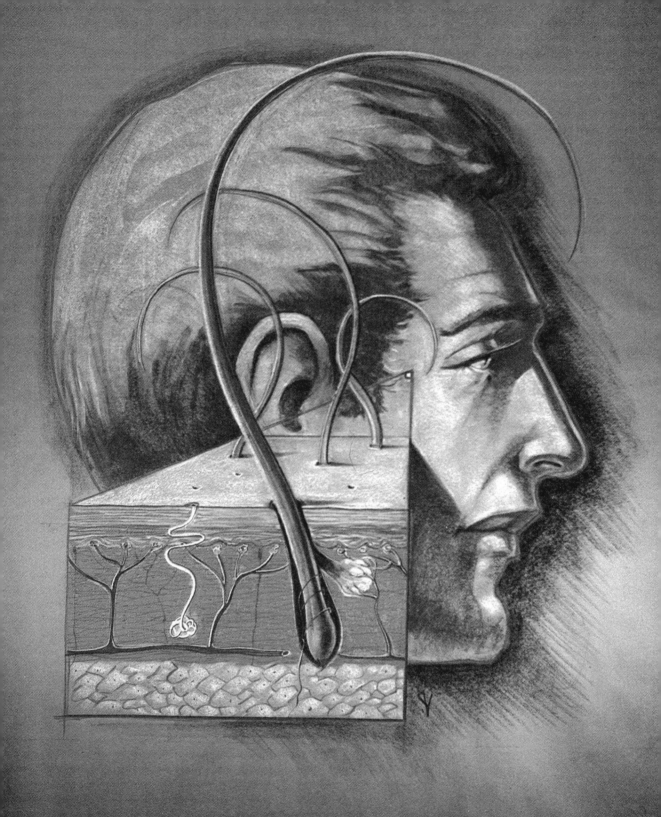

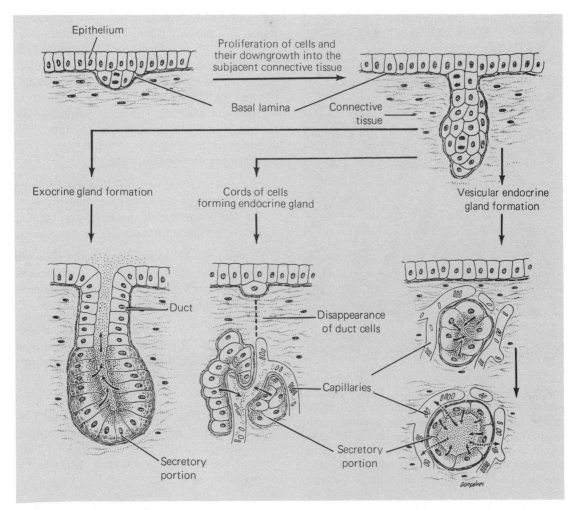

Figure 4-2 Formation of glands from covering epithelia. Epithelial cells proliferate and penetrate into connective tissue. They may or may not maintain contact with the surface. When contact is maintained, exocrine glands are formed; when contact is not maintained, endocrine glands are formed. The cells of these glands can be arranged in cords or follicles. The lumens of follicles accumulate large quantities of secretion; cells of the cords store only small quantities in their cytoplasm. (From *Histology*, 8th Ed., by A. W. Ham. Copyright © 1979 by J. B. Lippincott Co. Reprinted by permission.)

CONNECTIVE TISSUES

Connective tissues are found throughout the body. As their name implies, the principal function of these tissues is binding the body parts together. They form a framework for the internal organs; they also perform a variety of other functions, ranging from protection against injury to storage of fat.

A fundamental difference between connective tissue and epithelial tissue can be seen in their cellular compositions. Epithelial cells are directly adjacent to one another, separated only

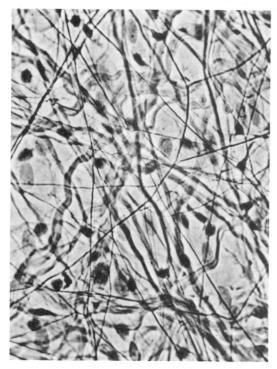

Figure 4-3 Fibrous connective tissue. Whole mesentery spread on a microscope slide. The preparation was stained by the Weigert method for elastic fibers and photographed under the phase contrast microscope. The thin, taut filaments are elastic fibers that branch and form a woven network. Collagen fibers are the thick and wavy structures. 200×. (From *Basic Histology*, 4th Ed., by L. C. Junquiera and J. Carneiro. Copyright © 1983 by Lange Medical Publications. Reprinted by permission.)

by a very small amount of intercellular substance called **matrix.** On the other hand, connective tissue contains few cells, and these are widely separated. The intercellular matrix is relatively abundant and usually determines the physical characteristics of a given connective tissue (Figure 4-3).

Connective Tissue Cells

In their embryonic stage, typical connective tissue cells are large and star shaped, with many projections called *processes*. These cells are *fibro-blasts*, and they arise from an early embryonic tissue, the **mesenchyme.** This tissue develops into many different forms, including blood cells and muscle. As the connective tissues develop, the cells lose their star-shaped appearance and become widely separated by large amounts of intercellular material. The adult connective tissue cell, which is responsible for the formation of fibers, is the *fibrocyte*. Other types of cells found in connective tissue are:

1. *Histiocytes or macrophages.* These cells move about through the connective tissue, ingesting foreign materials, bacteria, and cellular debris (phagocytosis).

2. *Plasma cells.* These small, irregular cells are associated with the formation of antibodies, an important part of the body's defense against foreign substances.

3. *Mast cells.* These are located near blood vessels and are involved in the production of heparin, an anticoagulant. They are also important in the production of histamine in allergic reactions.

4. *Blood cells.* The white blood cells, such as lymphocytes, monocytes, and neutrophils, are often present in connective tissue, where their function is to destroy bacteria by phagocytosis.

5. *Fat cells.* These specialized cells store fats and oils. Microscopically, a fat cell resembles a signet ring because the stored fat pushes the nucleus and the cytoplasm to one side of the cell (Figure 4-4).

Connective Tissue Fibers

The fibers characteristic of connective tissue are found within the intercellular matrix. There are three general types:

1. *Collagenous fibers.* These white fibers contain **collagen,** an albumin that is the principal structural protein of the body. This protein is synthesized by fibrocytes and is the major constituent of skin, ligaments, tendons, carti-

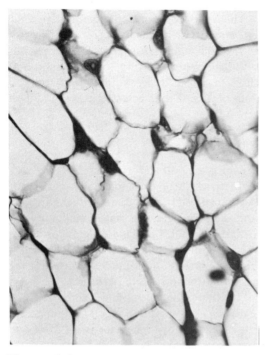

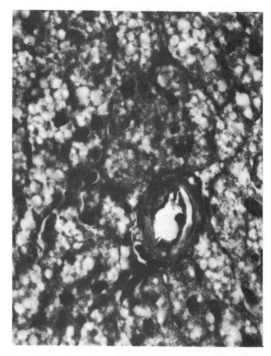

Figure 4-4 Photomicrographs of unilocular adipose tissue. (Left): Tissue in formative stage. (Right): Fully formed (mature) tissue. H and E stain, 320×.

(From *Basic Histology*, 4th Ed., by L. C. Junquiera and L. Carneiro. Copyright © 1983 by Lange Medical Publications. Reprinted by permission.)

lage, and bone. The fibers occur in bundles and are relatively nonelastic.

2. *Elastic fibers.* These fibers are yellow and may occur singly or in bundles. They are highly elastic and branch to unite with other fibers. Elastic fibers are both larger and straighter than collagenous fibers.

3. *Reticular fibers.* These short and very thin fibers branch freely, forming a cobweb network called a **reticulum.** These fibers are nonelastic and are usually found forming the internal framework of glands.

Types of Connective Tissue

Loose Connective (Areolar) Tissue

Loose connective tissue is made up of cells separated by a semifluid matrix. Within this matrix is an irregular network of white collagenous and yellow elastic fibers (Figure 4-3). The intercellular substance of areolar tissue allows the tissue to stretch without damage to the tissue structure. This tissue is the most abundant in the body. It is found beneath epithelial membranes and around blood vessels, nerves, and the ducts of glands. Areolar tissue is associated with maintaining the body's water balance, since it can absorb and hold a considerable amount of water. Because it contains many macrophages and white blood cells, this tissue also provides a second line of defense against bacteria.

Adipose Tissue

Areolar tissue beneath the epidermis is commonly filled with loosely packed fat cells held together in bundles by collagenous and elastic fibers (Figure 4-4). Adipose (fat) tissue is found primarily as a kind of padding around the joints, as soft pads between the organs, around the kidney and the heart, and in the yellow marrow of long bones. Its fat cells provide a reserve food supply and also insulate the body against heat loss.

Dense Connective Tissue

Many parts of the body require a stronger type of connective tissue to enclose, restrain, or separate functioning structures. Such dense tissue can be identified by its characteristic cells, which are concerned primarily with producing collagenous and elastic fibers. In the strong, tough fibrous tissue, there are many more collagenous than elastic fibers. These are arranged as cords or sheets, resulting in a tissue that is pliant but has little elasticity. Since fibrous tissue contains few blood vessels, it may take a long time to heal when injured or strained.

The basic forms of dense connective tissue are *tendons*, which attach muscle to bone; *ligaments*, which connect the bones that form joints; *aponeuroses*, which are thin, tendinous sheets attached to flat muscles; and *fasciae*, the thin sheets of tissue that cover muscles and hold them in place.

Cartilage

Cartilage is a modified form of connective tissue in which cartilage cells (chondrocytes) lie in small capsules, the **lacunae,** which are surrounded by an irregular matrix made up of fibers and/or a gel. Cartilage develops from the embryonic mesenchyme. Its adult form varies in the proportion and kinds of fibers, depending on the function it is to perform.

Hyaline cartilage, the most common form of cartilage, is a modified areolar tissue with a white or glossy appearance (Figure 4-5). It may contain any number of unevenly dispersed lacunae. Hyaline cartilage forms the skeleton in

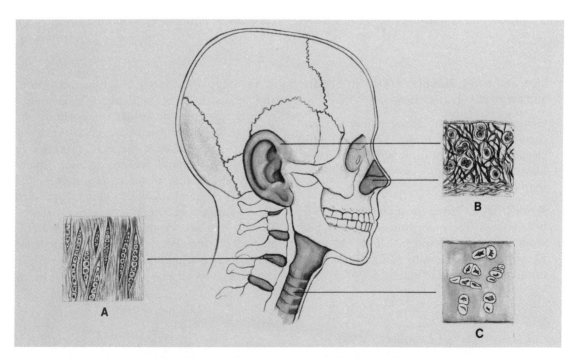

Figure 4-5 Schematic diagram showing the three types of cartilage. **A**, Fibrous, found in the intervertebral disks. **B**, Elastic, found in the ear and nose. **C**, Hyaline, found in the tracheal rings.

the embryo and covers the articulating surfaces of bones in the joint cavities. It also provides the structure for the nose and the connections of the ribs to the breast bone, and, in the respiratory tract, forms the ringlike trachea and bronchi.

Elastic cartilage contains many elastic fibers and is found wherever cartilage is required to move, such as in the epiglottis and the external ear (Figure 4-5).

Fibrous cartilage is less rigid than hyaline but contains heavy bundles of collagenous connective tissue (Figure 4-5). It is found in the intervertebral disks, which absorb shocks between the vertebrae of the backbone. Fibrous cartilage also reinforces the hyaline articular cartilages at the knee and hip.

Cartilage is not supplied with nerves and rarely contains blood vessels; therefore it has very poor healing properties. The perichondrium, a moderately vascular fibrous membrane, covers and nourishes cartilage, except in the articular regions.

THE INTEGUMENT

Most of the body's contacts with its environment occur through the skin, or **integument.** As discussed earlier, living cells must be surrounded by fluid. The surface of the skin in contact with air is made up of dead cells that form a protective covering for the inner, living cells.

Structure of Skin

The skin is made up of two distinct layers: the epidermis (cuticle) and the dermis (corium). These layers are illustrated in Figure 4-6. Beneath the skin is the subcutaneous tissue, which connects the skin to the underlying structures.

Epidermis

The tissue of the **epidermis** is stratified squamous epithelium. On its external surface there is a network of ridges caused by the arrangement of the tissues that lie beneath. The ridges

on the palmar surfaces of the fingers and hands and on the soles of the feet provide increased resistance to objects, making it easier to walk, to grasp objects, and to perform other motions involving contact with the environment. Each individual has distinct ridge patterns, such as fingerprints, and these can be used as a means of identification.

The epidermis is composed of four layers, or *strata:*

1. STRATUM GERMINATIVUM This is the deepest of the four strata and contains several layers of cells. One of its functions is to provide cells to the outer layers of the skin. As the cells of the stratum germinativum multiply and grow, cells previously formed are pushed upward toward the surface.

The stratum germinativum determines the thickness of the skin. It is the only layer of the epidermis that contains living cells capable of division, leading to the development of the outer protective layers. This layer also contains the greatest number of pigment cells, **melanocytes,** in the skin.

Melanin, a pigment that ranges in color from yellow to black, is found in all epidermal layers in dark-skinned people and in the deeper layers in light-skinned people. When the skin is exposed to ultraviolet radiation, the amount of melanin is increased and its color is darkened— a protective mechanism against radiation that causes tanning. Melanin is formed in the melanocytes by the action of an enzyme, tyrosinase, on an amino acid, tyrosine. When the enzyme is not present, the skin has no pigment *(albinism).* Skin blotches, such as freckles, are irregular patches of melanin.

2. STRATUM GRANULOSUM This is the layer of the epidermis directly above the stratum germinativum. Here the cells, as they progress upward, undergo a chemical degeneration of the nucleus. This leads to the death of the cell, which gives it a granular appearance microscopically. In this layer, the cells also change physically, becoming flattened or scalelike.

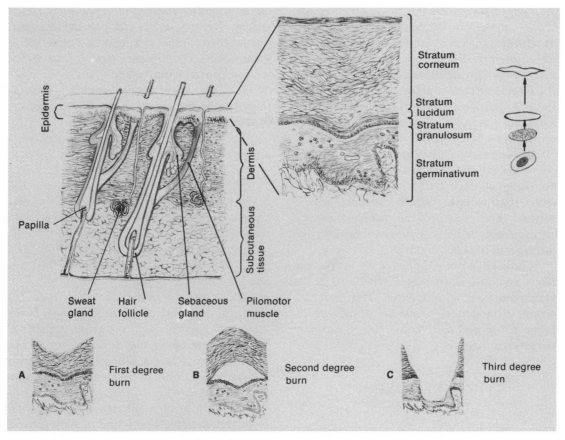

Figure 4-6 Diagram of a section of the skin to show its structure. The nucleated cell produced by the stratum germinativum dies (granulates) as it is forced outward to become the dead, scaly stratum corneum. The number of layers of the epidermis affected by the three types of skin burns is also shown. **A,** Only corneum cells are involved in first degree burns. **B,** Damage to the upper three layers occurs in second degree burns, forming a blister between layers 3 and 4. **C,** A third degree burn involves all epidermal layers and, therefore, usually requires a skin graft to replace the stratum germinativum.

3. STRATUM LUCIDUM In this next layer, the scalelike cells form a shield. They become hard, or keratinized. The process of keratinization of a cell involves the conversion of protoplasm into **keratin,** a hard, hornlike protein.

4. STRATUM CORNEUM This outermost horny layer of the epidermis consists of hard, scalelike cells that shield the body against the environment.

Natural sunlight acts directly on the cells of the skin and subcutaneous tissues to generate both pathological and protective responses. The chief protective response to ultraviolet light is tanning. The most familiar example of a pathological response is sunburn. In susceptible individuals exposed over many years, sunlight also causes a particular variety of skin cancer. Ultraviolet wavelengths in the narrow band from 290 to 320 nm cause the skin to redden within a few hours of exposure. Investigators generally agree that the inflammatory reaction, which may persist for several days, results either from a direct action of ultraviolet photons on small blood vessels or from the release of toxic compounds from damaged epidermal cells. The toxins presumably diffuse into the dermis, where they damage the capillaries and cause reddening **(erythema),** heat, swelling, and pain. A number of compounds have been proposed as the offending toxins, including serotonin, histamine, and bradykinin.

Sunburn is largely an affliction of industrial civilization. If people were to expose themselves to sunlight for 1 or 2 hours every day, their skin's reaction to the gradual increase in erythemal solar radiation that occurs during late winter and spring would provide them with a protective layer of pigmentation for withstanding ultraviolet radiation of summer intensities.

Immediately after exposure to sunlight the amount of pigment in the skin increases, and the skin remains darker for a few hours. The immediate darkening probably results from the photo-oxidation of a colorless melanin precursor and is evidently caused by all the wavelengths in sunlight. After a day or two, when the initial response to sunlight has subsided, melanocytes in the epidermis begin to divide and to increase their synthesis of melanin granules, which are then extruded and taken up into the adjacent keratinocytes, or skin cells. Concurrently, accelerated cell division thickens the ultraviolet absorbing layers of the epidermis. The skin remains tan for several weeks and offers considerable protection against further tissue damage by sunlight. Eventually the keratinocytes slough off and the tan slowly fades. (In the U.S.S.R., coal miners are given suberythemal doses of ultraviolet light daily on the theory that the radiation provides protection against the development of black-lung disease. The mechanism of the supposed protective effect is not known.)

Dermis

The **dermis,** or corium (true skin), is the skin's inner layer. It is highly sensitive and contains numerous blood and lymph vessels, nerves, glands, and the hair follicles. Elevations on its surface, called **papillae,** create the ridges that protrude above the epidermis. The papillae also contain a network of tactile corpuscles, or nerve endings, that are responsible for the sensation of touch (Figure 4-6).

The papillary and lower regions (reticulum) of the dermis also contain bundles of fibers that form an elastic connective tissue that can be stretched and extended. This property becomes evident when parts of the body become swollen, as in pregnancy. The skin stretches and becomes smooth and shiny. As an individual ages, the skin contains fewer elastic fibers and loses much of its extensibility, producing wrinkles.

Subcutaneous Tissue

The subcutaneous tissue is beneath the dermis. This tissue layer contains many fat cells, which provide insulation against the loss of body heat and also store food. The subcutaneous layer also contains blood and lymph vessels, nerves and nerve endings for the sensations of pressure and temperature, sweat glands, and hair follicles.

Blood Supply to the Skin

The arteries that supply blood to the skin form a network in the subcutaneous tissue and send branches to the papillae, hair follicles, and glands. The capillaries are so numerous that, when distended, they are capable of holding a large portion of all the blood contained in the body. When the skin is exposed to heat, or when the body temperature rises, the small arteries of the skin dilate, thus allowing more blood to flow through them **(vasodilation)** and permitting more heat to leave the body. On the other hand, when the skin is exposed to cold, the arteries constrict **(vasoconstriction)**; the blood flow through them reduces, thus preventing loss of body heat.

When blood vessels are damaged by disease or injury, there is a decrease or even a total loss of blood to an area of tissue, causing the death of the tissue cells. This cell death can result in open sores or ulcers. *Decubitus ulcers* (bedsores) occur when patients spend prolonged periods in one

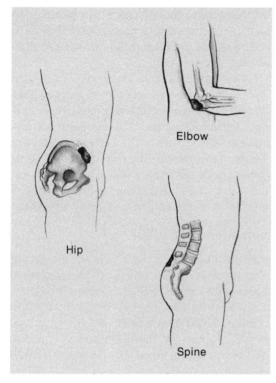

Figure 4-7 Areas of decubitus ulcer formation.

position (Figure 4-7). Patients requiring extended bed care may develop ulcers because the pressure of their body weight on the skin around a bony prominence (for example, the elbow, hip, or spine) cuts off the blood supply. Continual stimulation or manipulation of these areas is necessary to help return the blood supply to normal and to prevent further complications.

Skin Appendages

Sweat Glands

The sweat glands are simple tubules. Their lower ends lie in the deeper part of the dermis or in the subcutaneous tissue; their upper portions end in small openings in the skin's surface. The principal functions of the sweat glands are to help regulate internal body temperature and excrete a watery solution of so-

dium chloride, which contains traces of waste products such as urea and phosphates. Their pores (openings) are numerous and act as an emergency mechanism when it is necessary for the body to lose heat through evaporation or perspiration.

Sebaceous Glands

The sebaceous, or oil, glands are the most numerous of the skin appendages. These glands are found everywhere on the surface of the body except on the palm of the hand and the sole of the foot. They are generally associated with hair follicles, and their ducts usually open into the sides of the follicle. The glands themselves are located in the dermis. Their oily secretions, called **sebum,** lubricate the hair and horny layer of the skin, keeping them soft and pliable. Persons with underactive sebaceous glands develop dry hair and skin; those with overactive glands have oily hair and skin. When the ducts of these glands become blocked, infection follows—a common cause of blackheads **(comedome).** Too much sebum on the skin leads to seborrhea; the excess sebum may collect as dandruff. Acne is a chronic inflammation of the sebaceous glands.

Ceruminous Glands

The ceruminous glands are modified sweat glands found only in the skin of the passages leading into the ear (external canal). These glands secrete a waxy substance, **cerumen,** which helps to protect the eardrum and keeps the skin of the outer ear canal soft. Excessive amounts of cerumen can accumulate in the ear, impairing hearing or causing an irritating "cracking" sound when dry and loose.

Hair

Hair is another modification of the epidermis. Its **follicles,** deep indentations in the dermis, are formed by a downward growth of epidermal cells into the dermis, where germinal cells divide and are pushed upward until they eventually die and become keratinized. Each hair consists of a portion projecting above the skin's surface, the *shaft*, and a portion below the surface, the *root* (Figure 4-6). The root is embedded in the follicle.

A hair usually has a central core, the *medulla*, which is surrounded by an outer layer, the *cortex*. Hair color depends on the amount of melanin in the cortex and medulla. White hair has no pigment at all. The hair follicle is supported and controlled by an arrector muscle that, when stimulated, either straightens the follicle, causing the shaft to project upward, or pinches the dermis, resulting in "goose flesh." Flattening and bending of the hair follicle are related to curly hair; cylindrical, straight hair follicles are related to straight hair. Contrary to common belief, cutting or shaving hair does not stimulate its growth. Only stimulation of the epidermal cells in the hair follicle can cause hair to grow faster—or at all.

Nails

Nails are horny modifications of the corneal part of the epidermis. They protect the dorsal surfaces of the fingers and toes. The germinal layer beneath the nail is the nail bed. Nail growth occurs in the epithelium of the proximal portion of the nail bed. As a nail grows, it moves forward and becomes longer, but never loses its attachment to the proximal germinal layer. The crescent or "moon" at the base of the nail is called the *lunula*. It is generally overlapped by the soft cuticle.

Functions of the Skin

Protection

The skin acts as a barrier to prevent bacteria and other foreign organisms from invading the body. Harmless bacteria reside on the skin, but if a person's habits of cleanliness are poor, **pathogenic** (disease-causing) bacteria may invade the tissues, leading to infection.

Fluid Retention

The skin serves as a waterproof covering that prevents excessive water loss when the body is exposed to a dry environment. Perspiration and sebum form a film over the body, preventing dehydration and the absorption of chemical substances injurious to the tissues.

The importance of maintaining a normal water balance can be shown by a study of burns. Exposure to excessively high temperatures may produce local or general damage. Local damage can occur in one of three degrees (Figure 4-6):

- First degree burn—an erythema or slight reddening of the skin
- Second degree burn—blistering
- Third degree burn—destruction of the whole thickness of the skin

Burns result in a rapid loss of fluid from the underlying tissues and in a severe upset of the body's internal water balance. The resultant threat to life depends as much on the size of the area burned as the burn's severity. The "rule of nines" gives the relative distribution of the total body surface area: head and neck, 09%; anterior trunk, 18%; posterior trunk, 18%; upper extremity, 18%; lower extremity, 36%; and perineum, 01%. When a patient dies within a day or two after being burned, the death is due to shock (systemic collapse). Death occurring later is due to absorption of **toxins** (poisonous substances) from dead tissues.

Sensation

The importance of the skin's role in providing sensory awareness of the environment and its changes cannot be minimized. The nerve endings of the skin, which are responsive to touch, pain, pressure, and temperature changes, supply the brain and the nervous system with information that enables the body to adjust to its constantly changing surroundings.

Absorption

At one time, the skin was thought to be impermeable to all substances. Now, however, we know that gases (oxygen, carbon dioxide, nitrogen, and ammonia) readily penetrate the skin and travel through the sweat and sebaceous glands.

Electrolytes do not penetrate human skin in appreciable amounts; they are held back not by the stratum corneum (which acts as a gross sieve), but by an electrical double layer of H+ and OH⁻ ions that exist at the junction between the cornified and noncornified layers of the skin.

Some of the substances capable of penetrating the skin are of great practical importance. The fat-soluble vitamins A, D, and K, as well as carotenes, are readily absorbed through skin. So are various steroid hormones, including estrogens, progesterone, testosterone, and deoxycorticosterone. Various phenolic compounds are also readily absorbed through the skin. One of them, methylsalicylate (oil of wintergreen), is a common ingredient of liniment. When rubbed onto the surface of the skin, it rapidly penetrates to the sore muscles on which it acts.

Salts of lead, tin, copper, arsenic, bismuth, antimony, and mercury can penetrate through skin and in some cases may cause accidental poisoning. The salts themselves are not lipid soluble, but become so when they combine with fatty acids of the sebum at the skin surface.

Gases such as oxygen, carbon dioxide, hydrogen sulfide, nitrogen, ammonia, and hydrogen cyanide are also readily absorbed through the skin. (Fortunately, carbon monoxide is an exception and does not penetrate the skin.) There is some evidence that the passage of oxygen through the skin provides for a small part of the human body's oxygen needs, but the amounts of oxygen and carbon dioxide that can be transmitted through the skin are far below those required for human respiration.

Vitamin D Production

In the presence of sunlight or ultraviolet radiation, one of the sterols (7-dehydrocholesterol) that is found within the skin is changed to vitamin D_3 (cholecalciferol). Vitamin D_3 assists in the absorption of calcium and phosphate from ingested food and reduces the amount of phos-

phate excreted by the kidneys. In this way, vitamin D_3 is important in maintaining the calcium and phosphate levels of the body.

Excretion

In addition to perspiration's fluid retention and cooling effects, small amounts of nitrogenous waste products (urea) and sodium chloride leave the body via perspiration. The quantity and content of the excretions are regulated by the changing needs of the body.

One Square Inch of Skin

In the following description, the complexity of the skin is evident:

Human skin is not only skin deep. In fact, it is among the body's most complex organs. Of its three main layers, only the paper-thin epidermis is normally visible. Beneath the epidermis is the dermis, and below that is the subdermis. In a square inch of skin, you will find: 20 blood vessels; 65 hairs and muscles; 78 nerves; 78 sensors for heat, 13 for cold, 160–165 for pressure; 100 sebaceous glands; 650 sweat glands; 1300 nerve endings; and 19,500,000 cells.

The sweat glands do double duty, helping to eliminate wastes and cool the body. On a hot day, the skin can release up to 2500 calories of heat—enough to boil 6 gallons of water.

The body's largest organ, the skin measures about 21 square feet in an average adult. It accounts for 15 percent of total body weight and provides a protective shield against bacteria and viruses. It also absorbs shocks that might otherwise damage the bones and internal organs.*

CLINICAL IMPLICATIONS

Skin Grafts

Tissue transplantations involving skin, certain organs, and dismembered limbs have become increasingly successful in recent years owing to continued research in the fields of hematology

*"Significa" from *Parade* Magazine, October 1982, by Irving Wallace, David Wallechinsky, Amy Wallace.

(particularly in antigens) and plastics. Kidney and blood vessel transplants are now relatively common, although not always successful; severed arms and legs have been reconnected to the bodies of accident victims, again with limited success. Skin grafts have proved invaluable. Sometimes the skin is obtained from the individual who receives the graft (*autotransplant*); sometimes from another person (*homotransplant*); or from an animal (*heterotransplant*).

A method routinely used to determine the success or failure of a skin transplant or of the reimplantation of a limb is the Lewis triple reaction:

1. *Red line.* When the skin is stroked with a dull instrument, the normal capillary response shows a red line.

2. *Red flare.* A more forceful stroke demands a deeper response involving some of the connective tissue. The normal reaction to such a stimulus is a widened red area, or **flare.**

3. *Red wheal.* Continued forceful stroking of the skin area involved will induce a response of the vascular and nervous systems. This response results in a raised welt or **wheal,** rounded elevations with red edges and pale centers, appearing temporarily.

When there is a negative response to any of these tests, it is likely that the transplant did not "take" and that it will eventually die.

Research on skin grafting with artificial skin is being pursued in efforts to make skin grafting techniques more effective. One synthetic skin is made from collagen fibers—obtained from cowhide—and a major polysaccharide of cartilage obtained from sharks. This compound is covered with a sheet of medical-grade silicone rubber that serves as a barrier to infection and fluid loss and provides mechanical strength when the graft is sutured into place.

After the graft is in place, mesodermal cells begin to migrate into it and to produce more collagen, synthesizing a new dermal layer, or *neodermis.* Epidermal cells also begin to grow

inward from the edge of the graft. The synthetic skin, meanwhile, is slowly broken down. After 20 days, the silicone layer is removed, and epidermis from elsewhere on the patient's body is transplanted in its place. This procedure is much less traumatic than an autograft because no dermis is moved. New epidermis grows back over the source area in 7 to 10 days. Eventually, patients are able to distinguish heat and cold and experience pain. The main difference between the new skin and the rest of the patient's skin is that it does not have hair follicles or sweat glands.

In another recently developed technique, skin for transplants is first minced and grown on a base of pigskin, then grafted (pigskin and all) onto the injured site. The pigskin dries up and sloughs off, while living skin covers the damaged area.

Wound Healing

The first step in wound healing follows the initial trauma immediately. Bleeding into the wound results in the formation of a clot, which plugs the gap and provides a temporary mechanical protection. If the injury involves only the skin, it is rapidly repaired by the multiplication of cells in the deep layer of stratified squamous epithelium. A small defect may be entirely covered within 48 hours. In deeper and more extensive wounds involving underlying tissues, fibroblasts from the connective tissue begin multiplying within hours and migrate from the edges toward the central portion of the wound (Figure 4-8). Vascular buds develop and grow across the wound, forming new capillaries. These, together with the growing connective tissue, form granulation tissue, which gradually builds up on the wound surface. The granulation tissue has a characteristic pebbly appearance. As the fibroblasts proliferate and produce collagenous fibers, the wound is entirely filled with scar tissue.

Maintaining good general health is important for promoting rapid wound healing. In the case of extensive injuries, particularly burns,

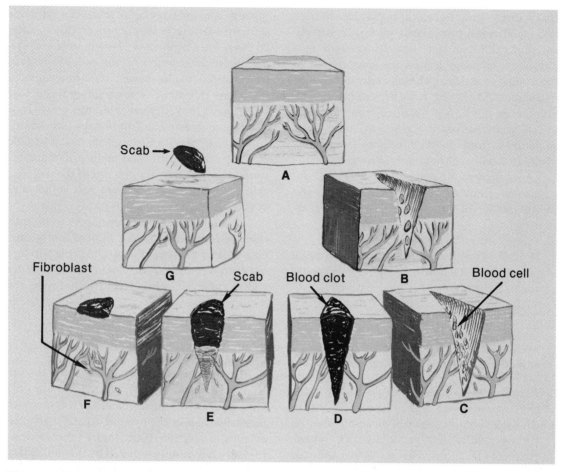

Figure 4-8 If normal skin (**A**) is injured deeply (**B**), blood escapes from dermal blood vessels (**C**), and a blood clot soon forms (**D**). The blood clot and dried tissue fluid form a scab (**E**), which protects the damaged region. Later, blood vessels send out branches, and fibroblasts migrate into the area (**F**). The fibroblasts produce new connective tissue fibers, and when the skin is largely repaired, the scab sloughs off (**G**).

considerable blood plasma may be lost, along with its proteins and electrolytes. This upsets the body's fluid balance, and the balance must be restored. Proteins are needed in ample amounts for building new tissues. Vitamins play a key role, both in healing and in preventing infections.

Tumors

Normal cells of the organs and tissues of the body are controlled strictly by the laws of growth. The liver, for example, or a bone increases in size not through growth of the individual cells of which it is composed but by a process of continual division and multiplication of these cells. Abnormal conditions may develop, however, which cause masses of cells to grow and proliferate without control, leading to the development of a **tumor,** or **neoplasm** (Figure 4-9*A*). This lack of control is particularly marked in malignant tumors **(cancer).** Thus cancer is essentially a disorder of cell growth and reproduction, and the cells of the malignant tumor are modified normal cells. These modifications are accompanied by loss of the cells' specialized functions, increase in size, and uncontrolled growth into neighboring tissues **(metastasis).** We know that cellular growth is a chemical process, and it would appear that the abnormal cellular growth must reflect some basic change in this process. That is why most cancer research is being done on a cellular-tissue level and why many cancer patients are being treated by chemotherapy.

Tumors may be subdivided into two broad categories: **benign,** nonpathogenic growths; and **malignant,** pathogenic growths that, if untreated, eventually will kill the patient, regardless of their location.

Benign tumors found in epithelial and connective tissues include:

1. *Papilloma.* This is an epithelial tumor found in the skin, mouth, and urinary bladder.

2. *Adenoma.* This epithelial tumor resembles a gland in its structure. It is found near the breast and near the thyroid gland.

3. *Fibroma.* This is a tumor of fibrous tissue. It may be found anywhere in the body, but most often is directly under the skin.

4. *Lipoma.* This soft, fatty tumor is found in the neck and on the shoulders, back, and buttocks, where fat deposits are abundant.

5. *Chondroma.* This tumor affects cartilage, and is found primarily at the ends of bones where some cartilage still persists.

Malignant tumors may be classified into two broad groups: carcinoma and sarcoma. (Both are included under the common name cancer.) **Carcinoma,** a tumor of epithelial tissue,

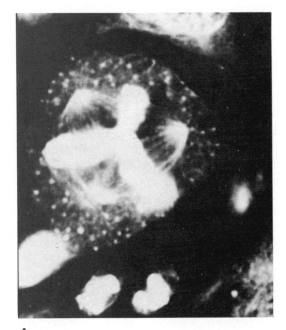

A

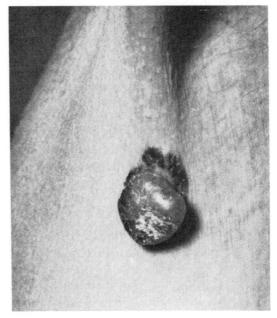

B

Figure 4-9 **A,** Abnormal tripolar mitosis occurring in malignant melanoma. (From *Pathology* by W. A. Anderson. St. Louis: C. V. Mosby, 1966. Reprinted by permission.) **B,** Malignant melanomas on posterior axillary fold. (From *Manual of Skin Diseases*, 4th Ed., by G. Sauer. Philadelphia: J. B. Lippincott, 1980. Reprinted by permission.)

is the most common, and its malignant cells are usually spread through the lymphatic system. These tumors may develop in the skin or in glandular organs such as the breast or the liver.

A **sarcoma** is a malignant tumor of fibrous connective tissue and can be found in any part of the body. The principal difference between sarcomas and carcinomas is that sarcomas tend to be localized and do not spread throughout the body. One common form of sarcoma is the *melanoma,* a malignant tumor originating in the melanocytes (Figure 4-9B).

The final diagnosis in suspected cancer—the decision whether the neoplasm is benign or malignant—must be made through laboratory studies of the tumor tissues and cells. When a tumor is removed by a surgical procedure, its type is determined by examination in the pathology laboratory. In other instances, a small sample of the diseased tissue is removed and studied; this process is known as **biopsy.** Rational treatment may then be recommended by the physician or surgeon.

Skin Diseases

The skin can be affected by diseases in varied ways. Skin infections are caused by common skin disorders or **dermatoses,** including contact reactions, reactions due to allergies and metabolic insufficiencies, and drug-related problems. Skin infections are also caused by viruses, bacteria, fungi, and parasites. Neoplasia or uncontrolled growth of certain cells can result in the formation of tumors of the skin.

Common Dermatoses

Inflammation of the skin is called **dermatitis.** *Contact dermatitis* is an acute or chronic inflammation that results from direct skin contact with chemicals or other irritants (such as poison ivy) (Figure 4-10). People with this condition have previously been sensitized to the antigen and may not produce a skin inflammation.

Eczema (atopic dermatitis) is a chronic noncontagious superficial inflammation of the skin

that manifests itself by redness (erythema), blisters **(vesicles),** and papular lesions (Figure 4-11). Eczema, an allergy, is caused by contact to a variety of chemicals, detergents, soaps, and certain foods. Psychological and emotional disturbances may aggravate an attack of eczema just as they bring on bouts of asthma and other allergic responses.

Urticaria (hives) is an inflammatory skin reaction to an allergen. The lesions are wheals that develop most often at pressure points (such as under tight clothing), but they may appear anywhere on the skin or mucous membranes. The lesions are extremely pruritic (itchy). Common causes are foods (shellfish, port, strawberries, wheat, eggs, milk, and chocolate), drugs (antibiotics, salicylates, metals, and vaccines), molds, insect bites, and emotional disturbances. The allergic response causes most cells to release histamine, which, in turn, causes vasodilation and increased cell membrane permeability, resulting in edema and itching due to the irritation.

Acne (vulgaris) is an inflammatory disease occurring in areas where the sebaceous glands are the largest, most numerous, and most active. It is most frequently seen during puberty, resulting from the hormonal changes that occur then, and affects both boys and girls equally. The increased levels of estrogen and testosterone stimulate not only growth but glandular activity. The sebaceous glands increase their secretions of sebum, which is released through hair follicles. The primary lesion is the comedone, or blackhead (exposure to air causes the black color). It may leave only a few dilated pores or superficial pustules, but prompt treatment prevents any scarring from occurring (Figure 4-12). Blackheads should not be squeezed or picked because broken skin offers a portal of entry to bacteria that are always present on the skin surface. Once **pyogenic** (pus-producing) bacteria enter the skin, pus forms, resulting in a pimple or pustule. Squeezing the pimples spreads the infection. In persistent cases, bacterial and chemical by-products can cause irritation and destruction of epider-

2. Cilia
3. Both goblet cells and cilia are found in columnar and pseudostratified epithelial tissues

D. Glands
1. Endocrine glands
   a. Internal secretion
   b. Ductless
   c. Example—thyroid
2. Exocrine glands
   a. Ducts for external secretion
   b. Example—salivary glands

III. Connective tissues
A. Function—support body parts and bind them together
B. Structure
1. Few cells, widely separated
2. Intercellular matrix
3. Fat cells—storage
C. Connective tissue cells
1. Fibrocytes—mesenchyme
2. Histiocytes—phagocytosis
3. Plasma cells—antibody production
4. Mast cells—heparin production
5. Blood cells—phagocytosis
6. Fat cells—storage of oils
D. Connective tissue fibers
1. Collagenous—white fibers, non-elastic
2. Elastic—yellow, highly elastic
3. Reticular—short and thin
E. Types of connective tissue
1. Loose (areolar)
   a. Most abundant in body
   b. Supports and maintains body's water balance
2. Adipose—passing around joints
   a. Fat storage bundles
   b. Reserve food supply
3. Dense connective tissue
   a. More collagenous fibers than elastic fibers
   b. Examples—tendons, ligaments, aponeuroses, fasciae
4. Cartilage
   a. Hyaline

   (1) Modified areolar tissue with glossy appearance
   (2) Forms skeleton of embryo, covers bones in joint cavities, forms nose and rings in respiratory tract
   b. Elastic
   (1) Many elastic fibers
   (2) Forms epiglottis and external ear
   c. Fibrous
   (1) Many collagenous fibers
   (2) Forms intervertebral disks

IV. Skin (integument)
A. Structure
1. Epidermis—stratified squamous epithelium
   a. Stratum germinativum—deepest of four epidermal layers
   (1) Produces the skin cells
   (2) Contains melanocytes that produce melanin (skin pigment)
   b. Stratum granulosum—layer above stratum germinativum
   (1) Cells undergo chemical degeneration, forming granules
   (2) Cells appear flattened and are dead
   c. Stratum lucidum—second layer from the outside
   (1) Made up of hard (keratinized) dead cells to form a shield for protection
   d. Stratum corneum—outermost horny layer of dead cells
2. Mechanism of sun-tanning
3. Dermis—true skin
   a. Inner layer of tissues
   b. Blood and lymph vessels
   c. Nerve endings—sense of touch
   d. Hair follicles
   e. Elastic connective tissue
4. Subcutaneous layer
   a. Fat cells

b. Deep nerve receptors for sensations of pressure and temperature
c. Sweat glands
d. Hair follicles

B. Blood supply
1. Network of capillaries
2. Vasodilation—increased blood supply and heat loss
3. Vasoconstriction—decreased blood supply and heat loss
4. Blood deficiency to cells—ulceration

C. Skin appendages
1. Sweat glands
    a. Tubules
    b. Regulate body temperature
2. Sebaceous glands
    a. Oil—sebum
    b. Lubricate skin and hair
    c. Chronic inflammation—acne
    d. Too much sebum—seborrhea
    e. Blockage of sebaceous glands—blackheads
3. Ceruminous glands
    a. Found only in ear canal
    b. Secrete cerumen
    c. Cerumen softens eardrum
4. Hair
    a. Dermal outgrowth
    b. Color depends on quantity of melanin in cortex and medulla of shaft
    c. Supported by arrector muscles
5. Nails—horny modification of epidermis

D. Functions of skin
1. Protection
2. Fluid retention
    a. Skin burns—first, second, and third degree
    b. "Rule of nines"
3. Sensation
4. Absorption
5. Vitamin D production
6. Excretion

V. Clinical implications
A. Tissue transplants
1. Skin graft
    a. Lewis skin reaction—red line, red flare, red wheal
2. Synthetic skin
B. Wound healing
C. Tumors
1. Benign
    a. Nonpathogenic growth
    b. Examples—papilloma, adenoma, fibroma, lipoma, and chondroma
2. Malignant
    a. Pathogenic
    b. Carcinoma—cells spread through lymphatic system
    c. Sarcoma—fibrous connective tissue growth; localized
3. Biopsy—diagnostic tool
D. Skin diseases
1. Common dermatoses
    a. Contact dermatitis
    b. Eczema
    c. Urticaria
    d. Acne
    e. Seborrhea
    f. Psoriasis
2. Viral skin infections
    a. Herpes simplex
    b. Herpes zoster
    c. Verruca vulgaris
3. Bacterial skin infections
    a. Impetigo
    b. Folliculitis

STUDY QUESTIONS AND PROBLEMS

1. Define a tissue. What are the four basic types of tissues?
2. What is epithelium? How is it different from other tissues?
3. If columnar epithelium is found in the intestine, what is its function?
4. Why do we have cilia and goblet cells in the respiratory tract?
5. Differentiate structurally between an endocrine and an exocrine gland. Give examples.
6. What is the principal function of connective tissue?
7. List at least five different types of cells found in connective tissue, and describe their functions.
8. What is collagen? What is its function?
9. How does one form of cartilage differ from another?
10. Locate various types of cartilage in the body.
11. What is adipose tissue, and what is its function?
12. Discuss the structure and functions of skin.
13. Where does hair grow from? Why do some people have white hair?
14. What kinds of substances can penetrate unbroken skin? Which cannot?
15. Discuss the role of vitamin D in relation to calcium and phosphate.
16. Differentiate among the three degrees of skin burn.
17. What three tests compose the Lewis triple reaction? For what is the reaction used?
18. Why does a skin graft have to include at least part of the stratum germinativum?
19. List the steps in wound healing.
20. Name the two general categories of tumors.
21. Name three examples of benign tumors, and tell where each is found.
22. What is the major difference between a malignant and a benign tumor?
23. How does carcinoma generally spread?
24. Define dermatitis, and give at least three examples of common dermatoses.
25. What is the difference between cold sores and shingles?

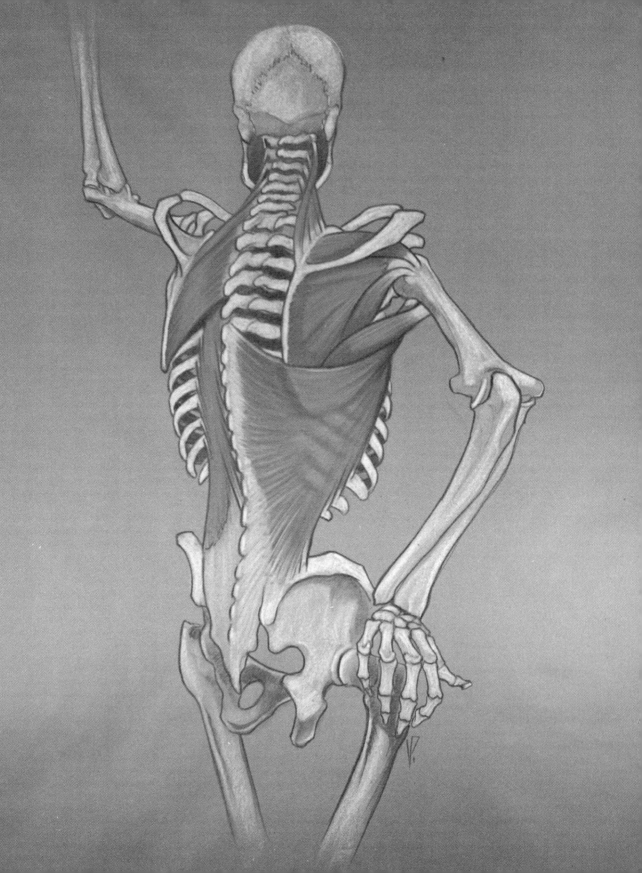

The Skeletal System

LEARNING OBJECTIVES

- List the functions of the skeletal system

- Describe the microscopic structure of bone

- Identify the gross anatomy of a bone

- Compare how membranous and cartilaginous bone develop

- Explain the function of vitamin D, parathormone, alkaline phosphatase, and calcitonin in relation to bone growth

- Distinguish between the axial skeleton and the appendicular skeleton, and name the components of each

- Identify the bones that make up the pectoral and pelvic girdle

- Draw and label a vertebra, and explain the differences in structure between the various types of vertebrae

- Explain why the atlas and axis are constructed the way they are

- Differentiate between fontanel and suture

- List and define the types of fractures

- Describe how a bone fracture heals

- Differentiate between a benign and malignant bone tumor

- Explain the process of performing an internal fixation, and indicate why it is necessary

IMPORTANT TERMS

appendicular	diaphysis	fracture	ossification	sinus
skeleton	eminence	hematopoietic	osteoblast	spine
atlas	endochondral	kyphosis	osteoclast	suture
axial skeleton	endoskeleton	lamina	osteocyte	thoracic
axis	epicondyle	lordosis	osteogenic	thorax
calcification	epiphysis	lumbar	perichondrium	trabeculae
callus	exoskeleton	malleolus	periosteum	tubercle
canal	facet	medullary	process	tuberosity
cervical	fontanel	meninges	ramus	tumor
condyle	foramen	myeloma	scoliosis	vertebra
crest	fossa	osseous	sesamoid	

The skeleton, the bony supporting structure of the body, is made up of 206 bones of many different shapes and sizes (Figure 5-1). In humans and other vertebrates, the skeleton is covered by soft tissues, such as muscle and fat, and therefore is called an **endoskeleton.** Many invertebrate animals (for example, lobsters, crayfish, and clams) are protected by another type of skeleton—an **exoskeleton,** which, as the name suggests, is on the outside of the soft tissues.

The endoskeleton has five principal functions:

1. *Protection.* The hard bony tissue provides internal soft tissues with a defense against possible trauma. The skull, for example, protects the brain; the vertebral column protects the spinal cord. The rib cage and the bones of the pelvis protect the organs contained in the thorax and the lower part of the abdomen.

2. *Support.* Most of the body's structures are soft, pliable, and delicate. The skeleton's rigid framework provides a means of suspension for body systems.

3. *Muscle attachment.* Muscles are attached to bone by tendons. The rigidity of bone, combined with the elasticity and contractility of muscle, makes the arms and legs strong levers, tools, or weapons that can be used for work, locomotion, or defense.

4. *Reservoir for minerals.* The **osseous** (bony) tissues of all parts of the skeletal system provide the body with a vast depot of mineral salts such as calcium and phosphates.

5. *Hematopoiesis.* During late fetal life and continuing in some bones through adult life, the red marrow of long bones, the ribs, the sternum, and the pelvic bones produces blood cells—a process called **hematopoiesis.**

It is important to remember that the skeleton is composed of living tissues. Its materials are constantly being dissolved and passed into the bloodstream. They are, in turn, constantly replenished by other tissues and from food.

FORMATION OF BONE

Osseous tissue develops from mesenchyme, embryonic connective tissue that is characterized by branching cells in a fluid matrix (see Chapter 4). The development of bone involves the transformation of mesenchyme into fibroblasts (embryonic fibrils) and collagenous and elastic fibers. This tissue may develop into any one of

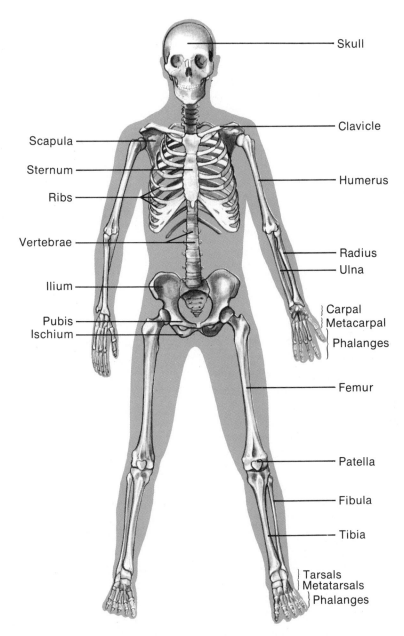

Skull

Clavicle

Scapula

Sternum

Ribs

Vertebrae

Ilium

Pubis

Ischium

Humerus

Radius

Ulna

Carpal

Metacarpal

Phalanges

Femur

Patella

Fibula

Tibia

Tarsals

Metatarsals

Phalanges

Figure 5-1 Human skeleton, anterior aspect.

the several types of mature connective tissue. However, when blood vessels are introduced, their rich contents stimulate mitosis; the cells increase in size, change from an irregular to a cuboidal shape, and then may be classified as bone-forming cells, or **osteoblasts** (Figure 5-2). The osteoblast deposits a mucoprotein substance that incorporates the fibers, giving the resultant osteoid tissue a rubbery, elastic consistency.

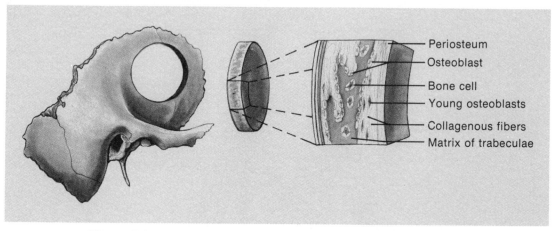

Figure 5-2 Developing intermembranous bone in the skull of the fetal cat.

Any cell that has the property of being able to lay down a matrix that calcifies without dying in the process is an osteoblast—a "bone builder." The precise origin of bone-forming cells is still controversial; their function, **ossification,** is much better understood.

Bone formation begins early in embryonic development. It occurs by two methods: intramembranous ossification and cartilaginous (or endochondral) ossification.

Intramembranous Ossification

The bones of the skull are formed by intramembranous ossification. In the embryo, one of the layers covering the developing brain is fibrous and is composed of fibroblasts and collagenous fibers. In the areas where bone is to be formed, the fibroblasts differentiate into osteoblasts, which then lay down an intercellular substance for the deposition of calcium. From this matrix, ossification progresses in radiating columns of calcified collagen fibers, between which the osteoblasts construct bone (Figure 5-2). Initially, there are several sites of ossification in membranous bone. Eventually, most of these sites will fuse together to form the main bone.

The ossification of the skull is not complete at birth, but continues through early childhood until the structure has developed into a strong bony case to protect the brain. Before ossification, the "soft spots" are called **fontanels.** They will be discussed in a later section.

Cartilaginous (Endochondral) Ossification

Most of the skeleton is formed by cartilaginous ossification, which involves a gradual replacement of hyaline cartilage by osteoblasts. As discussed in Chapter 4, cartilage is a specialized type of fibrous connective tissue that forms most of the temporary skeleton of the embryo. A skeleton of hyaline cartilage, a flexible semi-transparent substance, is fully formed by the second month of embryonic life. It is enveloped in a layer of dense fibrous connective tissue called the **perichondrium.** However, in the third fetal month bone begins to form directly on the shaft of the cartilaginous "model." The perichondrial layer develops into a fibrous connective tissue membrane covering the bone called the **periosteum** (Figures 5-2, 5-4, and 5-8). **Calcification,** the hardening of the skeletal tissue by calcium deposits, begins at this time.

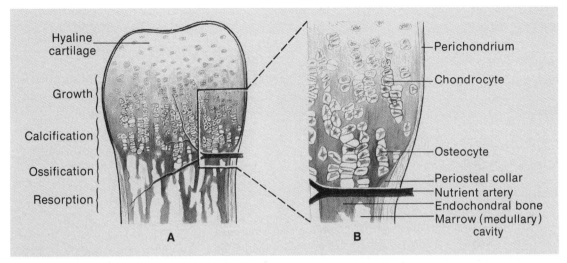

Figure 5-3 Endochondral ossification in longitudinal sections through the zone of epiphyseal growth of a fetal bone. **A**, Diagram of the zones of development, from unmodified cartilage to resorption by osteoclasts. **B**, High-power section of the zone of calcification, illustrating the conversion of a soft hyaline cartilage into hard bone through the replacement of chondrocytes by osteocytes and solidifcation of the extracellular matrix.

As the process of calcification proceeds, the cartilage cells swell and degenerate, ultimately leaving a linear arrangement of empty capsules (lacunae). Blood vessels then penetrate the periosteum, bringing osteoblasts to lay down a matrix of collagen and mucoprotein, two types of protein that serve as supporting substances. Through this process, the osteoblasts mature into **osteocytes** (Figure 5-3).

The growth of a bone continues as it becomes thicker at the outer periosteum and longer as more cartilage is replaced by osteocytes. These changes make it necessary for the cavity within the bone to increase in size, and bone material must be removed for this to occur. **Osteoclasts,** large cells that can dissolve bone, are responsible for this process, enlarging the bone's internal cavity while the osteoblasts continue to deposit new bone material outside. Thus the depth of the bone remains constant throughout most of a person's normal life while the bone is growing. Continued osteoclast activity with simultaneous gradual inhibition of osteoblasts occurs in some diseases and in old age.

This leads to a reduction in the depth of the bone, and, eventually, results in brittleness and fractures.

Within the spaces of bony structures are areas where blood cells reproduce, mature, and enter the circulation. These areas are commonly called the *bone marrow,* or *medullary cavities* (Figures 5-3 and 5-8).

In long bones (for example, those of the arms and legs) ossification proceeds in two directions. It begins at the shaft of the bone (**diaphysis**) and moves toward the head of the bone (**epiphysis**). At the same time, ossification occurs in the opposite direction, from the epiphysis toward the diaphysis (Figure 5-4). The epiphyseal cartilage persists until early adulthood, and continued growth of this layer by cell division provides for the lengthwise growth of the bone. Although skeletal growth begins early in fetal life, it is not complete until the person's eighteenth to twentieth year. When growth is complete, the epiphyseal cartilage cells cease to multiply (Figure 5-5) and are transformed into bone tissue.

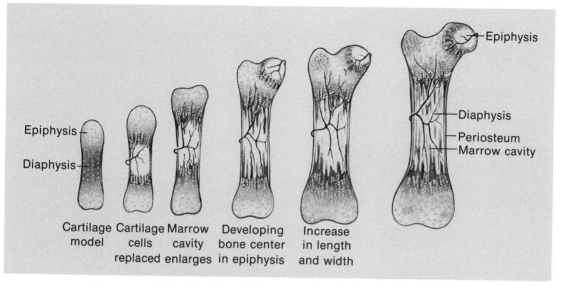

Figure 5-4 Diagrammatic representation of the development of a long bone.

Chemical Influences on Bone Formation

The final stage in the formation of bone occurs when the osteoblasts draw upon mineral ions that diffuse from the blood vessels into the *osteoid,* the intercellular matrix of young bone.

Other chemical processes are necessary to the development of bone. Vitamin D and alkaline phosphatase are two important compounds necessary for the absorption and availability of the mineral substances essential for the deposition of bone. *Vitamin D* is thought to be the chief agent in the absorption of minerals through the intestines. Insufficient quantities of the vitamin may lead to loss of calcium in the feces. *Alkaline phosphatase,* an enzyme found in many tissues, breaks down calcium salts and frees the all-important calcium and phosphate ions necessary for the hardening of bone.

The third important regulatory substance needed for normal bone deposition and growth is *parathormone,* a hormone produced by the parathyroid glands. Parathormone is responsible for regulating the calcium and phosphorus

levels of the blood, thereby maintaining homeostasis for bone growth.

Calcitonin (thyrocalcitonin), produced by the thyroid gland, has a direct stimulatory effect on bone formation and, indirectly, causes a decrease in blood calcium. Initially, it suppresses bone resorption by osteoclasts, preventing large amounts of calcium from returning to the blood. Next, it stimulates bone formation by osteoblasts and osteocytes, enhancing the uptake of the excess calcium from the blood. Calcitonin has its greatest effect in children, but probably plays only a minor role in the adult. To date, there have been no disorders resulting from calcitonin overproduction or deficiency.

Calcitonin is used in children and adults to treat Paget's disease, a disorder characterized by

Figure 5-5 Comparative sketches of baby and adult skeletons. **A,** The relation of size and shape of individual bones to total body size is comparable in the baby and the adult. **B,** The initial time of development and final growth period of some of the bones of the human skeleton.

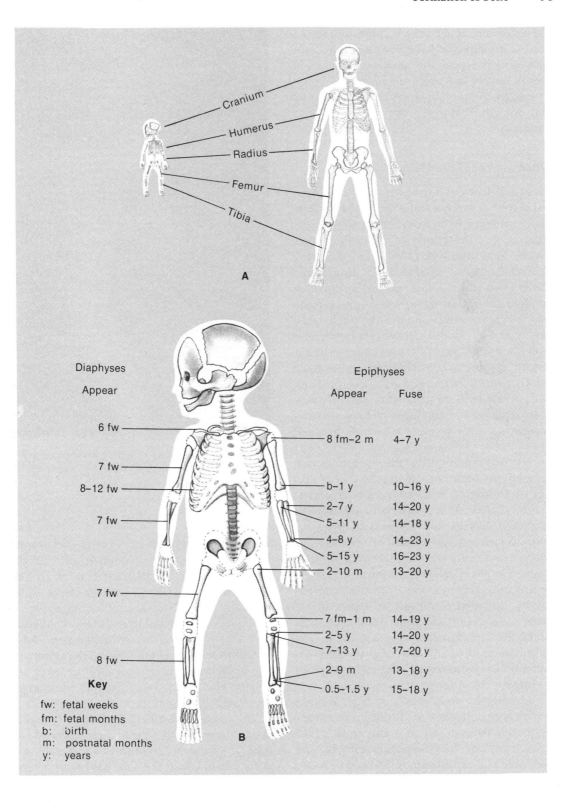

Cranium

Humerus

Radius

Femur

Tibia

A

Diaphyses

Appear

Epiphyses

Appear | Fuse

6 fw

7 fw

8–12 fw

7 fw

7 fw

8 fw

8 fm–2 m   4–7 y

b–1 y   10–16 y

2–7 y   14–20 y
5–11 y   14–18 y
4–8 y   14–23 y
5–15 y   16–23 y
2–10 m   13–20 y

7 fm–1 m   14–19 y
2–5 y   14–20 y
7–13 y   17–20 y

2–9 m   13–18 y
0.5–1.5 y   15–18 y

Key

fw: fetal weeks
fm: fetal months
b: birth
m: postnatal months
y: years

B

abnormal and rapid bone resorption and formation. Calcitonin blocks the bone resorption and thus also prevents the abnormal formation. Paget's disease can affect one or several bones. Untreated, the affected bones become deformed and so soft that they can be cut with a butterknife.

Disorders of Bone Development

An insufficiency of any of the chemical factors just described will generally result in a weakening or softening of bone. When the deficiencies occur during the period of ossification, they lead to the disease state called *rickets* (Figure 5-6). A similar deficiency state developing later in life is *osteomalacia*. Other metabolic disorders can lead to skeletal deformities. For example, hypofunction of the anterior lobe of the pituitary gland in its production of growth hormone (somatotropin, STH) may cause *dwarfism*, or stunted growth. Hyperfunction of the gland before maturity may cause *gigantism*, or excessive growth; hyperfunction after maturity may cause *acromegaly*, a condition characterized by overgrowth of the bones of the hands, feet, and face.

Hyperparathyroidism, the overproduction of parathyroid hormone, is usually due to a tumor of the parathyroid glands. It is characterized by *osteoporosis*, a condition in which bone resorption exceeds bone formation. Bone becomes very fragile. As a result of the increased osteoclastic activity and the inhibition of osteoblasts, the blood calcium becomes extremely high, causing kidney stones, depressed nervous activity, and irregular heart rhythms. In extreme cases, death will come from metastatic solidifying (precipitation) of calcium in the body tissues, especially the lungs. Hyperparathyroidism is corrected by surgical removal of the tumor.

Abnormal and uncontrolled cellular growth is characteristic of cancer. Primary cancer of the skeletal system arises from either bone or cartilage cells. Secondary cancers arise elsewhere and metastasize to the skeleton. For

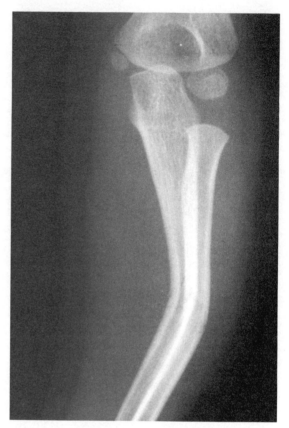

Figure 5-6 Deformity of a long bone produced by severe untreated rickets. (Courtesy of Salem Hospital.)

example, cancer of the breast often metastasizes to bone. The two most common forms of primary cancer are chondrosarcoma and osteogenic sarcoma. *Chondrosarcoma* is a slow-growing cancer of skeletal cartilage that is found mostly in males over the age of 40 years. It can be completely cured by surgery. *Osteogenic sarcoma* is most often found in persons under 40 years of both sexes. It is fast growing and usually fatal. Only 5% of patients who have it are alive 5 years after the initial diagnosis.

Osteomyelitis is an inflammation of the bone and marrow, usually due to staphylococcal bacteria. The bacteria enter through a wound, especially one in which the bone has been fractured and the ends have punctured the skin. The rich blood supply of bone provides an ideal environment for bacterial growth. This condition is easily treated with antibiotics, but is fatal if left untreated.

THE STRUCTURE OF BONE

Microscopic Structure

Collagen and minerals combine to form a homogeneous material that can be seen as an open meshwork of plates and bars in spongy, or *cancellous bone,* and as several relatively uniform concentric layers around each of the smallest blood vessels in dense, *compact bone.*

The basic unit of structure of compact bone is the osteon. It consists of a haversian canal and from four to twenty *lamellae,* which are bony plates arranged concentrically around the canal (Figure 5-7). The **canals** serve as passageways for nutrient materials to reach the living osteo-

cytes, which are located in lacunae, spaces within the lamellae. In addition to the haversian canals, compact bone also contains Volkmann's canals, through which blood vessels course (Figures 5-7 and 5-8). Volkmann's canals cross compact bone horizontally, maintaining a blood supply between the periosteum and the haversian canals. Delicate channels called *canaliculi* radiate from the haversian canals, connecting the separate lacunae, thus forming a complete circulatory system from the marrow cavity to the outer periosteum (Figure 5-7).

The epiphyses of long bones appear as a latticework of spongy cancellous bone (Figure 5-8). At first sight, the meshes of this network seem to be arranged in no apparent order; more careful examination, however, shows that the structural pattern of this bony tissue follows the lines of stress and strain to which the bone is exposed.

Bone strength is dependent on the structure and number of osteons. Areas of greatest strength have a large number of complete and

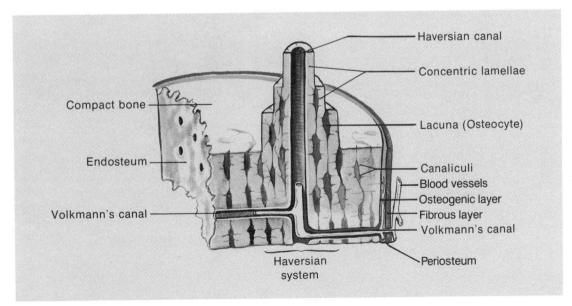

Figure 5-7 Three-dimensional sketch of a haversian system in compact bone. Haversian canals, Volkmann's canals, periosteum, and endosteum are concentric layers that make up the entire system.

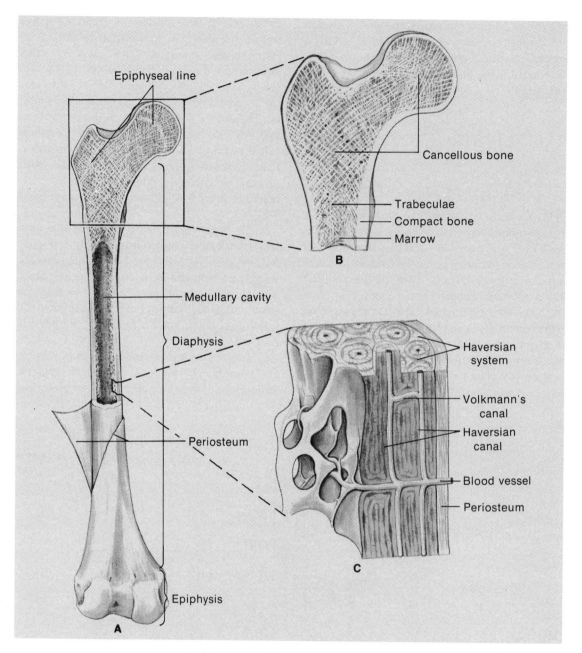

Figure 5-8 Schematic diagram of the structure of a typical long bone. **A**, Longitudinal section. **B**, Epiph-yseal section. **C**, Diaphyseal section.

partial osteons crowded together, and the supporting connective tissue, or **trabeculae,** are numerous. The collagen fibers in the lamellae are arranged diagonally, and the diameters of the haversian canals are small. The direction of the trabeculae is along the lines of stress. In contrast, the weakest areas have less densely packed osteons, and the trabeculae are few in number or absent. Within the osteons, the collagen fibers are arranged vertically along the axis of the osteon, and the haversian canals are large.

Gross Structure

A typical long bone consists of a shaft (the diaphysis) and two heads (epiphyses). The diaphysis is made up principally of compact bone in which lamellae are arranged to form a hard, solid mass. In the core of the diaphysis is the medullary cavity (Figures 5-3 and 5-8A), a large space filled with yellow bone marrow, a hematopoietic tissue.

Adult bone has two types of marrow: red and yellow. Red marrow is found in the vertebrae, sternum, ribs, and proximal epiphyses of long bones. Its major function is the production of blood cells and platelets. Yellow marrow, found in the cavities of long bones, is made up mostly of adipose tissue (fat cells). Under normal conditions, yellow marrow takes no part in red blood cell production in adulthood, but in certain abnormal circumstances, it may replace red marrow when the supply of the latter becomes depleted.

Outside, the bone is covered with periosteum. Nutrients and osteoid components are supplied to the bone by the large nutrient artery (Figure 5-3), as well as by many small blood vessels in the periosteum. These vessels permeate the compact bone and reach the medullary cavity through haversian and Volkmann's canals.

The marrow cavities of bone are lined with a membrane, the *endosteum*, which is basically a single layer of either osteoblasts or osteogenic cells (Figure 5-7).

TYPES OF BONES

Each bone is a distinct structural unit. However, all osteoid tissue has the same basic composition. Its pattern varies according to its functions, which are represented by the four basic types of bones:

1. *Long bones.* Long bones are the largest in the body. Included in this group are the prominent bones of the arms and legs. As previously described, each has a shaft and two ends. The general function of long bones is to act as levers. Examples are the femur of the leg and the humerus of the upper arm.

2. *Short bones.* These differ from long bones in that they are smaller and have less prominent ends. They are basically cuboid shaped and are composed of cancellous bone enclosed within a thin shell of compact bone. Their principal function is to provide strength. Examples are the carpals of the wrist and the tarsals of the ankle.

3. *Flat bones.* These are platelike and usually have two broad surfaces (tables) and narrow edges by which one flat bone articulates with another. Each flat bone is thin and is made up of two plates of compact bone that enclose between them a layer of cancellous bone. The general function of these bones is protection. Examples are the ribs, the scapula at the back of the shoulder, and the cranium.

4. *Irregular bones.* These are so named because of their blocklike, irregular shapes. Several surfaces may articulate with several other bones. Some of the bones of the skull are irregular, as are the vertebrae and the hip bones. The thinner parts of irregular bones consist of two thin plates of compact bone with cancellous bone between them. The general function of irregular bone is articulation.

A separate category of bone is **sesamoid bones,** which are seedlike. These bones develop in the capsules of joints or in tendons. They are small and round and are covered with cartilage or ligaments. Their function is to eliminate friction. An example is the *patella,* or knee cap.

THE SKELETON

Although the skeleton is a single functional entity, it may be divided for purposes of morphological description into an axial skeleton and an appendicular skeleton (Figure 5-9).

Axial Skeleton

The axial skeleton is composed of the eighty bones of the vertebral column, thorax, and skull.

Vertebral Column

The vertebral column, or "backbone," (Figure 5-10) is a long, flexible pillar that supports the trunk and neck in an upright position but also allows the body to make bending and twisting movements. In addition to these functions, the vertebral column protects the spinal cord and the roots of the spinal nerves (Figure 5-11).

When viewed from the lateral aspect, a normal vertebral column exhibits four curves (Figure 5-10). The thoracic and sacral curves are present at birth; the cervical curve appears when the infant begins to hold up the head. This usually occurs at about 3 months of age, and the curve becomes more pronounced when the child begins to sit up at about 9 months. The lumbar curve appears when the child begins to walk. These curves provide the vertebral column with the resilience and spring necessary for actions such as walking and jumping.

The vertebral column is formed of a series of twenty-six vertebrae, which are separated and cushioned by *intervertebral disks,* or *cartilages* (Figure 5-10). An inner canal is formed by

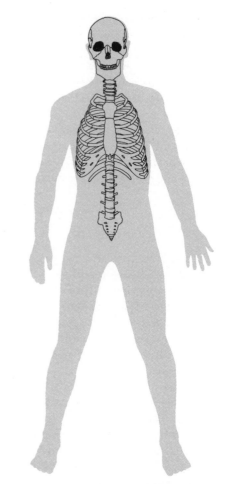

Figure 5-9 Axial skeleton. The bones of the head and trunk form the axial skeleton.

a succession of **foramina** (passages) in the individual vertebrae and by the ligaments and disks connecting them.

All the vertebrae are constructed basically the same, although different regions of the column have characteristic specializations. A typical vertebra (Figure 5-11) consists of five principal parts:

1. The *body* (or *centrum*) is the thick, disk-shaped anterior portion. Its upper and lower

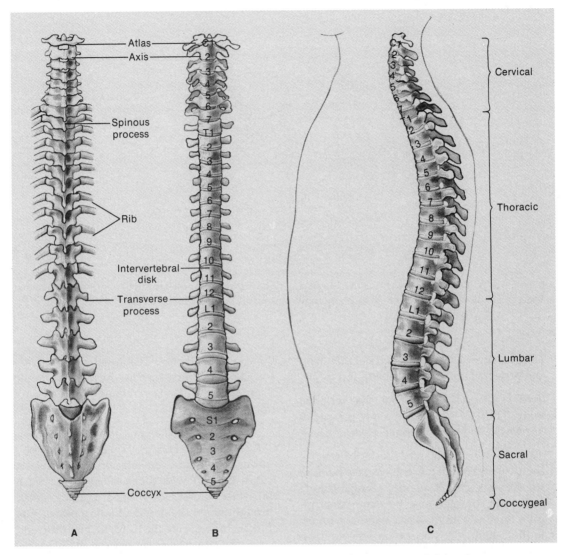

Figure 5-10 Vertebral column. **A,** Posterior aspect. **B,** Anterior aspect. **C,** Lateral aspect.

surfaces are roughened to allow for the attachment of intervening disks of fibrocartilage. Its anterior edge is pierced by small holes through which blood vessels enter to nourish the bone.

2. The *arch* projects dorsally from the body, enclosing a space (the neural canal) for the pas-

sage of the spinal cord. The arch bears three **processes** (projections) for muscle attachment: a spinous process, directed dorsally, and two transverse processes, one on either side of the vertebral canal or foramen.

3. Each vertebra has four *articular processes:* two superior and two inferior. These have

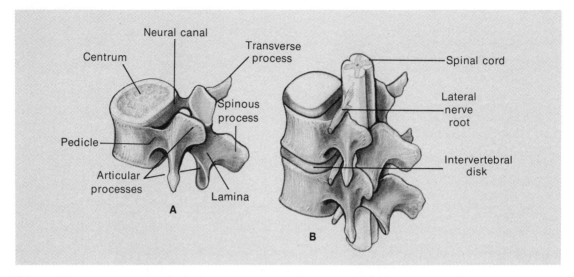

Figure 5-11 **A,** Diagonal view from the top of a typical verebra. **B,** Two vertebrae held in position by an intervertebral disk to show position of the spinal cord and its peripheral nerves.

smooth, curved surfaces for articulation with the vertebrae immediately above and below.

4. Two *pedicles* originate from the vertebral body. They are notched to allow passage of nerves to and from the spinal cord.

5. The *lamina* forms the roof (posterior wall) of the vertebral canal. It is the weakest point of the vertebra, and frequently a single lamina or several laminae must be removed surgically (laminectomy) after a traumatic injury to the column. This operation relieves pressure on the spinal cord and may aid in the fusion of two or more adjacent vertebrae.

The vertebral column, which extends the full length of the back, is made up of three groups of movable vertebrae and two groups of fixed vertebrae (Figure 5-10). The movable vertebrae include the seven **cervical** (in the neck), the twelve **thoracic** (in the thorax), and the five **lumbar** (in the loin). These remain separate throughout life. The five sacral and the four coccygeal vertebrae are fixed. In the adult, the sacral vertebrae have fused to form the *sacrum,* and the coccygeal vertebrae have united to form the *coccyx* (tailbone). The fused vertebrae are considered as one, which is why it is common to refer to the twenty-six vertebrae of humans, not thirty-three.

CERVICAL VERTEBRAE   These are the smallest vertebrae and are characterized by oblong bodies and very wide transverse processes. Each transverse process is pierced by a foramen through which the vertebral artery passes. The spinous processes of the third, fourth, fifth, and sixth cervical vertebrae are forked (bifid) to cradle the strong ligaments of the head.

The first two cervical vertebrae are unique. The first cervical vertebra, called the **atlas** (Figure 5-12), supports the head by articulating with the condyles of the occipital bone of the skull. This vertebra is a bony ring with no body and short, winglike transverse processes. Because of its articulation with the condyle, it is possible for you to rock your head back and forth on the atlas.

The second vertebra, the **axis** (Figure 5-13), has a small body and a projection of the centrum called the *dens,* or *odontoid process.* This

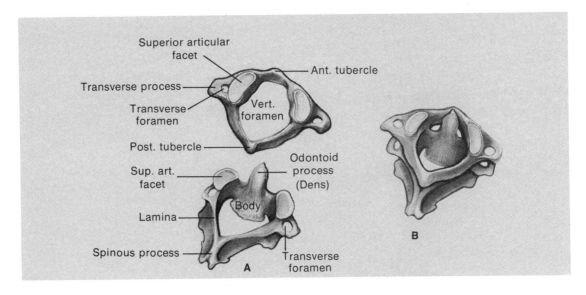

Figure 5-12 **A**, Atlas and axis, the first and second cervical vertebrae, seen diagonally from behind. **B**, Diagrammatic positioning of the dens of the axis up into the large vertebral formation of the atlas. This permits rotation of the atlas on the axis.

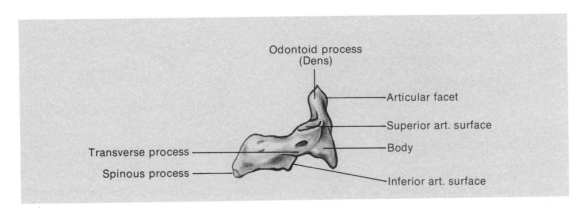

Figure 5-13 Axis, seen from the right side.

process passes through a canal in the first cervical vertebra to form the axis of rotation of the skull and the first cervical vertebra. As you turn your head from side to side, the atlas moves with the cranium, and these two turn on the axis, using the odontoid process as a center of rotation.

The seventh cervical vertebra has a very prominent spinous process with a **tubercle** (nodule) at its tip called the *vertebra prominens*, which can be seen and felt at the base of the neck.

THORACIC VERTEBRAE The twelve vertebrae of this group increase progressively in size from the base of the neck to the lumbar region. Thoracic vertebrae have two distinguishing features: a long, spinous process, pointed and directed downward; and six articular **facets,** flat surfaces that provide attachments for the ribs

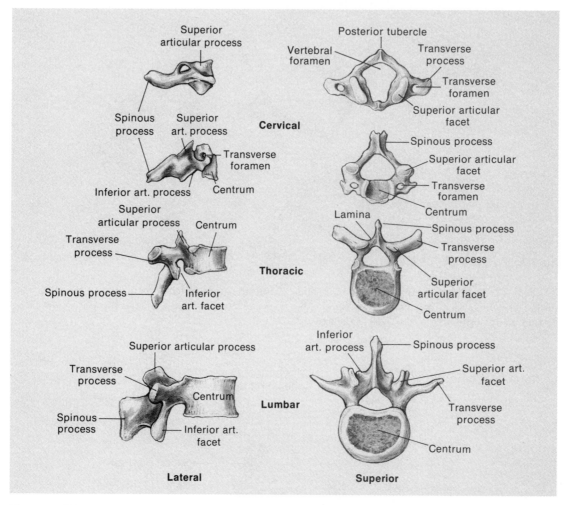

Figure 5-14 Atlas, lateral and superior aspects; a cervical, a thoracic, and a lumbar vertebra. Note the distinction between the processes and the facets: a process is a prominence or projection of bone; a facet is a small plane surface on a bone at the site where it articulates (joins) with another structure. A facet may or may not be situated on a process.

(Figure 5-14). There are three articular facets on each side of the vertebrae.

LUMBAR VERTEBRAE These largest and strongest of the vertebrae lack sharp spinous processes but have short, thin projections for attachment of the powerful muscles of the back.

SACRUM This triangular, curved bone forms the dorsal aspect of the pelvis and is wedged between the two hip bones (Figure 5-15).

COCCYX The coccyx articulates with the tip of the sacrum and is slightly movable. This property permits an increase in size of the birth canal during delivery. In some newborn infants, the coccyx may curve outward, but it can be removed surgically soon after birth without major discomfort.

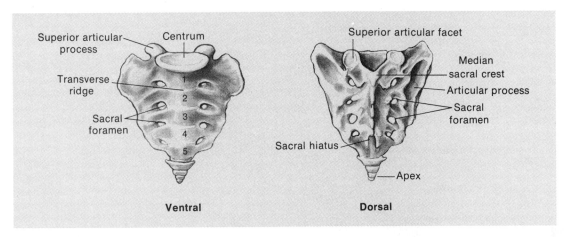

Figure 5-15 Sacrum and coccyx, dorsal and ventral aspects.

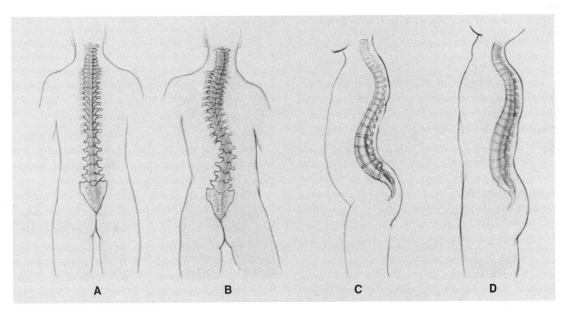

Figure 5-16 Dorsal and lateral positioning of the vertebral column. **A**, Dorsal view of a normal vertebral column. **B**, Scoliosis. **C**, Lordosis (swayback).

D, Lateral view of exaggerated posterior thoracic curvature-kyphosis (hunchback).

INJURIES AND DISEASES OF THE VERTEBRAL COLUMN The normal curves of the spine can become exaggerated as a result of injury, poor posture, or disease (Figure 5-16). When the posterior curvature is exaggerated in the thoracic region the condition is called **ky-**phosis (hunchback). An exaggerated anterior curvature in the lumbar region causes **lordosis** (swayback). **Scoliosis** (curvature of the spine) is a lateral curvature associated with other abnormalities such as herniation, osteoid tumors, and slipped intervertebral disks (Figure 5-16).

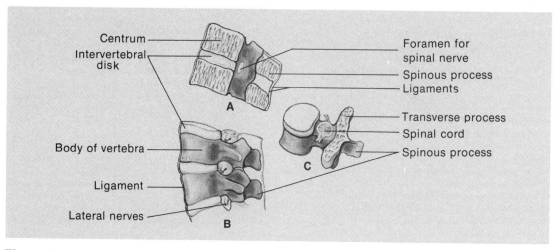

Figure 5-17 Vertebral column. **A**, Sagittal section through two vertebrae, showing the position of ligaments, intervertebral disks, and intervertebral fo-ramina. **B**, Lateral aspect. **C**, Posterosuperior aspect, showing the position of the spinal cord and lateral nerves.

The cartilaginous intervertebral disks between each pair of vertebrae act as cushions and shock absorbers (Figure 5-17). The vertebrae are bound together by strong ligaments, thick fibrous bands that strengthen joints. These are somewhat extensible, and the disks are compressible; therefore, movements of flexion, extension, abduction, and rotation are possible. When the spinal column is subjected to violent trauma, *fractures* (broken bones) and dislocations can result. The most common fracture is to the lamina of the vertebral body itself. In addition to such immediate shocks, the most dreaded complication is injury to the spinal cord, which sometimes may be completely severed. Such damage to the cord causes paralysis and loss of sensation below the point of injury. Thus, if the injury is high in the cervical region, respiration will no longer be possible, and death is inevitable.

The intervertebral disk is made up of a tough, fibrocartilaginous outer layer (*annulus fibrosus*) and a soft, putty-like interior material (*nucleus pulposus*). If the annulus becomes injured or punctured, the pressure within the disk is decreased, and the nucleus may protrude (Figure 5-18). Such a herniation can cause pressure on the spinal nerve associated with the affected vertebra, resulting in nerve root irritation (shooting leg pains) and possible compression fractures of the articular surfaces on the vertebrae above or below the injured disk.

Tuberculosis sometimes attacks the bodies of the vertebrae, practically eating them away. If the disease is untreated, the vertebrae eventually become so weak that they are crushed by the weight of the body parts above them, and a hunchback (see Figure 5-16) is produced. Such conditions have become relatively rare owing to the more effective treatment of tuberculosis in children.

Thorax

The bony structure of the **thorax,** or chest, is made up of four principal groups: the sternum, the costal cartilages, the ribs, and the bodies of the thoracic vertebrae. These twenty-five bones and cartilages are covered with muscles and skin; the floor of the thorax is formed by the

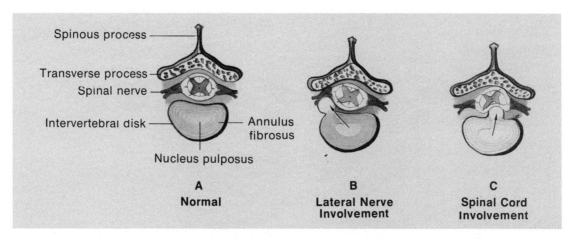

Spinous process

Transverse process
Spinal nerve

Intervertebral disk

Annulus
fibrosus

Nucleus pulposus

A
Normal

B
**Lateral Nerve
Involvement**

C
**Spinal Cord
Involvement**

Figure 5-18 Spinal vertebrae, superior aspect. **A,** Normal positioning of the spinal cord within the vertebral canal. **B** and **C,** Diagrams of ruptured intervertebral disks exerting pressure on the lateral nerve and spinal cord, respectively. Arrows indicate protrusion of the nucleus pulposus after injury to the annular rings surrounding it.

diaphragm. As a functional unit, the thorax protects and supports the heart and lungs, plays a leading role in respiration, and helps support the bones of the pectoral (shoulder) girdle. As mentioned earlier, red blood corpuscles are formed in the red marrow of the ribs and sternum. The thorax in the adult is cone shaped and has a broad muscular base.

STERNUM The sternum, sometimes called the *breast bone,* is centered at the front of the rib cage at the midline of the thorax (Figure 5-19). It is shaped like a sword with a short handle: the "handle" at the top is the *manubrium;* lying below is a long "blade," the body of the sternum (or *gladiolus);* then the tip, the *xiphoid process.* No ribs are attached to the xiphoid process. The manubrium and the gladiolus, however, are notched on either side to provide attachment for the first seven costal cartilages. The upper end of the sternum articulates with the clavicle and with the first rib on each side. The diaphragm, linea alba, and rectus abdominus muscles are attached to the xiphoid process.

COSTAL CARTILAGES At the front of the thorax, each of the fixed ribs is attached in a different manner (Figure 5-19). However, each ends some distance short of the sternum and is continued by a piece of hyaline cartilage, the costal cartilage. The first to the seventh pairs of costal cartilages attach to the sternum and contain small synovial cavities. These cavities disappear in old age. The eighth, ninth, and tenth costal cartilages do not reach the sternum; instead, each attaches to the cartilage above.

RIBS Twelve pairs of ribs form the bony portions of the thoracic walls (Figures 5-19 and 5-20). These are long, slender, curved bones. Posteriorly, they are attached to the thoracic vertebrae; anteriorly, to the costal cartilages. The first seven pairs attached to the sternum by separate costal cartilages are called *true ribs.* The remaining five pairs are *false ribs.* Of these, the eighth, ninth, and tenth pairs are attached by their costal cartilages to the costal cartilage immediately above; the eleventh and twelfth ribs have no cartilage attachments and are called *floating ribs.*

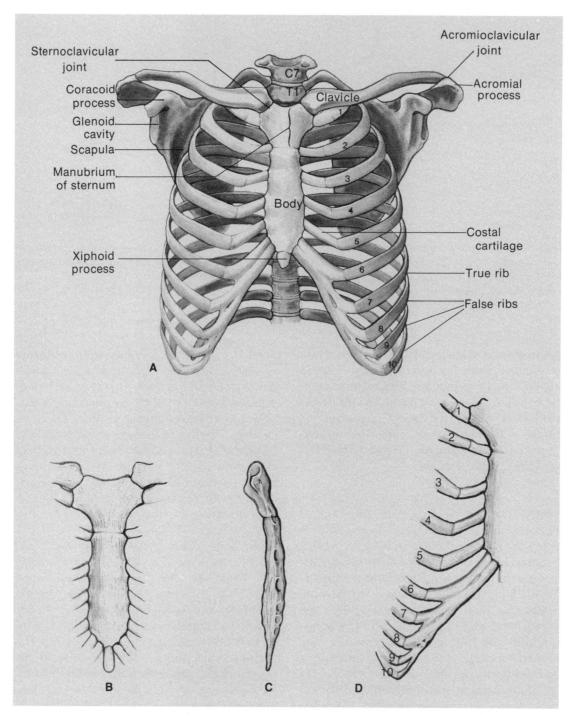

Figure 5-19 A, Rib cage, anterior aspect. **B**, Anterior surface of the sternum. **C**, Lateral view of the sternum. **D**, Costal cartilage, anterior aspect.

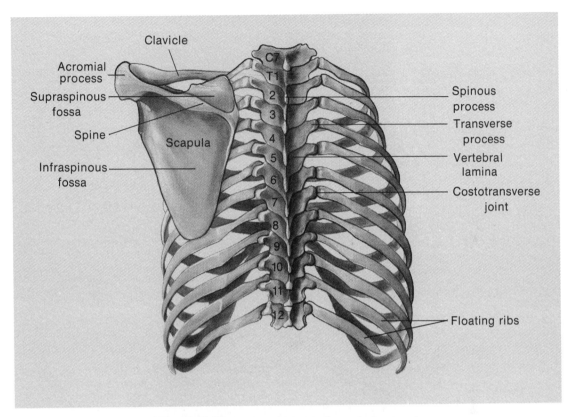

Figure 5-20 Rib cage and scapula, posterior aspect.

A typical rib has a *head*, the larger facet of which articulates with the body of the corresponding thoracic vertebra; a short *neck*; a protuberance from the neck, or **tuberosity**, which articulates with the transverse process of the vertebra; and a *shaft* (Figure 5-21). Except for the floating ribs, the upper ribs are shorter than the lower ones because the conical thorax is wider below than it is above. From their posterior vertebral attachments, the curved ribs slope at a downward angle as well as outward, thus increasing the size of the thoracic cavity, which is wider laterally than anteroposteriorly (front to back).

CLINICAL CONSIDERATIONS Fracture of one or more ribs is relatively common (Figure 5-21). Depending on the site and nature of the trauma, the rib or ribs will break—at the point of impact when hit by a small instrument with considerable force, or at the angle of the rib when a blow is applied over a large area on the front or back of the chest. In either case, the tips of the ribs will be driven inward, causing a decrease in thoracic space and an increase in pressure within the cavity. A free fractured rib tip may cause multiple abrasions and possible puncture of a lung. Immediate attention is required to prevent a possible collapsed lung.

Skull

The skull provides the skeletal structure of the face and head. It is composed of twenty-one sep-

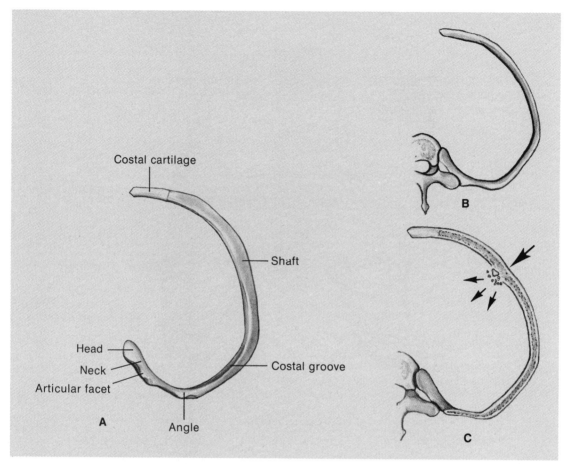

Figure 5-21 Ribs. **A**, Typical rib, superior aspect. **B**, Rib attached to the transverse process of a thoracic vertebra. **C**, Fractured rib.

arate bones, which are joined together and therefore move as a unit, plus one freely movable bone, the *mandible* (lower jaw). Eight of the skull's bones form the *cranium;* the remaining fourteen are facial bones. Except for the mandible and the ossicles of the ears, the bones of the adult human skull are immovable, united by **sutures,** or junction lines. The bones of the cranium enclose and protect the brain and its associated structures, the special sense organs.

CRANIAL SINUSES Within the bony structure of the cranium are cavities called *air sinuses* (Figure 5-22), which communicate with the nasal passageways. Four of these sinuses, the *paranasal sinuses*, communicate with each nasal cavity: the *frontal and ethmoidal sinuses* open into the nasal cavity; the *sphenoidal sinus* opens into the nasopharynx; and the *maxillary sinus* opens on the lateral walls of each nasal passage. These sinuses are lined with a mucous membrane that serves also as periosteum and is therefore called *mucoperiosteum.* This membrane is continuous with the mucous membrane that lines the air passages. Inflammation of the mucoperiosteum may extend into any one of the sinuses, causing

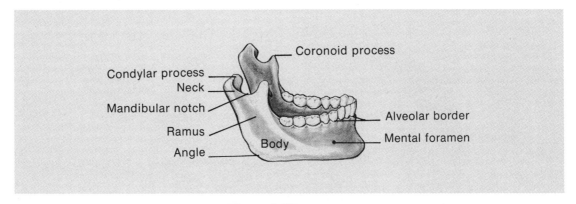

Figure 5-27 Mandible.

ations and abnormal conditions of these bones and the vomer are involved in some of the more common facial abnormalities.

The right and left *maxillae* are joined at the intermaxillary suture in the median plane of the bone to form the upper jaw (Figure 5-25). Fusion of the maxillae generally occurs before birth. When the two do not unite to form a continuous bone, a defect known as *cleft palate* (a depression of the tissue) occurs, usually associated with a *cleft* ("hare") *lip*. The maxillae also form part of the floor of the orbits, most of the roof of the mouth, the floor and lateral walls of the nasal cavity, and part of the wall of the nasolacrimal duct. Each maxilla consists of a body, which contains the large maxillary sinus, and four processes. The *alveolar process* supports the upper teeth, and the two *horizontal (palatine) processes* form the anterior and larger part of the hard palate.

The *mandible*, or lower jaw, is the largest bone of the face. It consists of a horseshoe-shaped body and two perpendicular portions, the *rami* (Figure 5-27). The lower teeth occupy the upper margin of the body. At the top of each ramus are two processes. The posterior process is the *condylar process;* it articulates with the temporal bone to form the temporomandibular joint. The *anterior coronoid process* attaches the

temporalis muscle and some of the fibers of the buccinator muscle. The deep depression between these two processes is called the *mandibular notch*. The *mental foramen*, just below the first molar tooth, serves as a passageway for the inferior dental nerve, a branch of the mandibular nerve. Branches of the nerve supply the molar and premolar teeth of the lower jaw. The mental foramen also provides a passage for blood vessels. The angle of the mandible is formed by the posterior border of the ramus and the lower border of the body.

At birth, the mandible consists of two parts that join at the symphysis in front to form one bone, usually during the first year of life. As the teeth erupt, the child chews, and the depth of the mandibular body increases, owing to the growth of the alveolar border. In the elderly person, the alveolar portion of the mandible ceases to grow. If the teeth are finally lost, the alveoli, or tooth sockets, become absorbed by the body of the bone, and the chin develops an angle, appearing more prominent.

LATERAL ASPECT OF THE SKULL   The lateral aspect includes parts of the frontal, parietal, temporal, occipital, and sphenoid bones (Figure 5-28).

The right and left *parietal bones* join to form

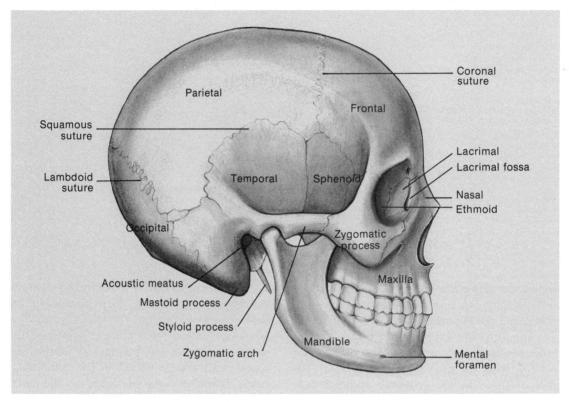

Figure 5-28 Skull, lateral aspect.

the greater part of the sides and roof of the skull. Their external surface is convex and smooth; the internal surface is concave and has **eminences** and depressions that accommodate the convolutions of the brain and the numerous blood vessels that supply the **meninges,** the membranous sheath of the brain (see Chapter 8). The two parietal bones join at the sagittal suture in the midline and at the coronal suture between the lines of articulation of the parietal and frontal bones.

The *temporal bones* like the parietal bones have right and left components and are at the sides and base of the skull. They enclose the internal structures of the ear and articulate with the mandible. Each temporal bone is made up of the squamous, petrous, mastoid, and tympanic parts. The *squamous portion* is the largest of the

four and is in the most superior position. It is a scalelike plate of bone at the side of the skull, and forms the temple. Projecting forward from its lower part is the *zygomatic process,* which forms the lateral part of the zygomatic arch, or cheek bone. Most of the interior surface of the squamous portion is occupied by the mandibular fossa, which receives the condyle of the mandible for articulation during chewing and speech.

The *petrous portion* of the temporal bone, shaped like a three-sided pyramid, contains the ear structures. On its cranial side there is a foramen, the *internal acoustic meatus,* which allows the passage of the facial, cochlear, and vestibular nerves. The petrous portion is located deep within the base of the skull between the sphenoid and occipital bones and contains both the inner ear and parts of the middle ear within its complex cavities. The *styloid process* projects downward from the inferior surface of the petrous portion of the temporal bone (Figure 5-28).

The *mastoid portion* is located behind and below the *external acoustic (auditory) meatus,* the external opening of the ear. The mastoid process projects downward from its inferior border (Figure 5-28). It is a rounded protuberance of the temporal bone that serves as a means of attachment for neck muscles. The mastoid portion contains air spaces called *mastoid air cells,* or *sinuses;* these are separated from the brain only by thin, bony partitions. Inflammation of the cells of the mastoid (*mastoiditis*) is not uncommon and when untreated is dangerous because

of possible invasion of the brain or the meninges by the causative bacteria or viruses.

The *tympanic portion* is a curved plate. The upper surface forms the anterior and inferior walls and part of the posterior wall of the external acoustic meatus—a 15 mm canal connecting the outer with the middle ear.

The *occipital bone* is situated at the back and base of the skull. It joins the parietal bones anteriorly at the lambdoidal suture. The inferior portion of the bone has a large opening, the *foramen magnum,* through which the spinal cord passes (Figure 5-29). It is at this level that the spinal cord joins the medulla oblongata of the brain. At the sides of the foramen magnum on the external surface are two rounded projections, or **condyles,** which articulate with the

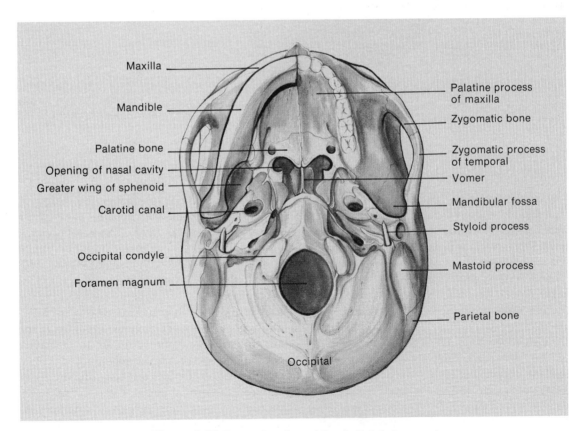

Figure 5-29 External surface of the skull, inferior aspect.

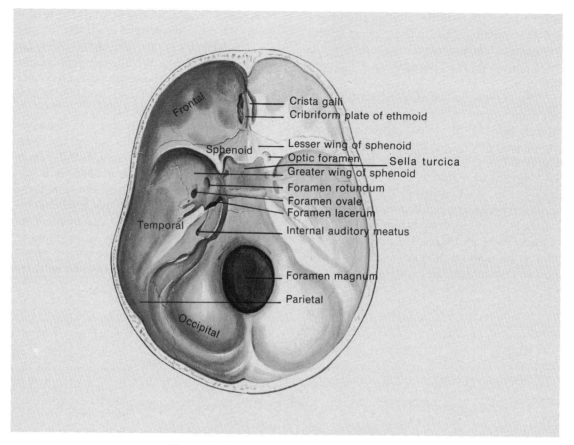

Figure 5-30 Base of the skull, superior aspect.

first cervical vertebra, the atlas. The atlanto-oc-cipital joint between the skull and the atlas per-mits backward and forward movements of the head. Other obvious projections from the occip-ital bone are the *external occipital crest* and the *external occipital protuberance*. These can be felt at the base of the neck and serve as a means for the attachment of muscles and ligaments.

INFERIOR ASPECT OF THE BASE OF THE SKULL   When the skull is viewed from below, the large foramen magnum, the openings of the nasal cavity, and the *palatine bone (hard palate)* may be seen (Figure 5-29). The bony roof of the mouth is made up of the hard palate and part of the maxillae and palatine bones. The hard pal-ate separates the mouth from the nasal cavity and also serves as the floor of the nasal cavity.

SUPERIOR ASPECT OF THE BASE OF THE SKULL   With the top of the cranium and its contents removed, the superior structures of the base of the skull may be studied (Figure 5-30). Once again the most obvious structure is the large foramen magnum. Anterior to the cavity is a wing-shaped bone, the sphenoid.

Anteriorly the *sphenoid bone* fills the space between the orbital plates of the frontal bone; posteriorly it separates the temporal and occipi-tal bones. It resembles a bat with its wings ex-

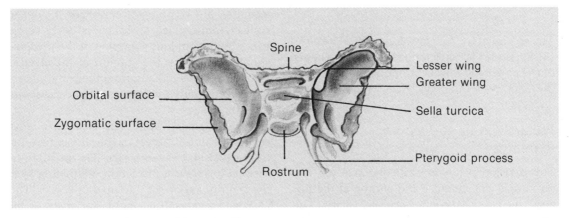

Figure 5-31 Sphenoid bone, superior aspect.

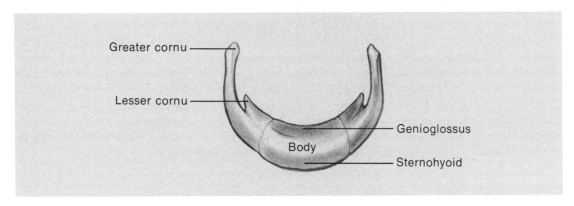

Figure 5-32 Hyoid bone, anterior surface.

tended and consists of a *body*, two *greater and two lesser wings* extending transversely from the sides of the body, and *two pterygoid processes*, which project downward (Figure 5-31). The body is joined to the ethmoid bone in front and to the occipital bone behind. It contains the sphenoidal sinuses, which communicate with the nasopharynx. The upper surface exhibits a depression in the form of a saddle, the *sella turcica*. The hypophysis (pituitary gland) lies in the depression. The undersurface of the body of the sphenoid forms part of the roof of the nose.

The last bone to be mentioned is the *hyoid* (Figure 5-32). This is an isolated, horseshoe-shaped bone that lies in the anterior part of the neck, a short distance above the larynx. It consists of a central part, the *body*, and two projections, the *greater and lesser horns* or *cornua*, which project backward. The hyoid bone is suspended from the styloid processes of the temporal bones and may be felt in the neck just above the laryngeal prominence (Adam's apple). It supports the tongue and provides attachments for muscles used in speaking and swallowing.

Appendicular Skeleton

The second of the two divisions of the skeleton, the appendicular skeleton, comprises the bones of the upper and lower extremities, pectoral girdle (shoulder), and pelvic girdle (hip).

Pectoral Girdle and Upper Extremity

This division of the appendicular skeleton, beginning at the shoulder and ending at the fingers, includes sixty-four separate bones. The first of these, the *clavicle* (collar bone) is long and has a double curvature (Figure 5-33). It is situated in a horizontal position at the upper and anterior part of the thorax, just above the first rib, at the root of the neck between the upper end of the sternum and the scapula (shoulder blade), to which it is attached and with which it articulates. In women, the curve of the clavicle is less pronounced, and the bone generally is smoother, shorter, and more slender than in men. The general function of the clavicle is to provide a prop for the shoulder. Because of its prominent position in the pectoral girdle, it is easily injured. Fractured clavicles, for

example, are common occurrences in violent contact sports. Such an injury is easily recognized by the obvious slumping of the involved shoulder. Without the support of the clavicle, the arm appears to hang from the body, since it is supported only by the scapula and the shoulder joint.

The *scapula*, a large, flat, triangular bone, covers the dorsal aspect of the thorax between the second and seventh ribs (Figure 5-34). Its *coracoid process* is a projection originating from the anterior surface of the superior border. This process provides attachments for some of the muscles that move the arm. If you reach over your left shoulder with your right hand, you can feel the spine of the scapula as it passes across the dorsal surface and ends in the large, flat *acromial process*, which forms the tip of the shoulder blade. The joint between the acromial process of the scapula and the lateral end of the clavicle is the only bony articulation between the upper extremity and the thorax. The projection of the acromial process can be thought of as the "roof" and the coracoid process as the upper medial "wall" of a fossa called the *glenoid cavity*, which acts as a socket for articulation with the humerus (Figure 5-35).

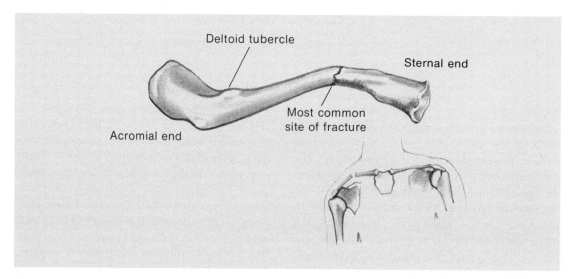

Figure 5-33 Right clavicle, inferior aspect. Inset demonstrates the "drooping" shoulder due to a fractured collar bone.

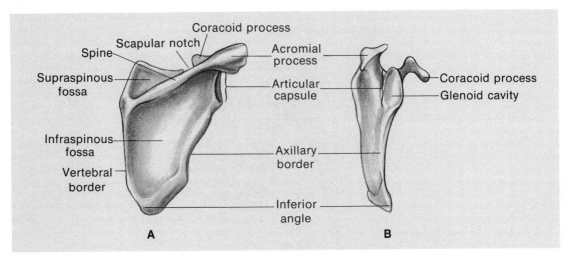

Figure 5-34 Right scapula. **A**, Posterior aspect. **B**, Lateral aspect.

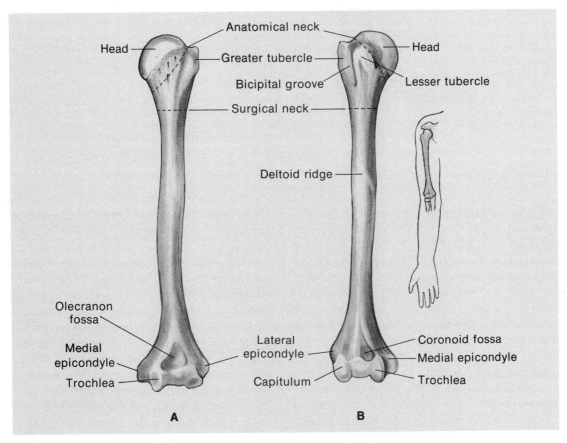

Figure 5-35 Right humerus. **A**, Posterior aspect. **B**, Anterior aspect.

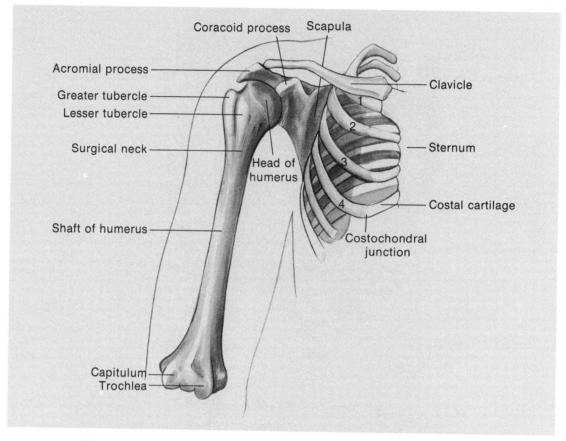

Figure 5-36 Relationship of the right humerus, scapula, and clavicle to the ribs.

The humerus of the upper arm is the longest and largest bone of the upper extremity (Figures 5-35 and 5-36). It extends from the shoulder to the elbow. The upper end of the humerus consists of a rounded *head* joined to the shaft by a constricted *neck* and two eminences, the greater and lesser tubercles, which provide insertions for many of the muscles of the upper extremity. There is a depression between the two tubercles, the *bicipital groove,* which receives the tendons of part of the biceps muscle. The portion of the neck above the tubercles is called the *anatomical neck;* the portion below is called the *surgical neck* because it is so often fractured. The head of the humerus artic-

ulates with the glenoid cavity of the scapula, forming a "ball-and-socket" joint (see Chapter 6).

The shaft of the humerus is long and slender. At its lower end (distal head) the humerus widens, developing two smooth areas. On the lateral side is the *capitulum,* which articulates with the head of the radius; on the medial side is the *trochlea,* which articulates with the ulna (Figure 5-36). Situated just above these articular surfaces are the *anterior coronoid fossa,* which receives the coronoid process of the ulna when the forearm is flexed, and the *posterior olecranon fossa,* which receives the olecranon process of the ulna when the forearm is extended.

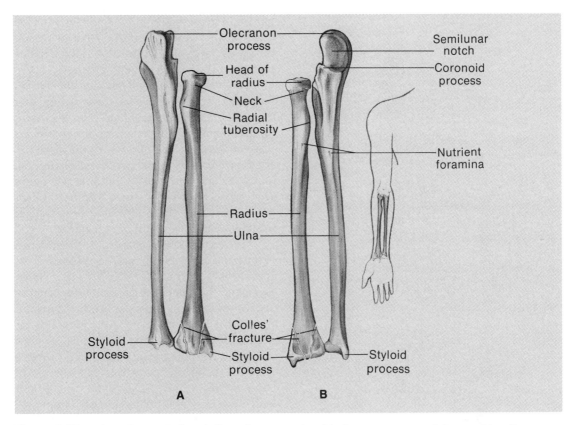

Figure 5-37 Right radius and ulna. **A**, Posterior aspect. **B**, Anterior aspect. Colles' fracture of the distal head is the most common injury to the radius.

Epicondyles project from the medial and lateral sides just superior to the condyles (Figure 5-25). These projections provide attachments for the tendons of both the ulna and the radius.

The *ulna* is the longer of the two bones of the forearm, and, as shown in Figure 5-37, is positioned on the medial side of the radius. Its shaft is triangular, and its enlarged upper end forms the *elbow*. There are two beaklike processes at this end: the *olecranon process*, which curves upward and forward; and the *coronoid process*, which, together with the olecranon process, forms a depression called the *trochlear*, or *semilunar notch*. This half-moon-shaped depression is the articular surface for the trochlea of the humerus. At the distal (lower) end of the ulna there is a small medial projection, the *styloid process*; at both ends of the bone are lateral facets that serve as areas of articulation with the radius.

The *radius* lies on the lateral or thumb side of the forearm (Figure 5-37). It joins the ulna by means of a thin, fibrous sheet, or *interosseous membrane*, which crosses the area between the shafts of the two bones. At the upper (proximal) end of the radius is a small, round head with a cuplike depression on its upper surface for articulation with the capitulum of the humerus. A prominent ridge surrounding the head allows it to rotate within the *radial notch* of the ulna. The head is supported by a constricted neck just superior to an eminence on the medial side of the

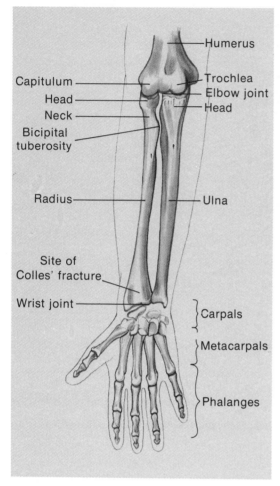

Figure 5-38 Bones of the right forearm and hand, anterior aspect.

hairline fractures and bone fragmentation because of the "jamming" of the wrist bones into the lower heads of both the ulna and the radius, thereby cracking their styloid processes.

The bones of the wrist are the *carpals,* which are arranged in two rows of four each (Figure 5-39). In the proximal row, beginning on the lateral side, are the *scaphoid (navicular), lunate, triquetrum,* and *pisiform.* In the distal row are the *trapezium (greater multangular), trapezoid (lesser multangular), capitate,* and *hamate* bones. The carpals are closely joined together by ligaments; however, these ligaments are arranged to allow a certain amount of movement.

The five *metacarpal bones* form the structure of the palm of the hand; each has a base, a shaft, and a head (Figures 5-38 and 5-39). The metacarpals are long and cylindrical, and their rounded distal ends form the knuckles. At their proximal ends, these bones articulate with the carpals and with each other; at their distal ends, they are connected with the proximal *phalanges* (fingers).

Each hand has fourteen phalanges, the bones of the fingers. Except for the thumb, which has only two phalanges, each finger has *proximal, middle,* and *terminal* (or distal) phalanges (Figure 5-39).

The Pelvic Girdle and the Lower Extremity

The pelvic girdle (sixty-two bones) provides support for the trunk and attachment for the bones of the legs. Although the bones of the lower extremity correspond in general to those of the upper extremity and bear a rough resemblance to them, their function is different. The lower extremities support the body in the erect position; therefore, they are more solidly built and their parts are less movable than those of the upper extremities.

The two *hip bones,* sometimes called the *innominate bones,* are the broadest in the body (Figure 5-40). Each is large and irregularly shaped and, when paired with the bone of the opposite side, forms the sides and the front wall of the pelvic cavity. Although fused in the

radius, the *radial tuberosity.* There are two articular surfaces at the lower end of the radius: the *styloid process,* which articulates with the scaphoid and lunate bones of the wrist; and the *medial ulnar notch,* which articulates with the ulna.

A specific percussion-type fracture of the lower end of the radius is called *Colles' fracture* (Figures 5-37 and 5-38). This fracture may occur in a fall when the person extends the hands to delay full body contact with the ground, thus putting too much stress on the lower third of the forearm. This kind of fall may also cause

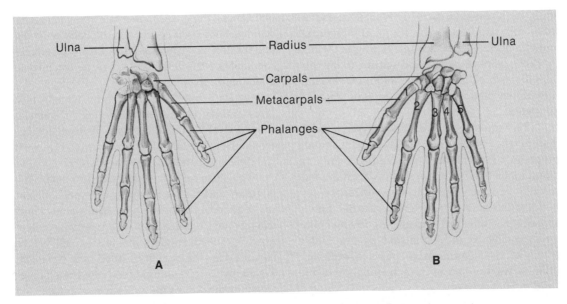

Figure 5-39 Bones of the right hand. **A,** Posterior aspect. **B,** Anterior aspect.

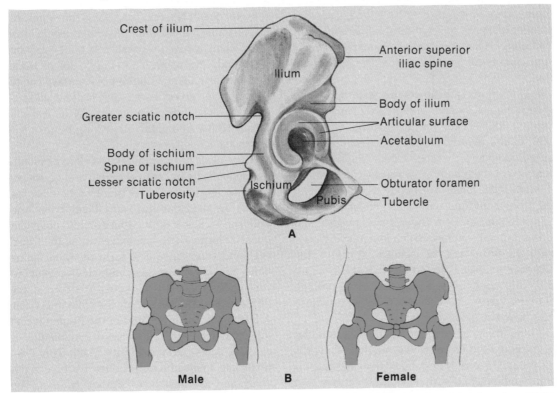

Figure 5-40 Hip bone. **A,** Right pelvic hip, lateral aspect. **B,** Comparison of proportion of male and female pelves. Note that the male pelvis is narrow and compact, whereas the female pelvis is broad and light to facilitate pregnancy and parturition.

adult, in the embryo each hip bone starts out as three separate bones: the ilium, the ischium, and the pubis (Figure 5-40A). A ring is formed by these pairs of bones, and within it are contained a part of the sigmoid colon, the rectum, the urinary bladder, and some parts of the genital system.

The female pelvis differs markedly from that of the male. These differences facilitate pregnancy and parturition (Figure 5-40B). In women, the pelvis is more shallow than in men, but it is relatively wider in every direction, especially laterally. The overall internal pelvic area of the female becomes greater during childbirth, owing to the expansion by the pubic bones at the semimovable pubic symphysis. This allows more room for manipulation of the child during delivery.

The *ilium* is the largest of the three pelvic bones; it forms the superior, broad, expanded prominence of the hip. At its crest are *anterior superior, posterior superior,* and *posterior inferior iliac spines.* These anatomic landmarks provide attachments for the muscles of the abdominal wall.

The *ischium* is the lowest and strongest bone of the pelvis. It extends downward from the *acetabulum,* the socket for articulation with the head of the femur, and expands into the large *ischial tuberosity,* which serves as a weight bearer when the body is in a seated position.

The *pubis* consists of a body, which unites at the frontal midline with its counterpart of the opposite side to form the *symphysis pubis,* and two **rami,** or branches, which join, respectively, with the ilium and the ischium. A cavity, the *obturator foramen,* created by the union of all three hip bones, provides space for the passage of major vessels and nerves to the lower extremities from the pelvic cavity.

The *femur,* or *thigh bone,* is the largest bone in the body, but its position is not that of a vertical pillar. Instead, it lies at a downward and inward angle from the hip joint. At the upper end of the femur is a rounded head with a constricted neck (Figure 5-41). There are also two

eminences, the *greater and lesser trochanters.* The femoral head itself articulates medially with the acetabulum to form the hip joint. Posteriorly the shaft of the femur has a ridge for muscle attachment, the *linea aspera.*

At its lower end, the femur widens into large eminences, the *lateral and medial condyles* (Figure 5-41). The *intercondyloid fossa* lies between these. Distally, the femur articulates with the tibia and patella at the knee joint.

The *sesamoid patella* (knee cap) is small, triangular, and flat. It occupies the space within the tendon of the quadriceps femoris muscle at the front of the knee joint. The patella, as noted, articulates with the femur (Figure 5-42), and is surrounded by large, fluid-filled sacs, or *bursae.* When the knee is straightened, the bone moves freely in support of its associated muscles.

The *tibia* (shin bone) lies at the front and medial side of the leg (Figure 5-43). Two bones form the lower leg, and the tibia is the larger. Its upper end expands into *lateral and medial condyles,* which are separated by the sharp intercondyloid fossa. The concavities of these condyles articulate with the tibial surface lateral to the inner ankle bone, the *medial malleolus.* The tibia also articulates at both of its extremities with the fibula, the other long bone of the lower leg.

The *fibula* of the calf, the slenderest bone in the body, lies parallel to the tibia and on its lateral side (Figure 5-43). At the upper end of the fibula, an irregular quadrant head articulates with the back part of the tibia but does not reach the knee joint. The lower extremity of the fibula is extended downward to form the outer ankle bone, the *lateral malleolus,* which is a pointed process just beneath the skin. At this end the fibula articulates with both the tibia and talus. A fracture of the lower end of the fibula involving injury to the lower tibial articulation is called a *Pott's fracture* (Figure 5-43). This fracture closely resembles the Colles' fracture of the wrist and is caused by similar trauma. Too forceful a strain or the sudden impact of too much weight from a fall will fracture the lower

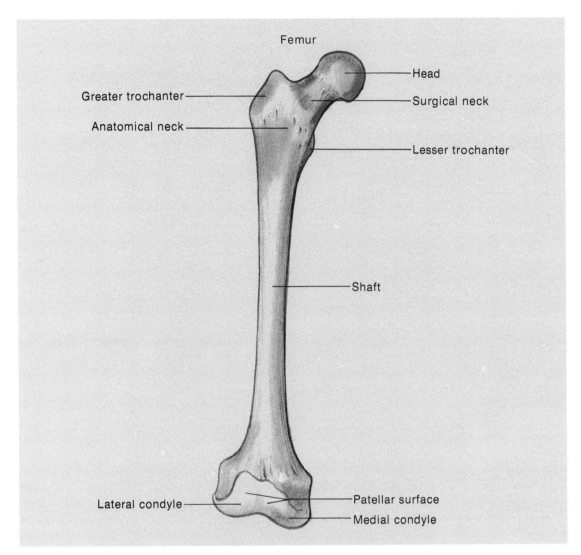

Figure 5-41 Right femur, anterior aspect.

third of the fibula or crack the lateral malleolus of the fibula and the medial malleolus of the tibia.

The *tarsus* is the general name for the seven tarsal (ankle) bones that make up the articulation between the foot and the leg (Figure 5-44). They are short and resemble the carpal bones of the wrist, although they are larger. Their names

are the *calcaneus, talus, cuboid, navicular, first cuneiform, second cuneiform,* and *third cuneiform.* The largest and strongest of the tarsal bones is the calcaneus (heel bone), which serves to transmit the weight of the body to the ground, forming a strong lever for the muscles of the calf (Figures 5-42 and 5-44).

The metatarsus, the bony structure of the

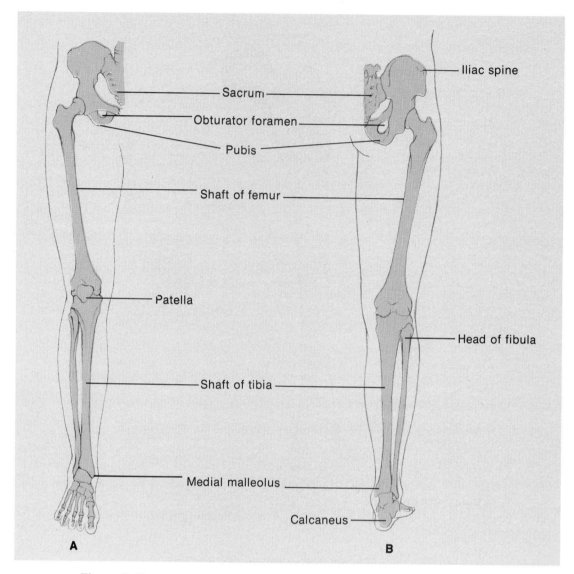

Figure 5-42 Bones of the right leg and foot. **A**, Anterior aspect. **B**, Posterior aspect.

sole and instep of the foot, is formed by five bones that closely resemble the metacarpal bones of the hand. Each metatarsal is long and consists of a base, a shaft, and a head (Figure 5-44*A*), and each articulates at one end with the tarsal bones and at the other with the first row of phalanges (toes). The shafts of the metatarsals provide attachments for the intrinsic muscles of the toes; the heads form the joints of articulation. The first metatarsal bears the body's weight and therefore is larger than the others.

The tarsal and metatarsal bones are arranged in the form of two distinct arches: one running from the heel to the toes on the inner

OUTLINE

I. Introduction
- A. Types of skeleton
  1. Exoskeleton
  2. Endoskeleton
- B. Functions of endoskeleton
  1. Protection
  2. Support
  3. Muscle attachment
  4. Reservoir for minerals
  5. Hematopoiesis

II. Formation of bone
- A. Embryonic tissue
  1. Mesenchyme
  2. Fibrous tissue containing fibroblasts
  3. Collagenous and elastic fibers
- B. Development of bone
  1. Blood vessels bring rich contents that stimulate cell division
  2. Osteoblasts form from previous mesenchyme
  3. Osteoblasts deposit mucoprotein to form a rubbery osteoid tissue
  4. Osteoblast
- C. Intramembranous ossification
  1. Bones of the skull
  2. Membranous covering of skull contains fibroblasts
  3. Fibroblasts change to osteoblasts, which then lay down a basis for calcium
- D. Cartilaginous ossification
  1. Replacement of preexisting cartilage by osteoblasts
  2. Hyaline cartilage forms embryonic skeleton
  3. Bone begins to form on cartilaginous model; covered by periosteum
  4. Blood vessels bring osteoblasts, which lay down collagen and mucoprotein
  5. Osteoclasts gradually dissolve old bone, while osteoblasts deposit new bone
  6. In long bones, ossification proceeds in two directions
  7. Bone growth ceases at about age 24 years
- E. Chemical influences on bone formation
  1. Vitamin D—absorption of minerals
  2. Alkaline phosphatase—releases calcium and phosphate ions from calcium salt solution
  3. Parathormone—regulates calcium and phosphorus levels in blood
  4. Calcitonin—stimulates bone formation
- F. Disorders of bone development
  1. Insufficiency—rickets, osteomalacia
  2. Hormonal dysfunction: gigantism, dwarfism, acromegaly
  3. Cancer—primary and secondary growths
  4. Osteomyelitis—inflammation of bone and marrow

III. Structure of bone
- A. Microscopic structure
  1. Compact bone—osteon
     - a. Haversian canals
     - b. Lamellae
     - c. Volkmann's canals
     - d. Canaliculi
  2. Cancellous (spongy) bone
- B. Gross structure (long bone)
  1. Two epiphyses
  2. Diaphysis
  3. Periosteum
  4. Medullary cavity
     - a. Yellow bone marrow—fat
     - b. Red bone marrow—red blood cells and platelets
  5. Endosteum

IV. Types of bones
- A. Long bones
  1. Function—as levers
  2. Examples—femur, humerus
- B. Short bones
  1. Same basic structure as long bones
  2. Function—strength
  3. Examples—carpals, tarsals

C. Flat bones
  1. Platelike
  2. No medullary cavity
  3. Function—protection
  4. Examples—cranium, ribs
D. Irregular bones
  1. Odd-shaped
  2. No medullary cavity
  3. Function—articulation
  4. Examples—vertebrae, hip bones, some skull bones
E. Sesamoid bones
  1. Seedlike bones found in joint capsules and in tendons
  2. Function—prevent friction
  3. Example—patella (knee cap)
V. Skeleton
  A. Axial skeleton
    1. Vertebral column
      a. Supports trunk and neck
      b. Protects spinal cord
      c. Four curves visible from lateral aspect
      d. 26 vertebrae separated by disks
        (1) 7 cervical
        (2) 12 thoracic
        (3) 5 lumbar
        (4) 1 sacral
        (5) 1 coccygeal
      e. Typical vertebra
        (1) Body
        (2) Arch
        (3) Four articular processes
        (4) Two pedicles
        (5) Lamina
      f. Vertebral injuries and diseases
        (1) Kyphosis
        (2) Lordosis
        (3) Scoliosis
        (4) Severed spinal cord—paralysis
        (5) Ruptured disk
        (6) Tuberculosis
    2. Thorax
      a. Sternum
      b. Costal cartilages
      c. Twelve pairs of ribs
        (1) seven true ribs
        (2) five false ribs (last two pair are floating)
    3. Skull
      a. Cranial sinuses
        (1) Air spaces in bone
        (2) Frontal and ethmoidal—nose
        (3) Sphenoidal—nasopharynx
        (4) Maxillary—nasal passages
        (5) Inflammation—sinusitis
      b. Cranial sutures
        (1) Coronal—side
        (2) Lambdoidal—rear
        (3) Squamous—side
        (4) Sagittal—middle
      c. Fontanels
        (1) Bregmatic—anterior
        (2) Occipital—posterior
        (3) Sphenoidal—anterolateral
        (4) Mastoidal—posterolateral
        (5) Premature fusion—microcephalus
        (6) Abnormally large quantity of spinal fluid—hydrocephalus
      d. Bones of skull (29)
        (1) Frontal—forehead
        (2) Ethmoid—nose
        (3) Nasal (2)—nose
        (4) Vomer—nose
        (5) Maxillae (2)—upper jaw
        (6) Mandible—lower jaw
        (7) Zygomatic (2)—cheek
        (8) Inferior conchae (2)
        (9) Lacrimal (2)—nose
        (10) Palatine (2)—mouth
        (11) Parietal (2)—lateral, superior
        (12) Temporal (2)—lateral, inferior
          (a) Squamous
          (b) Petrous
          (c) Mastoid
          (d) Tympanic
        (13) Occipital—posterior
        (14) Sphenoid—superior, infe-

rior
(15) Hyoid—anterior part of neck
(16) Middle ear ossicles (6)
B. Appendicular skeleton
1. Pectoral girdle and upper extremity (64 bones)
   a. Clavicle—collar bone
   b. Scapula—shoulder blade
   c. Humerus—upper arm
   d. Ulna—forearm, medial
   e. Radius—forearm, lateral
      (1) Colles' fracture
   f. Carpals (8)—wrist
   g. Metacarpals (5)—hand
   h. Phalanges (14)—fingers
2. Pelvic girdle and lower extremity (62 bones)
   a. Hip (innominate) bones
      (1) Ilium
      (2) Ischium
      (3) Pubis
   b. Femur—thigh, longest bone in body
   c. Patella—kneecap
   d. Tibia—shin, anterior

e. Fibula—shin, lateral to tibia
   (1) Pott's fracture
f. Tarsals (7)—ankle
g. Metatarsals (5)—foot
h. Phalanges (14)—toes

VI. Diseases and injuries affecting the skeletal system
A. Diseases
1. Osteitis fibrosa cystica
2. Osteomyelitis
3. Tuberculosis
B. Bone neoplasms
1. Benign—osteoma
2. Malignant—Osteosarcoma
   a. Multiple myeloma
C. Types of fractures and repair
1. Classification
   a. Complete—incomplete
   b. Simple—compound
   c. Depression—compression
   d. Various structural types
2. Repair
   a. Formation of bone callus
   b. Open or closed reduction
   c. Artificial prostheses
   d. Internal fixation

STUDY QUESTIONS AND PROBLEMS

1. Name five general functions of bone.
2. Differentiate between intramembranous and cartilaginous ossification.
3. Define osteoblast, osteocyte, and osteoclast, and give the role of each in the development of bone.
4. What roles do vitamin D, parathormone, alkaline phosphatase, and calcitonin play in the formation of bone?
5. What is rickets? Acromegaly? Osteomalacia? Osteoporosis?
6. What is a haversian canal?
7. Describe how the osteocyte gets nutrition from a vessel outside the bone.
8. What are the two major divisions of the skeleton?

9. Sketch a typical spinal vertebra and indicate why the lamina is its major point of weakness.
10. List the fontanels that are in the skull at birth, and describe microcephalus.
11. Differentiate among the curvatures of the spine. What do kyphosis, lordosis, and scoliosis mean?
12. List the cranial sinuses and give at least two reasons why they are important.
13. What are some of the differences between a male and a female pelvis?
14. Why is it important that we have an arch in the foot?
15. What is a bone callus?

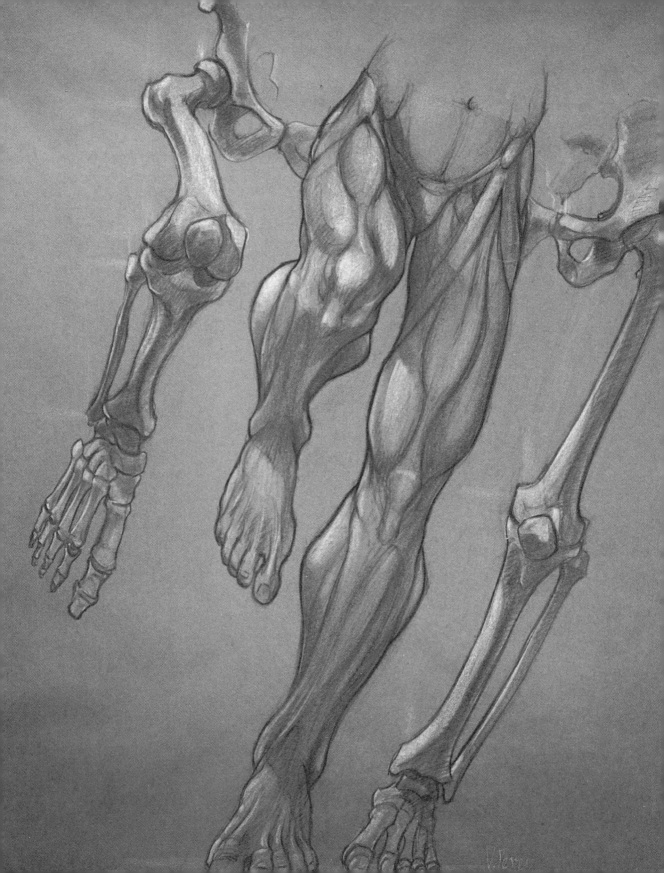

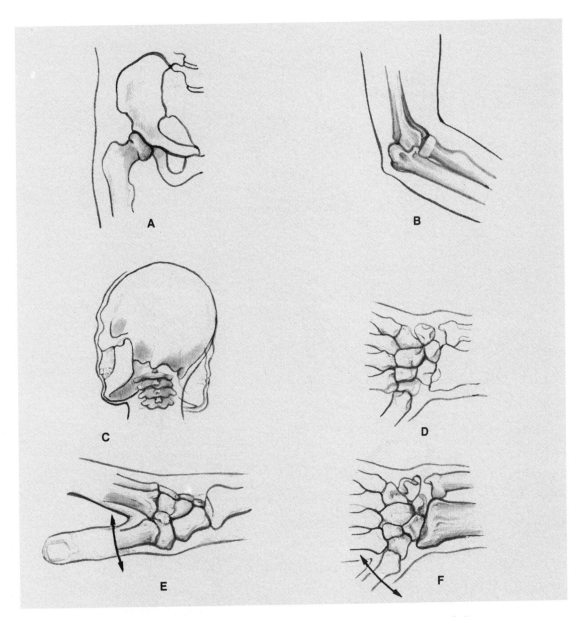

Figure 6-3 Types of diarthrodial joints. **A**, Ball-and-socket. **B**, Hinge. **C**, Pivot. **D**, Ellipsoidal. **E**, Gliding. **F**, Saddle.

- **Adduction**—the bone moves toward the midline (example: moving the arm or leg toward the midline and decreasing the angle between them)

- **Rotation**—turning or moving a bone on an axis without any displacement of the axis (example: one vertebra may pivot upon another while the plane of movement remains perpendicular to the long axis of the body)

- **Circumduction**—a combination of angular

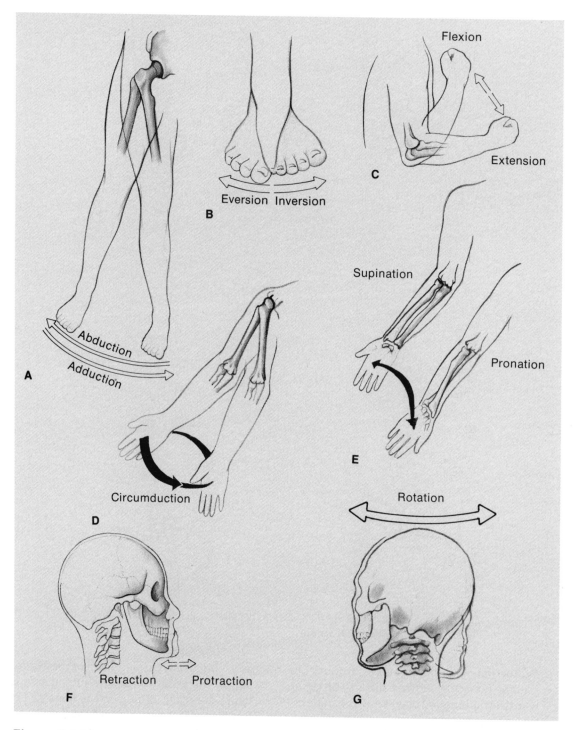

Figure 6-4 Movements of diarthrodial joints. **A,** Abduction–adduction. **B,** Eversion-inversion. **C,** Flexion–extension. **D,** Circumduction. **E,** Supination-pronation. **F,** Protraction–retraction. **G,** Rotation.

and rotary movements in succession (example: swinging the arms in a circle)

- **Supination**—the forearm is moved so that the thumb points away from the midline (example: when the forearm is made to turn the palm of the hand forward or upward, resulting in parallel positions of the radius and ulna)

- **Pronation**—the forearm is moved so that the thumb points toward the midline (example: turning the palm inward or downward so that the radius and ulna cross each other)

- **Inversion**—movement of the ankle joint so that the sole of the foot turns inward

- **Eversion**—movement of the ankle joint so that the sole of the foot turns outward

- **Protraction**—moving a part of the body forward (example: projecting the mandible)

- **Retraction**—moving a part of the body backward (example: withdrawing the mandible)

STRUCTURE OF SYNOVIAL JOINTS

Synovial joints have three distinct features: a layer of hyaline cartilage covering the articular surfaces of the opposing bones; an articular capsule composed of bands of fibrous tissue that unite the bones and encircle the joint cavity; and a synovial membrane that lines the articular capsule and secretes the lubricating fluid, synovia, into the joint cavity (Figure 6-5).

As previously discussed, the articular surfaces of the bones assume varying shapes according to their function. Most diarthrodial joints have congruent articular surfaces—opposed surfaces that correspond to each other in shape and curvature. Some joints, however, are incongruent, and this discrepancy is sometimes compensated by the interposition of an articular cartilaginous disk that serves to cushion the surfaces, as in the knee joint (see Figures 6-5 and 6-8). Other joints have disks that entirely separate

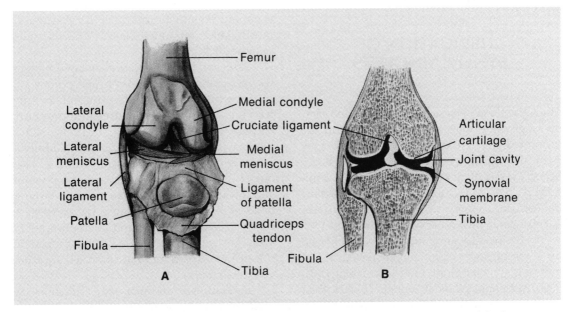

Figure 6-5 Right knee joint. **A,** Dissected from the front, with the patella resected and hanging down onto the tibia. **B,** Coronal section of the knee, exposing the joint cavity.

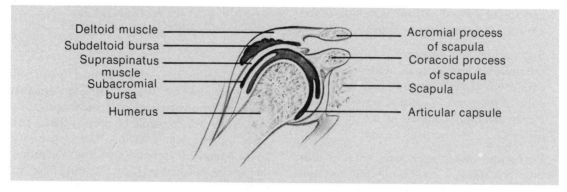

Figure 6-6 Coronal section of the shoulder, showing the relationship of the subdeltoid and subacromial bursae to the joint cavity.

the articular surfaces, as in the sternoclavicular articulation, or the joint margins may be provided with a fibrocartilaginous lip that tends to deepen the socket, as in the glenoid cavity of the shoulder (Figure 6-6). The surface area of most joints is increased by folds of synovial membrane, which may contain fat. These fat pads act to fill the spaces resulting from the separation of the articular surfaces as well as to reduce the friction generated by the opposing surfaces.

LUBRICATION OF SYNOVIAL JOINTS

The presence of synovial fluid in joints led early investigators to believe that human joints are lubricated much like machines are. We now know that this is not true. Rotating bearings utilize an even content of lubricating fluid to minimize friction. Most human joints do not rotate, but oscillate, thereby causing opposing surfaces to rub against each other and wear down.

In the healthy joint, lubrication is achieved by a combination of unique characteristics of both the synovial fluid and the articular cartilage. Although synovial fluid is thicker than water, a more significant property is its unique mucous character. This is due to the presence of hyaluronic acid, a mucopolysaccharide (a compound containing protein and complex sugar), which is transformed into large molecules called **mucin**. These mucin molecules, which are produced by the cells of the synovial membrane, distinguish joint fluid from plasma. Mucin molecules react with and cling to the cartilage surface. Because these molecules are larger than the outer cells of the cartilage, they do not penetrate the cartilage but stay on the surface, forming a slippery layer. Thus when two healthy joint surfaces move on each other it is not movement of cartilage on cartilage but of mucin layer on mucin layer. This is called *boundary lubrication.*

A second lubricating mechanism hinges on the fact that cartilage is elastic and porous and contains a large amount of extracellular fluid, which creates pressure (**hydrostatic pressure**). Because the elastic cartilage is compressible, whereas its water content is not, additional pressure created by movement squeezes water out of the cartilage in the area of the pressure and toward the surface because the underlying bone is impermeable. Because the cartilage is elastic, it forms a seal around the high pressure area. The additional surface fluid trapped by this process is used for lubrication. This is called *hydrostatic lubrication.*

In summary, normal joint function depends essentially on two types of lubrication—boundary and hydrostatic. These lubricating films produced by mechanisms can separate the two joint surfaces in the healthy joint under all reasonable conditions.

SUPPORTING TISSUES

One type of synovial membrane is frequently found outside the joint cavities where friction occurs. In these instances the membrane forms closed padlike sacs called **bursae,** which contain a fluid similar to synovia. The major function of these specially positioned bursae is to prevent opposing surfaces from rubbing together. Bursae are found between muscles, between muscle and ligament, and in any anatomical situation where muscle and tendon glide over bony prominences or ligaments (Figures 6-7 and 6-8).

Ligaments and muscles are also important to the function of joints because they provide added strength. **Ligaments** are fibrous bands

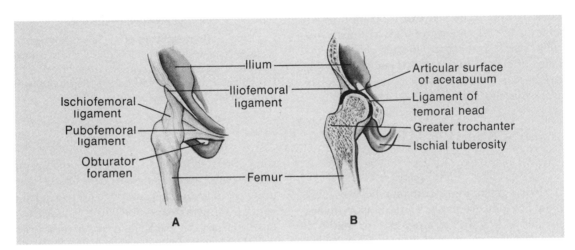

Figure 6-7 Right hip joint, anterior aspect. **A,** The articular capsule is held intact by massive groups of ligaments. **B,** Coronal section of the hip ball-and-socket joint.

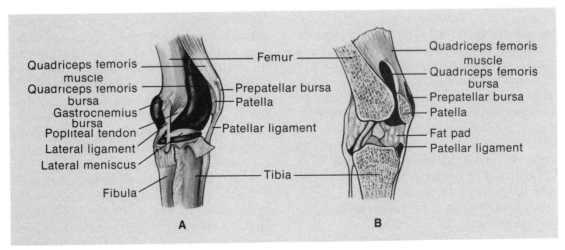

Figure 6-8 Knee, lateral aspect. **A,** Capsule of the right knee joint, indicating the position of the bursae. **B,** Sagittal section of the knee joint.

connecting bones. They may be incorporated into the articular capsule or be independent structures. Ligaments are essentially nonelastic, but they do afford enough flexibility to permit freedom of movement. They are usually arranged so that they remain taut when the joint is in its position of greatest stability.

Muscles, fibrous bands attached to bone that produce movement by contracting, provide an even more important mechanism for maintaining the stability of joints. They are stronger than ligaments, maintaining the articular surfaces in firm contact at every position of the joint during relaxation and contraction. In muscular paralysis a much greater than normal range of motion in the related joints is possible. In Chapter 7, we will take a more in-depth look at the muscles.

DISEASES OF THE JOINTS

Arthritis

Arthritis refers to any inflammatory involvement of a joint. *Suppurative arthritis* is an inflammation within a joint space due primarily to invasion of bacteria, most commonly of the coccus variety. This type of arthritis is more prevalent in middle and later adulthood. Any joint may be involved, but the most frequently affected are large joints such as the knee, hip, ankle, elbow, wrist, and shoulder. The synovial membrane becomes congested and thickens, and the joint fills with a thin, cloudy fluid containing white blood cells. As the disease process advances, the inflammatory alterations in the synovial membrane grow progressively more severe, and the fluid becomes a thick pus characteristic of this disorder. The resultant *synovitis* (synovial inflammation) may lead to ulceration and involvement of the underlying articular cartilage. Thus extensive destruction of the joint surfaces may result in fibrous scars that seriously hamper joint function.

Gouty arthritis is the chief manifestation of a metabolic disorder that results from an alteration in the metabolism of purine, a nitrogen

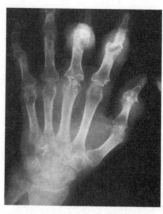

Figure 6-9 Gouty arthritis of the hand. Note extensive destructive changes, especially of the proximal interphalangeal joints. (From *Arthritis*, 8th Ed., by J. L. Hollander. Lea & Febiger, 1976. Reprinted with permission.)

compound. Uric acid, a product of purine metabolism, accumulates in the blood, and urate crystals are deposited in and around the joints, leading to destruction of the articular surfaces and possible permanent loss of joint function (Figure 6-9).

Osteoarthritis is a disease of the joints often seen in older patients. Articulations are subject to a great amount of attrition during normal function, and after years of use degenerative changes appear. Most studies reveal that all persons older than 45 years of age have mild and insignificant degenerative joint disease. These changes appear first in the weight-bearing joints of the spine and the lower extremities. The first pathological change is a softening of the cartilage in the central area of the articular surfaces where the stress of weight bearing is greatest.

The outstanding characteristic of *degenerative arthritis* is nature's attempt at remodeling. In an effort to equalize pressure in the altered joint, the body adds substance to the bony structures. These become lips, spurs (needlelike projections), and hard nodules. This remodeling can completely change the contours of the joint,

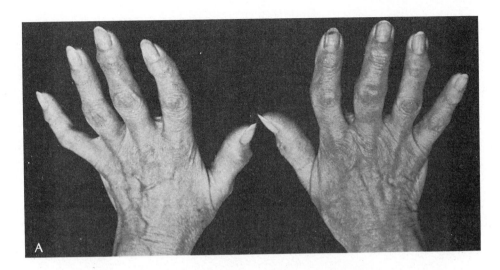

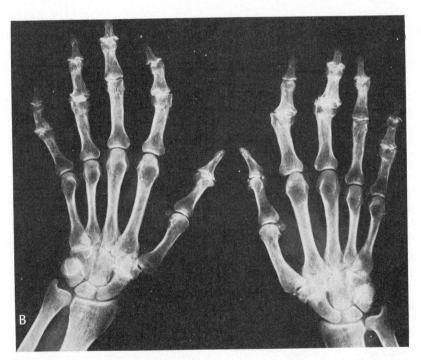

Figure 6-10 **A,** Primary osteoarthritis of the hands with distal interphalangeal joint involvement (Heberden's nodes). **B,** X ray of the same hand. (From *Arthritis*, 8th Ed., by J. L. Hollander. Lea & Febiger, 1976. Reprinted with permission.)

leading to grotesque deformities. The joints of the fingers often have small, hard nodules known as *Heberden's nodes* (Figure 6-10). Subse-quently a separation of the fibers occurs, fol-lowed by disintegration of the cartilage (Figure 6-11).

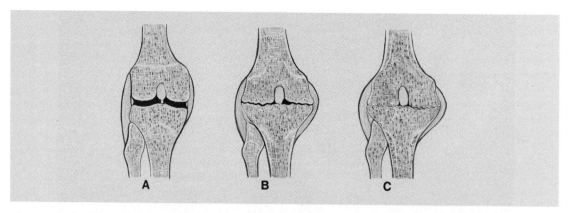

Figure 6-11 Osteoarthritis. Coronal sections of a developing degeneration from **A**, normal, to **C**, final fusion, when the two bones unite to become one. Note the increased width of the joint after fusion.

Rheumatoid arthritis is the most disabling form of chronic joint disease. The condition is systemic, and widespread involvement of surrounding connective tissue usually occurs. In its early stage this inflammation is localized in the joint capsule and synovia of small joints such as those of the hands and feet.

The basic disease process in rheumatoid arthritis is inflammation of the synovial tissue (synovium) secondary to an immunologic reaction. The inflamed synovium thickens, increases in size, and invades the cartilage. Cartilage is destroyed in part by physical invasion and in part by catabolic enzymes produced by the inflamed synovium. The destruction of the articular cartilage results in loss of the joint's normal lubrication mechanism. This in turn causes an increase in friction followed by fragmentation of the cartilage, which accelerates destruction of the joint. The diseased synovium also invades the surrounding joint-supporting tissues, leading to dislocation and deformity. The character of the joint fluid is also changed but, in the face of massive cartilage destruction, probably has little effect in most cases. If unchecked, the ultimate result is a totally destroyed joint (Figure 6-12).

After extensive damage to the cartilage has occurred, the synovitis lessens, and tough fi-

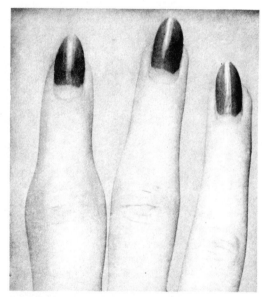

Figure 6-12 Second, third, and fourth fingers of a woman with rheumatoid arthritis; note the marked swelling in the interphalangeal joints. (From *Arthritis*, 8th Ed., by J. L. Hollander. Lea & Febiger, 1976. Reprinted with permission.)

brous tissue prevails in the joint. This tissue adheres tightly to the opposing joint surfaces, causing **ankylosis**—the inhibition or prevention of motion at the joint *(fibrous ankylosis)*. The

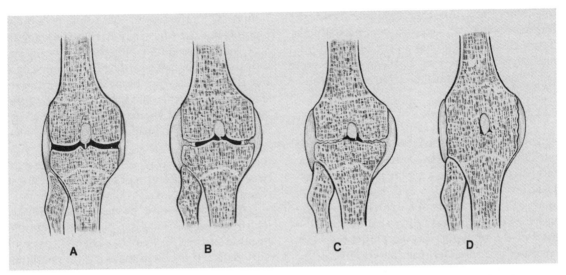

A B C D

Figure 6-13 Coronal sections through joint with developing rheumatoid arthritis. As the disease progresses from **A** to **D**, fibrous tissue between the bones becomes calcified and the two bones fuse into one, resulting in ankylosis, or obliteration of the joint.

fibrous tissue may then become calcified and change to osseous tissue, resulting in a firm bony union that obliterates the joint *(bony ankylosis)*, which is the final stage of the pathologic process (Figure 6-13).

Bursitis

Undue stress to or local infection of a synovial bursa may cause an inflammation called **bursitis.** Although this condition may develop in any of the surrounding bursae outside the joint, the more common locations are the subacromial bursa close to the shoulder joint, the prepatellar bursa near the knee joint, and the olecranon bursa at the point of the elbow. Pain and limited movement of the joint result. Usually there is a varying degree of **fibrositis** (inflammation of fibrous tissue) at the location of the involved bursa, and this worsens the stiffness, pain, and limitation of motion. If the condition is not corrected promptly, wasting of muscle may develop rapidly, and motion may be further restricted.

Inflammations of bursae about the ischial

tuberosities and the attachment of the Achilles tendon cause similar changes.

Bunion

A **bunion** is a deformity of the great toe, accompanied by the formation of a bursa at the metatarsophalangeal joint. It may develop as a result of pressure or friction from badly fitting shoes. The great toe is deviated laterally toward the second toe, and the joint is made unduly prominent.

Sprains and Ruptures

A **sprain** results when the joint is exposed to excessive external stress or pressure. The wrenching of the joint tears some of the fibers of a supporting ligament without displacing the bones or interrupting the ligament's continuity. Many sprains, such as those of the ankle, respond favorably to rest and support with straps, splints, or casts. Complete rupture of the supporting attachments can be much more serious. The strong ligaments and muscles of the thigh, for example, support the knee joint, and a twisting injury may tear the semilunar cartilage. Damage to this cartilage is a common athletic injury. Since cartilage has a poor blood supply, healing is rare. The torn cartilage or torn portion of the cartilage must be removed surgically.

Dislocation

A displacement of one of the bones of a joint or one of the parts that compose the joint is a more serious condition than a sprain; it is called a **dislocation.** The shoulder and knee joints are among the most vulnerable to dislocation. The socket of the scapula is rather shallow, and the head of the humerus may be pulled out of it, resulting in a shoulder dislocation. This sort of displacement can usually be reduced by pulling the arm away from the trunk and rotating it suitably. In the case of the knee, the most common injury is a crushing or tearing of one of the semilunar cartilages, resulting from a twisting injury. Not only is the condition extremely painful, but the knee may have a tendency to lock suddenly. Often surgery involving removal of the damaged cartilage is the only effective treatment. Jarring blows from the front or side can also damage the knee joint, tearing the ligaments that lash the bones together. Such injuries often leave football players and other athletes with a "trick knee," which may unexpectedly give way. An elastic bandage may be used to provide support to the joint; surgical repair may be necessary.

An incomplete or partial dislocation of a joint is called a *subluxation*. Dislocation or subluxation of the hip joint is a common type of birth defect. If it is detected early, the condition can be treated by immobilization of the hip joints in a brace or cast. This stimulates proper bone growth and formation of a fully functional ball-and-socket fit of the femur and hip (os coxae).

OUTLINE

I. Introduction
   A. Joint—articulation between two bones that allows movement
   B. Arthrology—study of joints
II. Classification of joints
   A. Synarthroses
      1. Fibrous
      2. Immovable
      3. Types
         a. Suture—cranial sutures
         b. Synchondroses—epiphysis of long bones
   B. Amphiarthroses
      1. Semimovable
      2. Types
         a. Symphysis—pubic bones
         b. Syndesmoses—fibrous tissue
   C. Diarthroses
      1. True joints—allow free movement
      2. Capsule lined with synovial membrane
      3. Synovial membrane secretes lubricating fluid—synovia
      4. Synovia fills joint cavity
      5. Types
         a. Hinge—elbow and knee
         b. Ball-and-socket—hip and shoulder
         c. Pivot—first and second cervical vertebrae
         d. Condyloid—radiocarpal joint
         e. Gliding—between carpal bones
         f. Saddle—thumb
III. Movement at synovial joints
   A. Extension
   B. Flexion
   C. Abduction
   D. Adduction
   E. Rotation

F. Circumduction
G. Supination
H. Pronation
I. Inversion
J. Eversion
K. Protraction
L. Retraction

IV. Structure of a synovial joint
   A. Articular cartilage
   B. Articular capsule
   C. Synovial membrane
   D. Supporting and protective structures
      1. Ligaments
      2. Muscles
      3. Bursae

V. Lubrication of a synovial joint
   A. Synovial membrane—secretes mucin composed of mucopolysaccharide (hyaluronic acid)
   B. Articular cartilage
      1. Porous
      2. Elastic
   C. Boundary lubrication—mucin clings to cartilage surface

D. Hydrostatic lubrication—water compressed out of cartilage

VI. Diseases of the joints
   A. Arthritis
      1. Suppurative—bacterial invasion
      2. Gouty—metabolic disorder
      3. Osteoarthritis—degeneration in middle or old age
      4. Rheumatoid—connective tissues as well as joints are involved
         a. Ankylosis—fusion of two bones into one
   B. Bursitis—stress or infection of synovial bursa
   C. Bunion—inward angulation of great toe and formation of bursa at metatarsophalangeal joint
   D. Sprains and ruptures
      1. Sprain—incomplete tear of some ligament fibers
      2. Rupture—disruption of tissue continuity
   E. Dislocation—displacement of one of the bones of a joint

STUDY QUESTIONS AND PROBLEMS

1. Describe the structure of a synovial joint.
2. List at least five different joints and indicate where they are found.
3. What is a diarthrodial joint? Can it be moved?
4. Where are the semimovable joints found?
5. Distinguish among synarthroses, amphiarthroses, and diarthroses with respect to structure of joint and degree of movement permitted.
6. What would happen if bone surfaces in movable joints were in direct contact, without intervening cartilage or fluid?
7. Which type of diarthrosis permits: (a) the freest movement? (b) movement in one plane only? (c) rotating movement? (d) sliding movement?
8. Distinguish between: (a) flexion and extension; (b) abduction and adduction; (c) protraction and retraction.
9. Define, by function, muscle, bone, joint, and ligament.
10. What is boundary lubrication? Differentiate boundary from hydrostatic lubrication.
11. What is arthritis?
12. Differentiate among the four types of arthritis described in the text.
13. Of what use are bursae? What is bursitis?
14. Distinguish among: (a) sprain, rupture, and dislocation; (b) arthritis, rheumatism, and bursitis; (c) rheumatoid arthritis and osteoarthritis.
15. What is a bunion? How may it result in rheumatoid or degenerative arthritis?

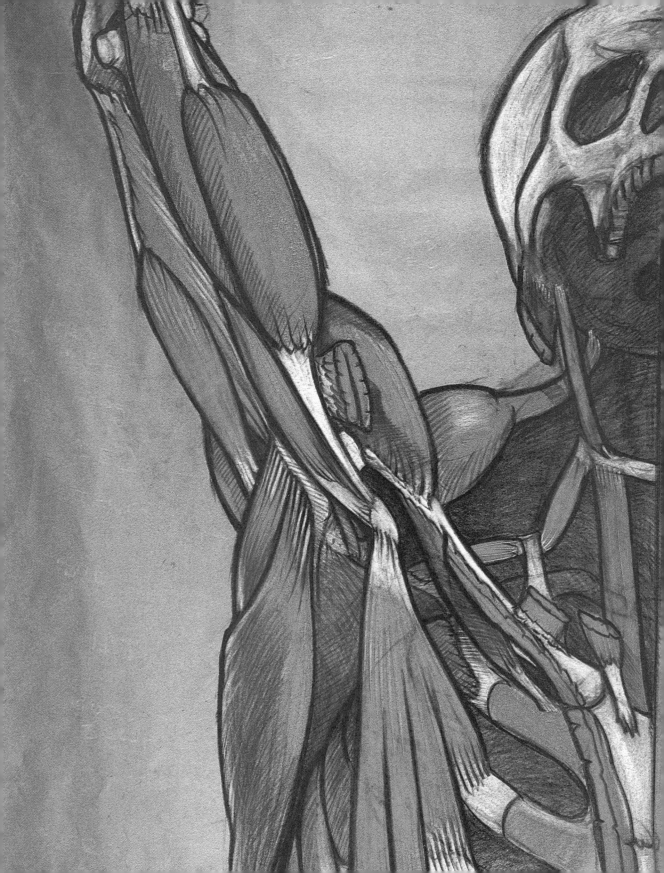

7

The Muscles

- Describe the structure and control of three different types of muscles

- Describe the microscopic anatomy of a contractile unit of muscle

- Explain the actomyosin complex theory

- Explain the biochemistry behind the three energy sources

- Detail a single muscle contraction

- Differentiate between tetany and fatigue

- Explain the difference between spastic and flaccid paralysis

- Describe the different body movements

- Analyze the major muscles according to their origin, insertion, and function

- Distinguish between isometric and isotonic contractions

- List how skeletal muscles are named

- State the all-or-none law

- Describe the oxygen debt phenomenon

- Distinguish between antagonistic and synergistic muscles

- Explain muscular dystrophy, and relate its genetic significance

- Diagram and label a neuromuscular synapse

- Identify conduction deficiencies resulting in muscle disorders

IMPORTANT TERMS

acetylcholine	creatine phos-	fibrositis	myosin	spasm
actin	phate (CP)	flaccid	myositis	stimulus
adenosine tri-	defibrillator	hernia	origin	synergist
phosphate (ATP)	dystrophy	hypertrophy	paralysis	tendon
antagonist	elasticity	insertion	peristalsis	tenosynovitis
aponeuroses	epimysium	isometric	phosphocreatine	tetany
asphyxiation	excitability	isotonic	response	tonus
atrophy	extensibility	motor end plate	sarcolemma	tropomyosin
cardiac	fascia	motor unit	sarcomere	troponin
cardiac muscle	fasciculus	myalgia	sarcoplasm	viscera
clonus	fatigue	myofibril	skeletal muscle	
contractility	fibrillation	myopathy	smooth muscle	

Motion, an essential function of the human body, is made possible by the special property of contractility in muscle tissue. In the broad sense of the term, motion includes not only the movements of the entire body or parts of the body in their physical environment, but also the internal movements of breathing and the beating of the heart. Movements of the alimentary (gastrointestinal) organs and glands as well as those of the blood and lymph vessels also must be considered.

MICROSTRUCTURE OF MUSCLE

In general, the cells and fibers of muscle tissue are specialized to possess the property of contraction. This process is accomplished by a delicate network of intercellular organelles, the muscle fibrils or **myofibrils,** which are complicated protein molecules that surround the nucleus and fill the muscle cell. When the myofibrils contract, the entire muscle fiber is made to contract or shorten.

The fibrils of muscle fiber are enclosed in a delicate but strong sheath, the **sarcolemma,** which also encloses the semifluid protoplasm of the fiber, the **sarcoplasm.** A fiber may have one or many nuclei, depending on its type and function. Individual fibers are arranged so that they lie parallel to one another, and are held together in bundles **(fasciculi)** by interlacing networks of areolar connective tissue, the **fascia.** The fibers are small, ranging from 0.001 to 0.1 mm in diameter (about the size of a human hair) and from 0.1 to 45 mm (about 2 inches) in length.

Types of Muscle

Three types of muscle can be identified by their structure, function, and location in the body: smooth muscle, skeletal muscle, and cardiac muscle.

Smooth Muscle

The fibers of **smooth muscle** are made up of long, slender, tapered cells grouped into thin sheets. This muscle is called smooth because of the heterogeneous arrangement of protein fibrils in its sarcoplasm. The fibrils are not grouped in large bundles as in other muscle types; therefore, they are not as readily visible under the microscope.

Each of the elongated, spindle-shaped cells of smooth muscle has a single, large central nu-

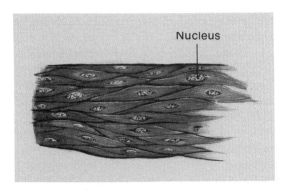

Figure 7-1 Smooth muscle. Each fiber has a single, central nucleus.

siderably. It is peculiar in its ability to be stretched without a change in tension. This property is exhibited by the urinary bladder and by the uterus during pregnancy.

All smooth muscle is termed *involuntary* because its innervation originates in the autonomic nervous system, which is not under conscious control by the brain. There are nerve endings about smooth muscle cells, but motor nerve connections, or end-plates, to the muscle fiber have not been found. It is probable that impulses are transmitted by a chemical agent or the mechanical pull exerted by one cell in contraction, which serves as a stimulus for other cells to contract.

cleus (Figure 7-1). The cells are usually arranged in layers surrounding viscera, forming an inner, thick, circular coat and an outer, thin, longitudinal coat. The rhythmic waves of contraction of smooth muscle that pass along the walls of a hollow organ, such as the intestine, are called **peristalsis.**

Visceral muscle is most often involved in slow, sustained contractions. A single contraction may take as long as a few minutes, and the muscle tissue is capable of being stretched con-

Skeletal Muscle

The fibers of skeletal muscle are constructed to fulfill their primary purpose of allowing skeletal movement. Each skeletal muscle can be thought of as a separate organ, since each is surrounded by its own sheath of connective tissue, the **epimysium.** Extensions of the epimysium form a tough, fibrous connective tissue, the **tendon,** which connects muscle and bone (Figure 7-2). Because the overall action of skeletal mus-

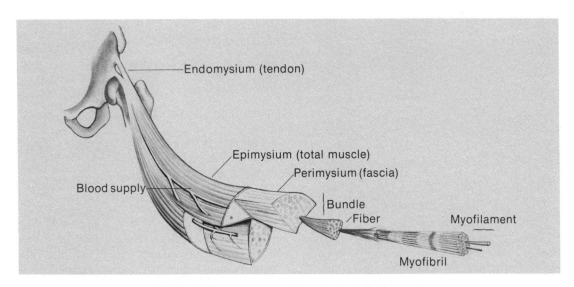

Figure 7-2 Cross-sectional detail of a skeletal muscle.

cles is directed by the higher brain centers, their action is voluntary, and they are therefore called *voluntary muscles.*

The striped or striated appearance of skeletal muscle fibers under the microscope has led to a further designation, *striated voluntary muscle.* Each fiber consists of a long, tubular cell containing many nuclei that are positioned at the periphery of the sarcoplasm beneath the sarcolemma. Microdissection reveals cross-striped myofibrils that are closely packed together and run lengthwise through the entire fiber. The sarcoplasm surrounds the fibrils, and has a soft, contractile consistency (Figure 7-3). Electron microscopic studies show that a myofi-

bril is composed of two distinct filaments, a thick filament (about 100 Å in diameter) and a thin filament (about 50 Å in diameter). (An angstrom, Å, is 1/10,000,000 of a millimeter.) These filaments are about 1.5 and 2.1 mm long, respectively. (A micrometer is 1/1000 of a millimeter.) Chemical analysis has demonstrated that the thick filament contains **myosin;** whereas the thin filament contains **actin.** Actin and myosin are large protein molecules ultimately responsible for muscle contraction (Figure 7-3). The arrangement of actin and myosin filaments contributes to the alternating light and dark areas on the striated muscle fiber. It is the overlapping of the filaments that provides the shaded

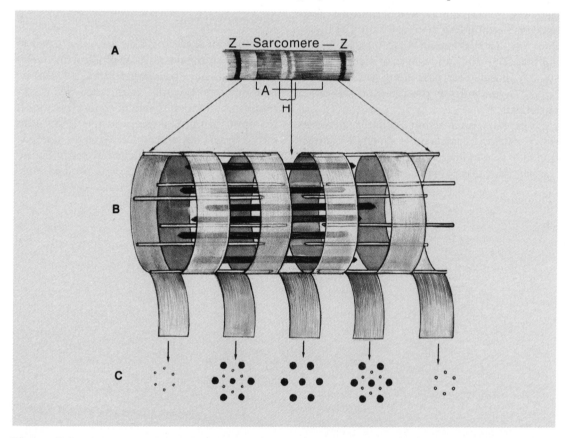

Figure 7-3 Three-dimensional transverse section through a skeletal muscle fiber. **A** and **B,** Each single muscle fiber is made up of myofibrils that appear as striations and are repeated in organized networks of bands called sarcomeres. **C,** The thick and thin fila-

ments are arranged in a hexagonal pattern, with a thick filament surrounded by many thin filaments, except in areas where they do not overlap one another.

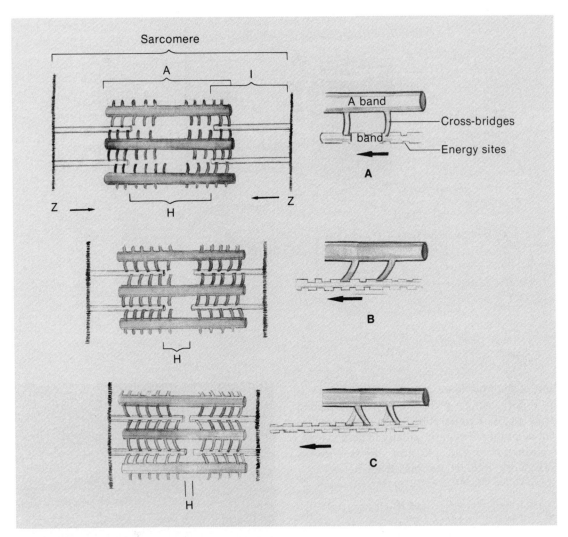

Figure 7-6 Contraction of a muscle. **A,** Sliding filament theory proposes that the A bands contain flexible cross-bridges that come in contact with energy sites on the more numerous I bands. **B** and **C,** With the availability of energy, the cross-bridges pull the active filament a short distance **(B)**, release it, and attach to another site **(C)**, resulting in a shortening of the H zone between the I bands: contraction.

energy (input) of a muscle that can perform work is less than 25 percent, the remainder being lost as heat or used to maintain normal body temperature. This quantity can be realized only when the muscle contracts at a moderate speed—neither too slowly nor at an excessively rapid rate.

Many features of muscle contraction can be demonstrated by a *kymograph,* a device used to elicit muscle twitches in the laboratory (Figure 7-7). This can be accomplished either by stimulating the nerve that conducts impulses to a muscle or by stimulating the muscle belly itself. The stimulus causes a single, sudden contrac-

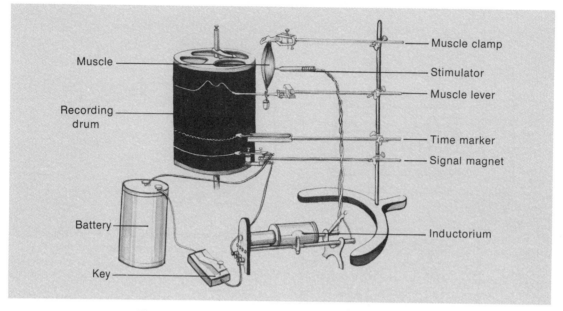

Figure 7-7 Kymograph used in analyzing muscle action.

tion of the muscle—a muscle twitch (Figure 7-8)—which can be recorded on the blackened drum of the kymograph. Succeeding twitches can be produced experimentally to demonstrate environmental and physical effects such as heat, work, cold, fatigue, and so forth.

Properties of Muscle Related to Contraction

The distinct properties of muscle related to contraction are excitability, contractility, extensibility, and elasticity.

Excitability

Excitability (irritability) is the property of being able to receive a stimulus and respond to it. Thanks to this property it is possible to stimulate a muscle directly with an electric shock (current) and cause it to contract or respond. This is of utmost importance in the treatment of paralysis due to motor deficiencies or to disorders caused by disease. A muscle that does not

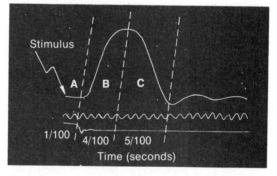

Figure 7-8 A single muscle twitch. **A,** Latent period of approximately 1/100 of a second after muscle receives an impulse. **B,** Contraction period: muscle works for 4/100 of a second. **C,** Muscle returns to normal size and shape during the relaxation period, 5/100 of a second.

respond over a prolonged period will degenerate, or **atrophy.**

Contraction does not occur immediately upon the presentation of an adequate stimulus. A finite period is necessary for the chemical processes resulting in a contraction to take place

(Figure 7-8). This period, approximately 1/100 second, is called the *latent period.* At its end, the muscle begins to shorten and thicken, a period of approximately 4/100 second called the *contraction period.* During the contraction period, tension is developed. After the work is done, the muscle returns to its normal size and shape and undergoes a relaxation period of about 5/100 second. As an example, the frog's gastrocnemius muscle takes about 1/10 second to respond to an adequate stimulus and to return to normal.

At this point, some definitions of terms are necessary. A **stimulus** is an excitant or irritant that causes a reaction, such as a movement. A stimulus can be one of the following:

- *Liminal (threshold)*—the weakest stimulus from a nerve fiber that will cause a response

- *Subliminal*—any stimulus of lesser intensity than the liminal or threshold

- *Maximal*—a stimulus that causes all the motor units of the muscle to respond, resulting in the greatest possible contraction of the muscle

- *Adequate*—a stimulus of the specific form of energy to which the nerve fiber is most sensitive

A **response** is the reaction, such as movement of a muscle, that results from an adequate stimulus.

Although a single subliminal stimulus cannot make a muscle respond, a series of subliminal stimuli applied to a muscle can elicit a response. This is spoken of as the *summation effect.*

Contractility

Contractility is the property of a muscle that enables it to change its shape. As previously stated, the conversion of potential energy (ATP) into energy for muscle contraction is accompanied by the loss of some energy in the form of heat.

When a muscle contracts **isometrically,** it remains the same length, and all the energy de-veloped escapes as heat, since no mechanical work is performed. When a muscle develops enough tension to overcome its work load, the muscle contracts and performs mechanical work. An **isotonic** contraction is one in which the tension within the muscle stays the same.

Examples of these types of contraction can be found in two types of exercise. In exercises using the principle of opposing forces, for example, pushing against a wall, no mechanical work is done; thus the muscle works only to expend a great deal of energy as heat. Contrast this isometric exercise with the popular weightlifting procedures, in which the muscle overcomes the opposing weight and contracts isotonically; the muscle does mechanical work as well as generates heat.

The basic unit of muscular contraction is the **motor unit,** which consists of many muscle fibers, all innervated by the same motor nerve fiber. These fibers are actually intermeshed so that a single impulse causes a contraction throughout the entire muscle body, not only in one area of the muscle. (See "Conduction of the Contractile Impulse," p. 168.)

Motor units are subject to two basic laws of contraction: the *all-or-none law* means the muscle fiber will either respond entirely to a stimulus or not respond at all. With the *law of tonicity* the healthy muscle is in a state of partial, continual contraction or **tonus.** Tonus is also described as the involuntary resistance to stretch (stretch reflex).

These two laws might seem at first to conflict with each other, but in reality they are complementary in their actions. The all-or-none principle applies strictly to the individual fiber, whereas the law of tonicity applies to the entire muscle, that is, to hundreds of fibers. The fundamental mechanism whereby muscle tone (tonus) is produced is not fully understood, but physiologically tone is known to be due to nerve impulses.

Tonus in skeletal muscle is created by the stretch reflex. It is completely dependent on the central nervous system for maintenance. Muscles that control the skeleton are stretched over the joints. One example is the extensor muscles used for maintaining posture. By means of tonic contraction, posture is maintained for long periods with little evidence of exhaustion. This effect is achieved mainly by a mechanism in which different groups of muscle fibers contract in relays, giving alternating periods of rest and activity to given groups of fibers. In humans, the retractors of the neck and extensors of the back are among the muscles exhibiting the highest degree of tonus.

Muscle tone is often reduced during illness or when prolonged bed rest is necessary. Muscles can lose their tone in several ways. When its nerve is cut, injured, or diseased, a muscle does not receive a constant flow of impulses and thus loses its tone. It soon becomes **flaccid** (soft) and then atrophies. This process is called the *atrophy of denervation.*

Tone may also be lost by immobilization or inactivity of a part, such as in bedridden people or when limbs are held in a cast for long periods. The normal flow of nerve impulses to such muscles is reduced, and atrophy follows. This is called *atrophy of disuse* and is not as serious as atrophy of denervation, since the nerve connection remains normal, and the muscle can recover when it is again normally active.

Tonus in smooth muscle differs from that of skeletal muscle in that it is dependent on the peripheral rather than the central nervous system (see Chapters 8 and 9). Smooth muscle tonus can be influenced markedly when drugs are introduced into the tissue fluid that bathes the muscle fiber. The response to a particular stimulus may occur as a rhythmic contraction, an increase in tonus, or a decrease in tonus. Drugs such as epinephrine (adrenaline) and sympathin are known to stimulate a greater activity of the muscle; other drugs, such as strong concentrations of acetylcholine and curare, lead to a general loss of tone and muscle activity.

Cardiac muscle tone is also independent of the central nervous system. Because of its special nature, it will be discussed in greater detail later in this chapter.

Extensibility

Extensibility is the property that allows the muscle to be stretched or extended. All types of muscle can be stretched when force is applied. For example, this property makes possible the increase in size of the urinary bladder as urine accumulates. If the bladder could not distend, a very uncomfortable dribbling of urine would occur instead of the usual periodic emptying.

Elasticity

Elasticity is the ability of a muscle to return to its original form when a stretching force is removed. Provided muscle damage (rupture) has not occurred, on removal of the weight the muscle will return to its normal length. Since most muscles of the body are generally somewhat stretched, the two properties of extensibility and elasticity are of value in that they tend to keep the muscle in continuous readiness for contraction.

Some Phenomena of Contraction

Skeletal muscles contract quickly and relax promptly. In contrast, visceral (smooth) muscle contractions develop slowly, are maintained for some time, and fade slowly. The contraction of a skeletal muscle is the result of a stimulus discharged by the nerve fibers innervating it. If a single contraction were to be analyzed, it would show brief periods of pause, contraction, and relaxation, as discussed previously. In general, the stronger the stimulus (up to a certain maximum), the stronger the contraction. Muscles do their best work at body temperatures of about 37°C (98.6°F). If the temperature is raised or lowered, the muscle will respond accordingly in the strength and duration of its contraction (Figure 7-9). If the temperature is raised much above 37°C, the muscle loses excitability and becomes functionally depressed very rapidly—a condition called *heat rigor,* a permanent short-

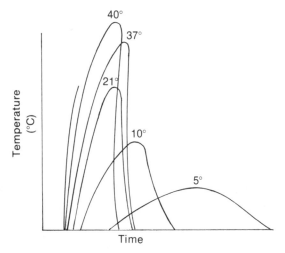

Figure 7-9 The effect of temperature on muscle action. The lower the temperature, the longer the contraction, but the strength of the contraction is lessened.

ening of the muscle fibers. This effect of temperature is understood and put to use by athletes. Before strenuous muscular activity during physical contests, they use "warm up" exercises or wear jackets to keep their bodies at a constant temperature. This prevents the body from cooling off and thereby keeps the muscle from contracting less efficiently.

As discussed earlier in relation to isotonic contraction, the development of tension enables a muscle to overcome resistance and, under proper conditions, to perform such mechanical work as lifting a weight. The amount of mechanical work done is determined by multiplying the weight of the load by the height to which the load is lifted. The result expresses the work done by the muscle. When a muscle contracts without a load, or when the load is too heavy to be lifted, no mechanical work is done.

Tetany occurs when a number of stimuli are applied to a muscle in rapid succession, so that little time is offered for relaxation between the successive contractions (Figure 7-10). There is a degree of fusion of the twitches; the greater the frequency of the stimuli, increasing from three to eight impulses per second, the more nearly complete the fusion (*incomplete tetany*), until at sixteen impulses per second the recording shows no evidence of individual twitches. This steady state of contraction is called *complete tetany*. The tension developed during tetany may be four or more times as great as that during a single twitch.

A single muscle fiber contraction would be of little use in coordinated movement. Those muscles that enable us to stand, walk, run, or bend maintain a state of physiological tetany for

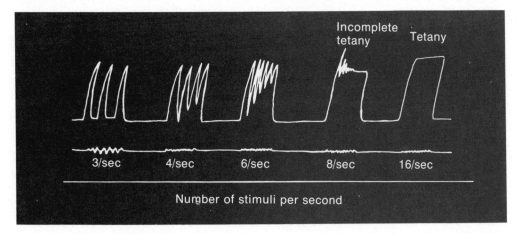

Figure 7-10 Tetanic contraction. As the number of impulses to a muscle is increased, the rest periods become less until finally the muscle cannot rest: tetanic contraction.

a long period without ill effect. The condition of tetany can apply, however, only to a portion of a muscle rather than to the entire muscle. The slow flexing of a muscle is a coordinated movement involving many fibers in a state of contraction, and is much slower than a muscle twitch.

Tetany is closely related to muscle tone. In the maintenance of muscle tone, only a few motor units are involved at one time, so that no actual movement occurs. If all the motor units in use contracted and relaxed at the same time, the result would be a series of jerks instead of a smooth, coordinated contraction. Such a condition, called **clonus,** sometimes occurs in response to forcible extension of some muscle.

A **spasm** is a sudden involuntary contraction of a skeletal muscle. If it persists, it is called a *tonic spasm,* or *cramp.* The characteristic "charley horse" is a spasm due to an overexertion of the muscle of the thigh. Most cramps or spasms are painful.

In muscles undergoing contraction, the formation of carbon dioxide and lactic acid, both toxic waste products of cellular metabolism, first causes an increase in the irritability of the muscle. But, if the muscle is continuously stimulated, the strength of contraction becomes progressively less until the muscle refuses to respond (Figure 7-11). This is true **fatigue** and is

caused by the toxic effects of waste products that accumulate during exercise. Added factors are the loss of nutritive materials and the inability of the muscle to store sufficient oxygen to compensate for strenuous exertion. In fatigue not only does the muscle gradually fail to contract, but it also fails to relax (Figure 7-11). This explains why muscle fatigue is sometimes confused with a cramp or spasm.

Exercise stimulates circulation, thereby bringing fresh blood to the tissues and altering the local pressure in the muscle. Exercise increases the size, strength, and tone of muscle fibers. Massage and passive exercise (as in hydrothermy machines) may, if necessary, be used as substitutes for active exercise. Although physical exercise is desirable to aid the body's metabolism, the continued use of fatigued muscles is harmful if their energy supply (glycogen) is exhausted and they must utilize the protein of their own cells. Normally the sensation of fatigue protects us from such extremes.

CONDUCTION OF THE CONTRACTILE IMPULSE

We can control with great precision the force applied by our muscles. We can perform such delicate actions as picking up a tiny insect or,

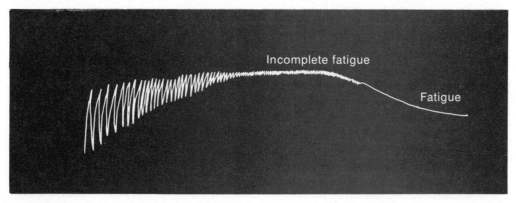

Figure 7-11 Fatigue. During stimulation at intervals of 1 second, the strength of contraction of this muscle diminished until it could not respond at all: fatigue.

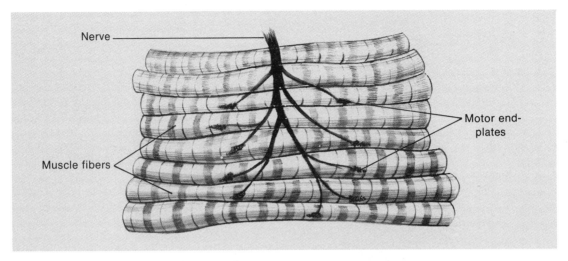

Figure 7-12 Innervation of skeletal muscle.

using a powerful grip, lift a hundred-pound weight. This gradient of force is regulated by the nervous system. The muscle fibers contract in response to impulses from the central nervous system, to which every skeletal muscle is attached by means of a branch of a nerve fiber. The impulse proceeds down each nerve fiber branch, ending at the neuromuscular junction, where it acts to trigger the muscle fiber and cause contractions. As discussed earlier, a nerve fiber and its associated muscle fibers together form a motor unit. In the large muscles of the legs, a single motor nerve may serve as many as one hundred muscle fibers (Figure 7-12).

At one time it was thought that the nerve impulse crossed to the muscle fibers in the same way that an electrical spark jumps across a gap. It is now known that the arrival of nerve impulses at the motor end-plate (Figure 7-13) causes the production of **acetylcholine,** a chemical that alters the muscle cell membrane like an electrical shock. The wave of excitation (stimulation) produced spreads along the muscle fiber in each direction until the muscle fiber contracts.

Acetylcholine is the mediator or carrier at all junctions involving muscle and nerve.

Within a fraction of a second, acetylcholine is destroyed by the enzyme, cholinesterase, which prevents the reexcitation of the muscle fiber until the arrival of another impulse. If the breakdown of acetylcholine into acetyl and choline did not occur, the chemical would keep the synaptic bridge intact and allow continual contractions, which could result in exhaustion of the muscle.

The synthesis of acetylcholine at the end-plate involves a second enzyme, choline acetylase, which is found in large quantities at nerve endings where acetylcholine is the mediator. Choline acetylase catalyzes the production of acetylcholine from the available acetyl and choline compounds. It remains in an inactive state until an impulse from the motor nerve activates it again.

Events at the neuromuscular junction can be modified in many ways, such as by disuse or drugs. Curare, a South American Indian poison, acts to bind acetylcholine and block transmission of the impulse across the neuromuscular junction, thus preventing the contraction of muscles. This knowledge has led to the controlled use of curare and associated drugs in conjunction with anesthesia for various types of

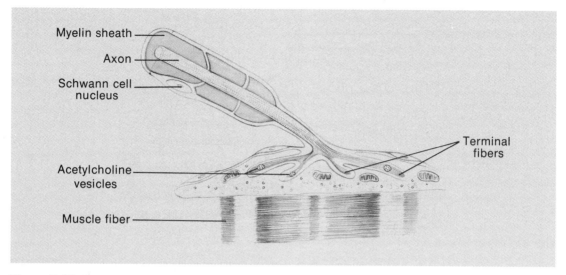

Myelin sheath

Axon

Schwann cell
nucleus

Terminal
fibers

Acetylcholine
vesicles

Muscle fiber

Figure 7-13 Cross-section of a motor end-plate on a skeletal muscle, showing the actual connection between the muscle fiber and the terminal branch of the nerve fiber.

surgery where muscular relaxation is desired. Curare has also been useful in the emergency treatment of children who develop muscular spasms of the throat (croup), preventing strangulation or suffocation in such patients.

Succinylcholine also blocks transmission at the myoneural junction. Its action involves binding the cholinesterase and preventing the muscle fibers from returning to normal after they have contracted once.

Some complex organic fluorine compounds are potent inhibitors of cholinesterase. These compounds are the so-called nerve gases developed as military weapons. They can cause loss of muscle function (**paralysis**) and death from **asphyxiation,** an inadequate oxygen supply to the tissues.

Another group of substances affects the release of acetylcholine from the nerve terminals. Botulinus toxin, produced by *Clostridium botulinum*, blocks the release of acetylcholine, preventing excitation of the muscle membrane. Botulinus toxin is responsible for many deaths from food poisoning. This, again, is a weapon used in germ warfare.

Myasthenia gravis, a condition characterized by weakness of the skeletal muscles, is due to a blockage of impulses at the neuromuscular junction. The beneficial therapeutic action of anticholinesterase agents indicates that myasthenia is essentially a problem of competitive inhibition by a curarelike substance.

SKELETAL MUSCLES

Skeletal muscles are connected, for the most part, to the skeleton, and their principal function is to produce movement of one bone upon another. To accomplish this, they must have two attachments (Figure 7-14*A*): one to the bone that is fixed, called the **origin** of the muscle; the other to the bone to be moved called the **insertion** of the muscle. This does not mean that muscles attach only to bones. Some attach to other muscles as well.

As mentioned earlier, muscles are attached to bones by tendons, cords or bands of white fibrous connective tissue. **Aponeuroses** are

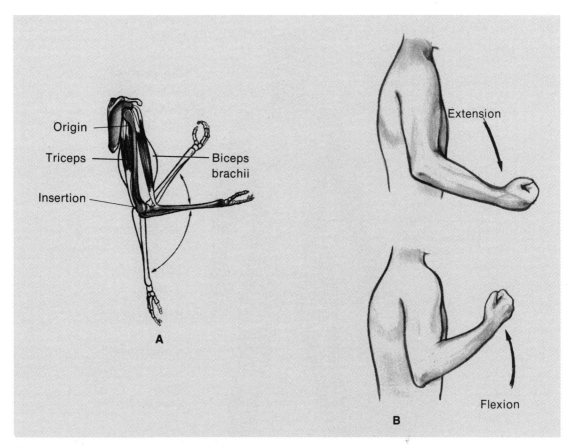

Figure 7-14 **A,** Skeletal muscle attachments. The origin is attached to the less movable bone, whereas the insertion is attached to the more movable bone. **B,** Two examples of muscle action.

sheets of connective or membranous tissue that connect muscle and the part it moves.

Muscles are named in various ways as follows:

- *Location*—examples: intercostal (between the ribs), brachii (arm), occipitalis (head)

- *Action*—examples: adductor, flexor, extensor

- *Shape or size*—examples: trapezius (trapezoid), maximus (largest), minimus (smallest), longus (long), brevis (short)

- *Direction of fibers*—examples: rectus (straight), transversus (across)

- *Number of heads of origin*—examples: biceps (two heads), triceps (three heads), quadriceps (four heads)

- *Points of attachment*—example: sternocleidomastoid (sternum, clavicle, and mastoid process of the temporal bone)

The motions produced by muscle contraction are described with reference to the standard anatomical position of the body—standing erect with arms straight down and palms directed forward (Figure 7-15). As described in Chapter 6, movements parallel to the longitudinal axis are flexion, or folding together, and ex-

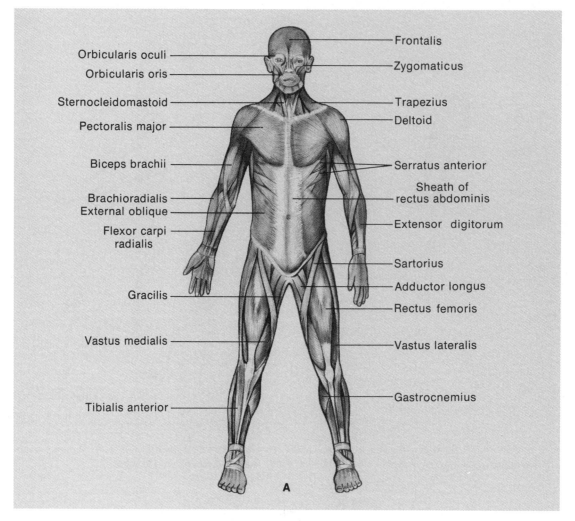

Orbicularis oculi

Orbicularis oris

Sternocleidomastoid

Pectoralis major

Biceps brachii

Brachioradialis
External oblique

Flexor carpi
radialis

Gracilis

Vastus medialis

Tibialis anterior

Frontalis

Zygomaticus

Trapezius

Deltoid

Serratus anterior

Sheath of
rectus abdominis

Extensor digitorum

Sartorius

Adductor longus

Rectus femoris

Vastus lateralis

Gastrocnemius

A

Figure 7-15 Muscles of the body. **A,** Anterior aspect.

tension, an unfolding movement in the opposite direction (Figure 7-14*B*). Abduction refers to movement away from the axis; adduction means toward the axis of the body.

Muscles that produce movement in directions opposite to one another are **antagonists** (Figure 7-14). Muscles that work together to accomplish a particular movement are **synergists.** Two muscles that bend an arm at the elbow are synergists with one another (for example, the biceps and brachialis flex the lower arm), but both are antagonistic to a muscle that straightens the arm (the triceps extends the lower arm).

Muscles of Facial Expression and Mastication

The muscles of facial expression enable one to express physically such feelings as disdain, pleasure, pain, disgust, fear, anger, and surprise (Figure 7-16 and Table 7-1). The muscles of

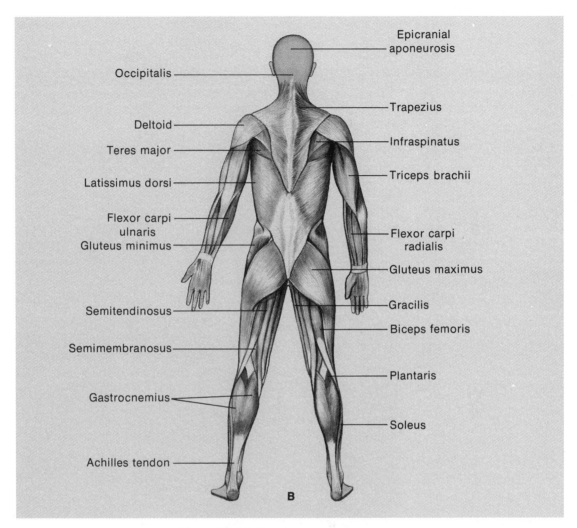

Figure 7-15 *Continued.* **B,** Posterior aspect.

mastication aid in the digestive process by providing power for chewing food before it is swallowed.

The epicranius consists of two separate muscular portions: the *occipitalis* and *frontalis* muscles covering their respective cranial bones. Connecting the two is a large sheet of fibrous tissue, the *galea aponeurotica* of the scalp. The epicranius as a whole raises the eyebrows and moves the scalp, thereby expressing surprise.

The *orbicularis oculi* encircles the eyelids and closes the eye (Figures 7-15*A*, and 7-16). It is a flat, elliptical muscle that arises from the nasal portion of the frontal bone and the frontal process of the maxilla. The fibers spread laterally, forming a broad, thin layer that occupies the eyelid, encircles the orbit of the eye, and spreads over part of the temple and downward on the cheek. Contraction of some of the fibers causes the eyelids to close gently, as in sleep. Contraction of most fibers creates a squinting effect, as when looking into the sun.

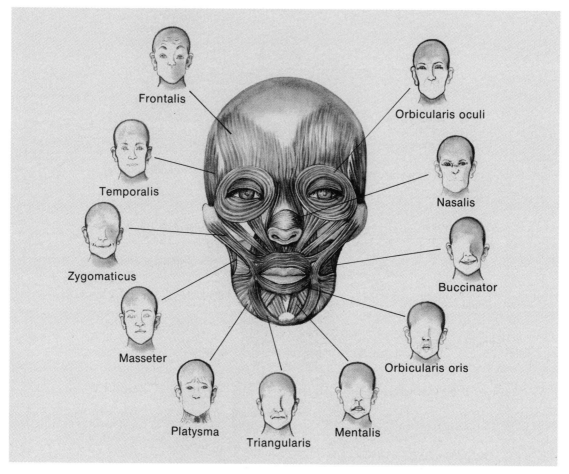

Figure 7-16 Muscles of facial expression and mastication.

The *orbicularis oris* encircles the mouth and functions in closing the lips (Figures 7-15A, and 7-16). Some of the fibers are derived from other facial muscles that are inserted into the lips; some pass in an oblique direction from under the surface of the skin through the thickness of the lips to the mucous membrane; others connect the maxillae and septum of the nose above with the mandible below. This muscle may cause protrusion or pouting of the lips.

The *zygomaticus* comprises major and minor parts and arises from the zygomatic bone (Figures 7-15A, and 7-16). The major part inserts into the skin at the angle of the mouth; the mi-

nor part inserts into the orbicularis oris muscle. As a whole, the zygomaticus pulls the angles of the mouth upward, backward, and laterally, as in smiling or laughing.

The *buccinator* is the muscle of the cheek (Figure 7-16). It arises from the maxilla and mandible, converges toward the angle of the mouth, and inserts into the orbicularis oris. This muscle draws the corner of the mouth laterally and flattens the cheek. It aids in swallowing, whistling, and blowing such wind instruments as the trumpet, from which it received its name. It compresses the cheek during mastication, keeping food under the pressure of the teeth.

Table 7-2   Muscles of the Head, Shoulder, and Arm

Muscle	Origin	Insertion	Action
Sternocleidomastoid	Two heads from sternum and clavicle	Mastoid process of temporal bone	Acting individually, each draws head toward shoulder of same side; acting together, they flex head forward
Semispinalis capitis	First six thoracic vertebrae; last four cervical vertebrae	Occipital bone	Extends and rotates head
Trapezius	Occipital bone; seventh cervical and all thoracic vertebrae	Acromial process of clavicle and spine of scapula	Draws scapula toward spine; rotates scapula to lift shoulder
Levator scapulae	Upper four cervical vertebrae	Vertebral border of scapula	Elevates scapula
Rhomboid major	Second to fifth thoracic vertebrae	Vertebral border of scapula	Elevates and retracts scapula
Rhomboid minor	Last cervical and first thoracic vertebrae	Vertebral border of scapula	Elevates and adducts scapula
Latissimus dorsi	Last six thoracic vertebrae; lumbar vertebrae; iliac crest	Anterior surface of upper humerus	Adducts, extends, and rotates arm medially
Pectoralis major	Cartilages of first six ribs, sternum, and medial half of clavicle	Crest of greater tubercle of humerus	Flexes, adducts, and rotates arm medially
Pectoralis minor	Cartilages of second to fifth ribs	Coracoid process of scapula	Pulls scapula forward and downward
Serratus anterior	Lateral surface of first nine ribs	Vertebral border of scapula	Draws scapula forward and laterally; slight rotation
Coracobrachialis	Coracoid process of scapula	Medial surface of humerus	Flexes and adducts arm; medial rotation of arms
Deltoid	Spine of scapula; clavicle; acromial process	Lateral surface (deltoid tubercle) of humerus	Abducts arm
Teres major	Lower angle of scapula	Anterior surface of humerus (lesser tubercle)	Adducts, extends, and rotates arm medially
Teres minor	Axillary border of scapula	Greater tubercle of humerus	Rotates arm laterally
Supraspinatus	Supraspinous fossa of scapula	Greater tubercle of humerus	Abducts arm and rotates it laterally
Infraspinatus	Infraspinous fossa of scapula	Middle facet of greater tubercle of humerus	Rotates arm laterally

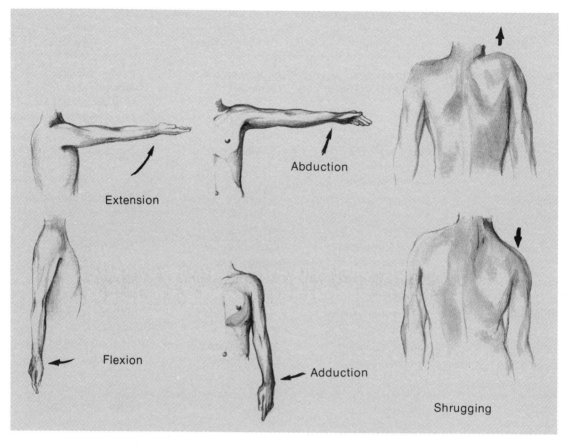

Figure 7-20 Movements at the shoulder.

curs in swimming. This muscle is the chief muscle used in downward blows (karate) or in lifting the body (climbing).

The pectoralis muscles form the anterior axillary fold (Figure 7-21). The *pectoralis major* originates on the clavicle, sternum, and first six rib cartilages and inserts on the greater tubercle of the humerus (see Figure 7-15A). This muscle adducts the arm, draws it across the chest, and rotates it medially. It draws the shoulder girdle forward and down, creating the flight motion so efficiently demonstrated by birds. The *pectoralis minor* is a thin, triangular muscle quite unlike the thick, large pectoralis major. The minor lies behind the major and arises from the costal car-

tilages of the second to fifth ribs. It inserts on the coracoid process of the scapula and pulls the scapula forward and downward, assisting in flight motion. Both the major and minor pectoralis muscles are removed surgically in a radical mastectomy for breast cancer.

The *serratus anterior* occupies the side of the chest and the medial wall of the axilla (armpit) (Figure 7-15A). It arises from the outer surfaces of the first nine ribs and the intercostals between them. The fibers pass upward and backward, inserting on the vertebral border of the scapula. The name of this muscle derives from the sawtooth appearance of its origins from several ribs. These origins on the chest wall permit

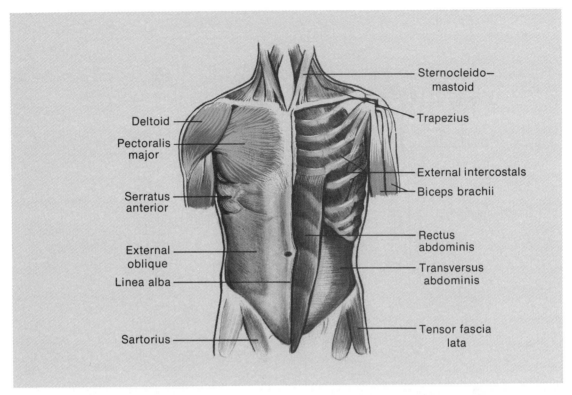

Figure 7-21 Muscles of the neck, thorax, and arm, anterior aspect. Some superficial muscles have been removed on the left side to permit exposure of the deep muscles.

it to move the scapula forward away from the spine (abduction) in a pushing motion.

The *coracobrachialis* is located at the upper and medial part of the arm. It arises from the coracoid process of the scapula and inserts on the medial surface of the humerus. This muscle carries the arm forward (flexion) and assists in its adduction.

The *deltoid* is a thick, powerful muscle that covers the shoulder joint. It arises from the clavicle, acromial process, and spine of the scapula and inserts onto the lateral (deltoid) tubercle of the humerus. Contraction of the fibers abducts the arm from the side of the body and raises it laterally, as when reaching for something sideways. The thickness of the muscle makes it a prime area for intramuscular injections.

The *teres major* is thick and flat (Figure 7-15B). It arises near the lower angle of the scapula and inserts on the anterior surface of the humerus (lesser tubercle). This muscle extends the arm, drawing it downward. It also helps to adduct the arm and rotate it medially.

The *teres minor, supraspinatus,* and *infraspinatus muscles* collectively form a tendon after the muscles have passed over the shoulder joint and insert on the greater tubercle of the humerus. The teres minor arises from the axillary border of the scapula; the supraspinatus and infraspinatus originate from their respective fossae in the scapula. These three muscles protect the shoulder joint and assist in the lateral rotation of the arm. The supraspinatus also aids the deltoid in abduction of the arm.

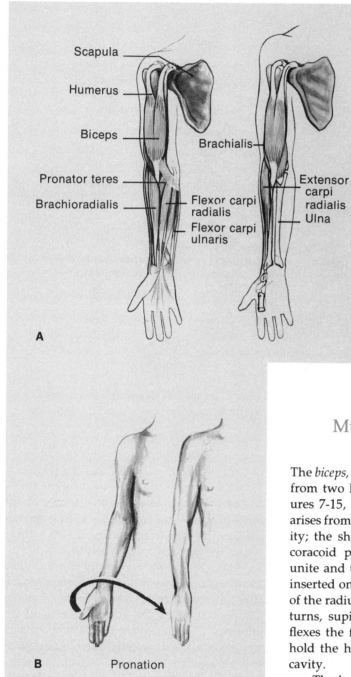

A, Muscles of the arm, superficial and deep.

- Scapula
- Humerus
- Biceps
- Pronator teres
- Brachioradialis
- Brachialis
- Flexor carpi radialis
- Flexor carpi ulnaris
- Extensor carpi radialis
- Ulna
- Supinator
- Flexor pollicis longus

B Pronation

Figure 7-22 **A,** Muscles of the arm, superficial and deep. **B,** Pronation of the arm.

Muscles of the Forearm and Hand

The *biceps,* or *biceps brachii,* of the arm originates from two heads, one long and one short (Figures 7-15, 7-22, and Table 7-3). The long head arises from the upper margin of the glenoid cavity; the short head arises from the tip of the coracoid process of the scapula. Both heads unite and terminate in a flat tendon, which is inserted onto the tubercle on the proximal end of the radius. As the biceps contracts, the radius turns, supinating the hand. This muscle also flexes the forearm, and its long head helps to hold the head of the humerus in the glenoid cavity.

The *brachialis* has its origin from the distal, anterior portion of the humerus and inserts onto the coronoid process of the ulna, covering the front part of the elbow joint (Figure 7-22). It is a strong flexor of the forearm.

Table 7-3   Muscles of the Forearm and Hand

Muscle	Origin	Insertion	Action
Biceps brachii	Short head from coracoid process; long head from scapula (glenoid cavity)	Tubercle on proximal end of radius	Flexes forearm; supinates forearm and hand; long head helps hold head or humerus in glenoid cavity
Brachialis	Anterior surface of humerus (distal part)	Coronoid process of ulna	Flexes forearm
Triceps brachii	Long head from axillary border of scapula; lateral and medial heads from posterior surface of humerus	Olecranon process of ulna	Extends forearm; long head extends and adducts forearm
Pronator teres	Medial epicondyle of humerus and coronoid process of ulna	Middle of lateral surface of radius	Pronates hand; flexes forearm
Supinator	Lateral epicondyle of humerus, ridge of ulna	Anterior and lateral surfaces of radius	Supinates hand and forearm
Flexor carpi radialis	Medial epicondyle of humerus	Base of second metacarpal	Flexes and abducts hand
Flexor carpi ulnaris	Medial epicondyle of humerus, proximal two-thirds of ulnar border	Pisiform bone	Flexes and adducts hand
Palmaris longus	Medial epicondyle of humerus	Fascia of palm	Flexes hand
Flexor digitorum superficialis	Medial epicondyle of humerus, coronoid process of ulna	Middle phalanges of fingers	Flexes middle phalanges
Flexor digitorum profundus	Shaft and coronoid process of ulna	Distal phalanges of fingers	Flexes distal phalanges
Flexor pollicis longus	Anterior surface of radius, coronoid process of ulna	Distal phalanx of thumb	Flexes thumb
Extensor carpi radius longus	Lateral supracondylar ridge of humerus	Base of second metacarpal	Extends and abducts hand
Extensor carpi ulnaris	Lateral epicondyle of humerus	Base of fifth metacarpal	Extends and adducts hand
Extensor pollicis longus	Lateral side of dorsal surface of ulna	Base of second phalanx of thumb	Extends and abducts thumb
Extensor digitorum	Lateral epicondyle of humerus	Common extensor tendon of each finger	Extends fingers and hand
Abductor pollicis longus	Posterior surfaces of ulna and radius	Base of first metacarpal	Abducts and extends thumb
Adductor pollicis	Principally second and third metacarpals	Proximal phalanx of thumb	Adducts and opposes thumb

The *triceps,* or *triceps brachii,* arises by three heads of origin: the long head from the axillary border of the scapula and the lateral and medial heads from the posterior surface of the humerus (Figure 7-15B). The muscle fibers terminate in a common tendon that inserts onto the olecranon process of the ulna. The triceps is the great extensor of the forearm and therefore is the direct antagonist of the biceps and brachialis.

The *pronator teres* has two heads of origin: the humeral, or larger head, and the ulnar (Figure 7-22). This muscle extends obliquely across the forearm and inserts as a flat tendon onto the lateral surface of the radius. Its primary function is to pronate the hand, but it also flexes the forearm.

The *supinator* originates on the lateral epicondyle of the humerus and the ridge of the ulna and is inserted onto the lateral margin of the radius (Figure 7-22). As its name implies, it supinates the hand and the forearm.

Movements of the wrist and fingers depend on several muscles of the forearm (Figures 7-22 and 7-23). The *flexor carpi radialis* and the *flexor carpi ulnaris* originate on the medial epicondyle of the humerus and aid in flexion at the elbow and in flexion of the hand at the wrist joint. The radialis abducts the hand; the ulnaris adducts the hand.

The *extensor carpi ulnaris* and *extensor carpi radialis* originate on the lateral epicondyle of the humerus and are synergistic with the flexor carpi muscles in their action across the elbow joint. More important, however, they extend the hand at the wrist through their insertions on the dorsal side of the second and fifth metacarpals at the ulnar and radial sides of the hand.

Muscles of the forearm also control flexion and extension of the fingers. The proximal joints of the fingers are flexed by the *flexor digitorum (profundus and superficialis)* muscles that arise from the medial epicondyle of the humerus and the coronoid process of the ulna. Extension of the fingers is carried out by the *extensor pollicis* muscles in conjunction with the extensor digitorum group of muscles (Figure 7-15A).

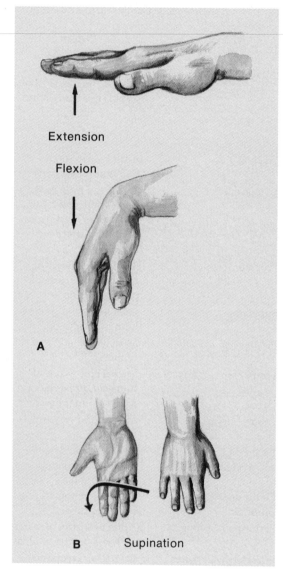

Figure 7-23 **A,** Flexion of the wrist. **B,** Supination of the hand.

The thumb and fingers can be abducted (fanned out) by contraction of the *abductor pollicis* (thumb) and *interossei dorsales* (dorsal metacarpal area) muscles. Contraction of the adductor pollicis and *interossei palmaris* (palm) muscles adducts the thumb and fingers (Figure 7-24).

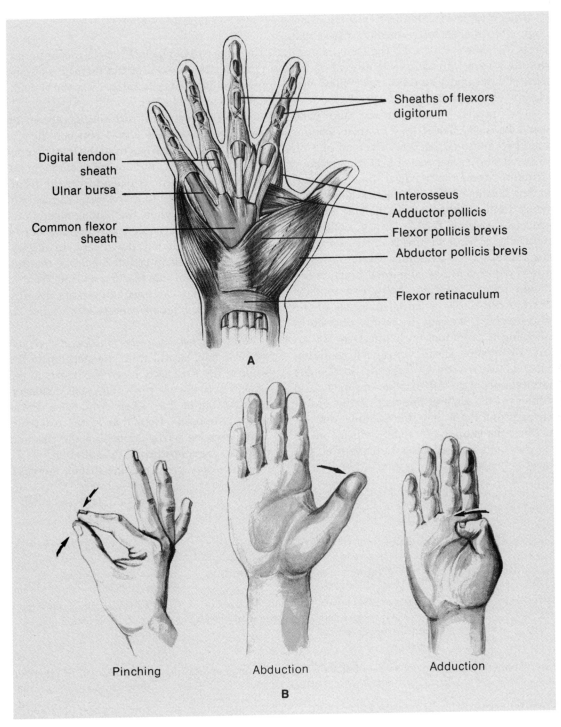

Sheaths of flexors digitorum

Digital tendon sheath

Ulnar bursa

Common flexor sheath

Interosseus

Adductor pollicis

Flexor pollicis brevis

Abductor pollicis brevis

Flexor retinaculum

A

Pinching

Abduction

Adduction

B

Figure 7-24 The hand. **A,** Palmar surface of the wrist and hand, showing the muscles and sheaths of the flexor tendons. **B,** Movements of the hand.

The *palmar aponeurosis*, or *palmar fascia*, is a tough fibrous sheet of connective tissue that covers the palm just beneath the skin. Its presence allows great pressure to be exerted on the palm without injury to the nerves, blood vessels, and tendons beneath.

The muscles of the forearm and hand would be ineffective if they or their tendons could become stuck together or rub on one another, causing a great deal of heat from friction. Therefore, numerous synovial (fluid-secreting) tendon sheaths encase muscles and tendons to reduce friction between tendons and adjacent structures (Figure 7-24A). Exact anatomical knowledge of these sheaths is a necessity when surgical repair of a tendon laceration or severance is required to restore normal function following injury. Suturing the wrong tendons together can result in partial paralysis of the parts involved. Occasionally an infection sets in following a laceration. Such infection can invade the tendon sheath, causing a condition called **tenosynovitis** (p. 195). An inadequate nerve supply or growth of adhesions may cause stiffness of the wrist or fingers. Stiffness or impaired use of the thumb, for example, can immobilize the entire hand, compromising one of the functions that separates humans from the lower animals.

Muscles of Respiration

There are two independent mechanisms for respiration: the muscles of the thoracic wall and the diaphragm and its antagonists, the abdominal muscles (Table 7-4).

The *external intercostal muscles*, eleven on each side, occupy the spaces between the ribs (Figure 7-21). Each arises from the lower border of a rib and inserts onto the upper border of the rib below. The muscles extend from the tubercles of the ribs behind to the costal cartilages of the ribs in front, where they end in membranes that connect with the sternum. The direction of the fibers is obliquely downward. In respiration, the external intercostals enlarge the thoracic cavity from side to side and from front to back when they contract and elevate the ribs. Thus these muscles are concerned with inspiration.

The *internal intercostal muscles*, also eleven on each side, extend from the sternum to the angle of the ribs. Each arises and inserts in the same way as the external intercostals; however, the direction of their fibers, while also downward, is opposite from that of the externals. These muscles act in depressing the ribs and therefore are concerned with expiration.

The *diaphragm* is a dome-shaped musculo-

Table 7-4 Muscles of Respiration

Muscle	Origin	Insertion	Action
Diaphragm	Xiphoid process; lower six costal cartilages and four ribs; upper lumbar vertebrae	Central tendon of diaphragm	Expands thorax; compresses contents of abdominal cavity
External intercostals	Lower border of each rib	Upper border of next rib	Draw adjacent ribs together (inspiration)
Internal intercostals	Lower border of each rib	Upper border of next rib	Draw adjacent ribs together (expiration)

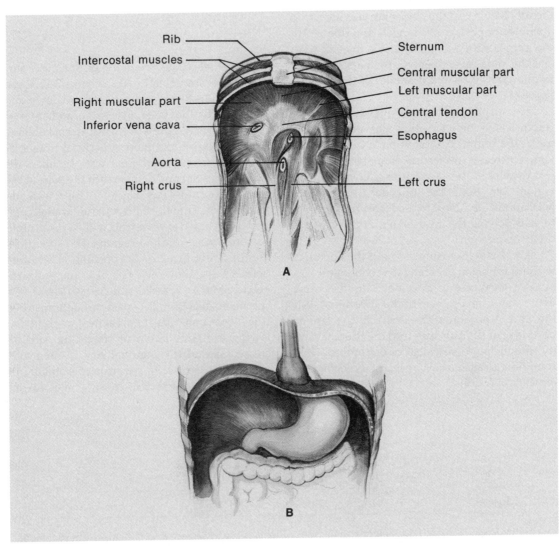

Figure 7-25 The diaphragm, **A,** Undersurface of the large muscle partition, showing the openings of digestive, nervous, and circulatory systems. **B,** Diaphragmatic or hiatus hernia.

fibrous partition that separates the thoracic and abdominal cavities (Figure 7-25A). It forms the convex floor of the thoracic cavity and the concave roof of the abdominal cavity. Its peripheral portion arises from the xiphoid process of the sternum. The costal fibers arise from the lower six costal cartilages and the lower four ribs on either side, and the lumbar fibers arise from the tendinous arches and bodies of the upper lumbar vertebrae. The fibers converge, to be inserted onto the central portion of the diaphragm, called the *central tendon.*

There are three large openings in the diaphragm: the esophageal opening, for passage of the esophagus and the vagus nerves; the aortic opening, for passage of the aorta and frequently the azygos vein and the thoracic duct as well; and the vena caval opening, for passage of the inferior vena cava and branches of the phrenic nerves.

The diaphragm is the principal muscle of respiration. When the muscular portion contracts, the central tendon is pulled downward, which increases the vertical diameter of the thorax. Because of this, a vacuum is created, sucking in air from the outside (inspiration). Relaxation of the diaphragm decreases the vertical axis, forcing the air back out of the thoracic cavity (expiration).

Abnormal stretching of one of the openings in the diaphragm, generally the esophageal, can cause a weakness to develop. This weakness may result in an opening in the abdominal wall **(hernia).** A portion of the stomach (the cardiac portion) can slip part way into the thoracic cavity, producing a great deal of discomfort. This condition is called a *diaphragmatic,* or *hiatus hernia* (Figure 7-25B).

Muscles of the Abdominal Wall

The *external oblique* originates on the outer surface of the lower eight ribs (Figure 7-15A and Table 7-5). Its fibers pass forward and down and insert on the iliac crest, iliac spine, and anterior rectus sheath (Figure 7-26). The external oblique is the strongest and most superficial muscle of the anterolateral abdominal wall. Its fibers are short and broad and do not reach the edge of the rectus abdominis (see Figure 7-21), which lies medial to it. Therefore the anterior wall is composed of a broad aponeurosis, which inserts into the linea alba (midline region). The superficial inguinal ring is a circular opening in the aponeurosis just above and lateral to the pubic tubercle. Major vessels and nerves (and the spermatic duct in males) pass through this ring. It is a common area for weakness to occur following excessive strain or stretching, and increased size of the inguinal ring in men may allow the intestines to pass from the abdominal cavity into the scrotum, causing marked dis-

Table 7-5   Muscles of the Abdominal Wall

Muscle	Origin	Insertion	Action
External oblique	Lower eight ribs	Iliac crest; anterior rectus sheath	Compresses abdomen; flexes and rotates vertebral column
Internal oblique	Inguinal ligament; iliac crest; lumbodorsal fascia	Lower three or four costal cartilages; pubic bone	Compresses abdomen; flexes and rotates vertebral column
Transversus abdominis	Lower six costal cartilages; iliac crest; lumbodorsal fascia; inguinal ligament	Linea alba; crest of pubis	Compresses abdomen
Rectus abdominis	Symphysis pubis; crest of pubis	Fifth to seventh costal cartilages; xiphoid process	Flexes vertebral column and pelvis

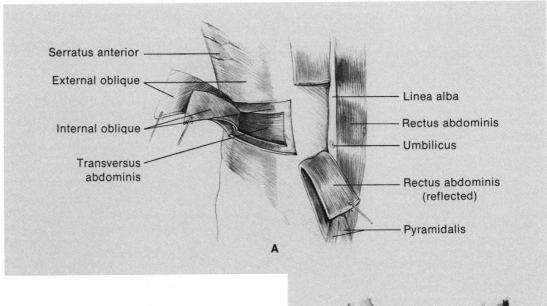

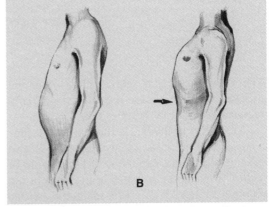

Figure 7-26 Muscles of the abdominal wall. **A,** The muscles have been reflected to illustrate their layered position in the formation of the abdominal aponeurosis. **B,** Abdominal contraction aids respiration.

comfort (Figure 7-27). This condition is called an *inguinal hernia*. If the ring subsequently closes, a strangulation of the intestines occurs resulting in an *incarcerated* or *strangulated hernia*. Immediate surgical attention is necessary because a gangrenous condition may result from the constricted blood supply to the intestine.

The *internal oblique* lies just beneath the external oblique (see Figure 7-26). It arises from the inguinal ligament, iliac crest, and lumbodorsal fascia. The fibers extend upward and forward and insert onto the lower three rib cartilages to form the aponeurosis that inserts onto the linea alba. The lower part of the aponeurosis inserts on the pubic bone.

The *transversus abdominis* is positioned deep in the lateral abdominal wall (see Figures 7-21 and 7-26). Its fibers arise from the lower six costal cartilages, lumbodorsal fascia, and inguinal ligament. Most of its fibers, as the name of the muscle suggests, have a horizontal (transverse) direction and form the aponeurosis that inserts onto the linea alba and the crest of the pubis. The deep inguinal ring is found in this muscle. It provides the internal opening for the inguinal

canal, whose outer opening is found in the external oblique muscle.

The external oblique, internal oblique, and transversus abdominis muscles, with their aponeuroses, form a wall or girdle for the abdomen. Their muscle fibers cross in three directions, acting to compress the abdominal contents. This action also assists in expiration.

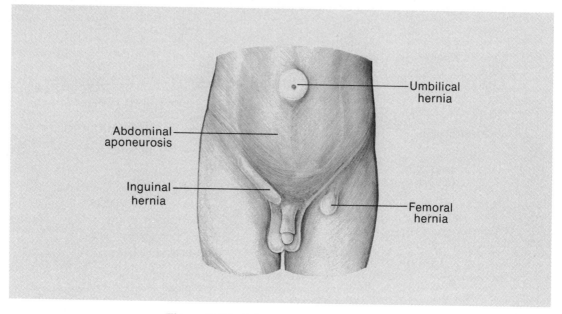

Figure 7-27 Abdominal hernias in the male.

When the diaphragm relaxes, the abdominal muscles contract; when the diaphragm contracts, the abdominal muscles relax.

The *rectus abdominis* is the last of the four major abdominal muscles (see Figures 7-21 and 7-26). It is a long, straplike muscle made up of vertical fibers and is situated at the front of the abdomen, where it is enclosed in the fibrous sheath formed by the aponeuroses of the other three abdominal muscles. Its fibers arise from the symphysis pubis and the crest of the pubis. The muscle inserts onto the costal cartilages of the fifth to seventh ribs and onto the xiphoid process of the sternum. Its major function is to flex the vertebral column and the pelvis.

When all the abdominal muscles are contracted, they assist in childbearing, defecation, urination, and vomiting. They also bend the thorax forward, and, when acting on one side only, bend the trunk of the body in that direction.

In addition to the inguinal hernia already discussed, other weaknesses of the abdominal wall include umbilical and femoral hernias. An *umbilical hernia* is a protrusion of the abdominal contents in the region of the umbilicus. A *femoral hernia* is an enlargement of the canal in the femoral region just behind the inguinal ligament. The inguinal canal is larger in males than in females; therefore, the inguinal hernia is more common in men and the femoral hernia more frequent in women.

Muscles of the Thigh

The ball-and-socket arrangement of the hip joint permits free movement in all directions (Figure 7-28 and Table 7-6). The thigh muscles have an important role in facilitating hip joint movement.

The *iliopsoas* arises from the transverse processes of the lumbar vertebrae and the iliac crest and fossa. Its fibers converge and pass along the edge of the pelvis to the front of the thigh, where they are inserted onto the lesser trochanter of the femur. The iliopsoas is actually a compound muscle made up of the iliacus and psoas major. The action of both is primarily to flex the

The four-headed *quadriceps femoris* covers the front and sides of the femur. Its parts are the *rectus femoris*, which arises from the ilium, and the *vastus lateralis, vastus medialis,* and *vastus intermedius,* which arise from the femur (see Figures 7-15A, and 7-29). The fibers of these four muscles unite at the lower part of the thigh to form a strong tendon inserted on the tuberosity of the tibia, which is closely applied to the patella (knee bone). The patella develops within the tendon. The quadriceps femoris extends the lower leg, aiding greatly in walking and running. The rectus femoris portion flexes the thigh.

The ankle is a hinge joint. In the foot, upward movement is called *dorsal flexion;* that is, the foot is bent toward the anterior part of the leg. Downward movement is called *plantar flexion* (extension): the foot is bent backward and downward away from the body. Inversion (turning the sole of the foot inward) and eversion (turning the sole of the foot outward) also take place at the ankle joint. The movement of the ankle depends on the following muscles of the leg.

The gastrocnemius and soleus, together known as the triceps surae, form the calf of the leg (see Figures 7-15B and 7-30). The *gastrocnemius* has two heads that arise from the medial and lateral condyles of the femur. The *soleus* arises from the head of the fibula and upper end of the tibia, underneath the gastrocnemius. Both muscle fibers are directed downward and are inserted into a common tendon, the *tendon of Achilles,* the thickest and strongest tendon in the body. These two powerful muscles are used extensively in walking, standing, and leaping, as they extend, or plantar flex, the foot at the ankle joint and flex the femur on the tibia.

The *tibialis posterior* arises from the posterior surfaces of the tibia and fibula, and inserts onto the undersurfaces of the tarsal bones (cuneiform, cuboid) and the second to fourth metatarsal bones. This muscle plays an important role in supporting the longitudinal arch of the foot. It also extends the foot at the ankle joint in conjunction with the soleus and gastrocnemius.

The *peroneus longus* is found on the lateral aspect of the leg. It arises from the lateral condyle of the tibia and the head and lateral surface of the fibula and inserts onto the first cuneiform and first metatarsal bones after crossing obliquely over the sole of the foot and terminating in a long tendon. This muscle extends and everts the foot and also helps to maintain the transverse arch.

The antagonists to these extensors are the *tibialis anterior* and the *peroneus tertius* and *brevis.* The tibialis anterior arises from the lateral condyle and upper portion of the tibia and inserts into the first cuneiform and the base of the first metatarsal bones (Figure 7-30). The peroneus tertius arises from the distal third of the fibula and inserts onto the base of the fifth metatarsal bone. These two muscles flex the foot dorsally. The tibialis anterior also inverts the foot at the ankle. The peroneus brevis is a short muscle that arises from the lower two-thirds of the fibula and inserts into the fifth metatarsal bone. The peroneus brevis acts with the longus to extend and evert the foot upon the leg. Nerve involvement of the paralyzing variety in the region below the knee generally affects the ability of the tibialis anterior to flex the foot. If this function is impaired, the foot tends to hang from the lower leg, a condition called *foot drop.* There are marked problems in walking or standing erect. A person with this abnormality would have to lift the affected leg high and "throw" the foot forward to take a step.

DISORDERS AFFECTING MUSCLES AND RELATED CONNECTIVE TISSUE

The major symptoms of disorders affecting muscles are paralysis, weakness, atrophy, pain, and spasm.

Myalgia means muscular pain. **Myositis** is inflammation of muscular tissue, especially of voluntary muscles. **Fibrositis** is inflammation

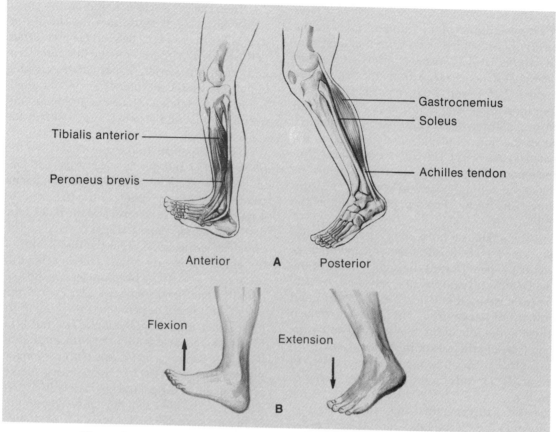

Figure 7-30 **A,** Superficial muscles of the lower leg. **B,** Movements of the foot.

of white fibrous tissue, as in muscle sheaths. Tendons, tendon sheaths, and fascia are also subject to inflammatory disorders. **Myopathy** is any muscular disease.

Paralysis

Examples of paralysis, a common pathological condition of muscle resulting from nerve injury, are known to all. Such muscles cannot contract normally and may, with disuse, show indications of atrophy. The failure of some muscles to contract may prove fatal: for example, after abdominal operations, the muscles of the

ileum (the distal portion of the small intestine) may become paralyzed, resulting in a potentially lethal obstruction. Paralysis of the muscles of respiration makes it necessary for a mechanical breathing device to be employed.

Spastic paralysis, a taut tetanic contraction, is experienced when a group of neurons connecting the brain and the spinal cord is injured. *Flaccid paralysis* (a limp, flabby form) can follow injury to the group of neurons connecting the spinal cord to muscle. If paralysis continues, the lack of use of the affected muscles will result in atrophy. In addition, muscles may undergo spontaneous atrophy of unknown cause that first appears in the lower limbs. Either a deficiency of potassium or too much potassium (potassium intoxication) can cause paralysis.

The enzyme transaminase is present in great concentrations in both cardiac and skeletal muscle. If this muscle tissue is destroyed, levels of transaminase in the blood will increase, providing a valuable test for determining the presence and extent of injury, especially in the heart. The test is used in the diagnosis of myocardial infarction.

Fibrillation

Sometimes muscle fibers will contract only individually or in small groups. This phenomenon, called **fibrillation,** gives the appearance of quivering. Immediate mechanical stimuli or electric shocks stop the muscle, allowing it time to start contracting again in rhythm. This is the procedure used to stop the heart from fibrillating.

Muscular Dystrophy

The **dystrophies** are actually a group of related degenerative muscle diseases that are apparently inherited, at least in part. Women are usually the carriers, but males are most often affected. The shoulder and hip muscles become enlarged as occurs in **hypertrophy,** but this is not a true hypertrophy because the muscle is gradually replaced by fat and connective tissue. A blood test will reveal elevated levels of creatine and abnormal amounts of transaminase.

Myasthenia Gravis

Myasthenia gravis, characterized by easy exhaustion of muscles, is a combined neuromuscular problem since its cause is an impairment of impulse conduction from nerve to muscle. It is thought to be a deficiency of the conducting neuronal transmitter, acetylcholine, or an increase in cholinesterase, the enzyme that readily inhibits acetylcholine activity.

Primary Fibrositis

Primary fibrositis is a disease of fibrous connective tissue. It may occur in acute attacks, causing painful stiffness of the neck or lower back (*lumbago*), or it may begin insidiously and become a nagging, chronic ailment affecting many parts. Its cause is unknown, but current research suggests that chemical changes may be the chief etiologic factor. The involved portions become stiff and painful. Irritation of the periarticular connective tissue and muscles leads to restriction of movement.

Tenosynovitis

Tenosynovitis is an inflammation of the tendon sheaths. It may interfere with free passage of the enclosed tendons, leading to dysfunction of the associated joints. Weakness and atrophy result if the infection becomes chronic. Since the joints are not diseased, no deformities result. When the infection is arrested, function returns to normal.

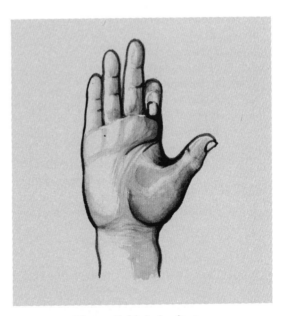

Figure 7-31 Index finger.

The area along the flexor tendon sheaths of the fingers is commonly affected. The finger frequently cannot be extended by the "frozen" flexor apparatus, and a "trigger finger" results (Figure 7-31).

Ganglion

A *ganglion* is a small cystic swelling that develops in connection with a tendon sheath close to a joint. The commonest location is the back of the wrist. The swelling is attached to the outer surface of the tendon sheath and eventually un-dergoes mucoid degeneration, with numerous small cysts forming in the area. Eventual impairment of wrist movement warrants surgical removal of the ganglia.

Clubfoot

Clubfoot is a congenital malformation of one or both feet in which the front part of the foot is inverted and rotated. It is accompanied by shortening of the Achilles tendon and contracture of the fascia in the sole of the foot.

OUTLINE

I. Introduction
   A. Motion
      1. Made possible by contractility of muscle tissue
      2. Includes breathing, movement, beating of heart, digestion, and visceral movements
II. Microstructure of muscle
   A. General structure
      1. Myofibrils
      2. Sarcolemma
      3. Sarcoplasm
      4. Fasciculi
      5. Fascia
   B. Types of muscle
      1. Smooth
         a. Long, slender fibers
         b. Fibers grouped in thin sheets
         c. Each fiber has a single, central nucleus
         d. Found in blood vessels and surrounding viscera
         e. Involuntary control
      2. Skeletal
         a. Voluntary control of the skeleton
         b. Surrounded by epimysium
         c. Each fiber cell has many nuclei

      d. Protein bands appear as striations on outside of fiber
         (1) Thick filament—myosin (A band)
         (2) Thin filament—actin (I band)
         (3) Arranged in rows that interlock (A band)
         (4) Fibers divided into segments (sarcomeres) by two lines
         (5) Availability of energy affords a contraction
      3. Cardiac
         a. Cells form network of branching fibers
         b. Involuntary
         c. Intercalated disks
         d. Many mitochondria to supply abundant energy
III. Chemistry of muscle tissue
   A. Makeup of skeletal muscle
      1. 75 percent water
      2. 25 percent solid material
         a. 20 percent protein (actin and myosin)
         b. 5 percent carbohydrates, lipids,

inorganic salts, and nonprotein nitrogenous compounds

B. Energy sources
   1. ATP—immediate
   2. CP—intermediate
   3. Glycogen—ultimate
C. Oxygen debt mechanism
   1. High energy derived from aerobic respiration
   2. Deficiency of oxygen to tissues prevents energy production
   3. Lactic acid accumulates
   4. Body rest resupplies the muscles with the needed oxygen to pay back the energy need

IV. Contraction of muscles
   A. Mechanism of muscle contraction
      1. Available ATP combines with actin
      2. Myosin, in the presence of calcium, acts as an enzyme to free energy from ATP
      3. Magnesium, tropomyosin, and troponin inhibit reaction between actin and myosin; myofibril remains relaxed
      4. Actomyosin complex theory
         a. Conversion of ATP to ADP binds actin and myosin together at energy bridges and sites, causing a shortening of the segments of a muscle fiber—a contraction
   B. Properties of muscles related to contraction
      1. Excitability
         a. Muscle receives an impulse and responds to it
         b. Muscle contraction
            (1) Latent period
            (2) Contraction period
            (3) Relaxation period
         c. Stimulus
            (1) Liminal
            (2) Subliminal
            (3) Maximal
            (4) Adequate
         d. Response—reaction

2. Contractility
   a. Permits muscle to change its shape
   b. Isometric contraction
   c. Isotonic contraction
   d. All-or-none law
   e. Tonicity (tonus)
   f. Atrophy
   g. Epinephrine
3. Extensibility
   a. Muscle can be stretched
4. Elasticity
   a. Muscle returns to its original size
C. Some phenomena of contraction
   1. Skeletal muscles contract and relax more quickly than smooth muscles
   2. Muscles do their best work at body temperature
   3. Tetany
   4. Spasm
   5. Fatigue
   6. Exercise stimulates circulation and strength in a muscle
D. Conduction of the contractile impulse
   1. Motor unit
      a. Nerve body
      b. Nerve fiber
      c. Motor end-plate
      d. Associated muscle fibers
   2. Chemical conductor
      a. Acetylcholine
   3. Acetylcholine cycle
      a. Activated in the end-plate by an impulse
      b. Diffuses into the muscle with the impulse
      c. A muscle contraction occurs
      d. Cholinesterase inactivates the acetylcholine
      e. Separate acetyl and choline radicals diffuse back into the end-plate
      f. Choline acetylase reestablishes acetylcholine, which awaits a new impulse

  4. Inhibitors
     a. Curare
     b. Succinylcholine
     c. Nerve gases
     d. Botulimus toxin
     e. Myasthenia gravis
V. Skeletal muscles
  A. How muscles are named
    1. Location—intercostal
    2. Action—flexor
    3. Shape—trapezius
    4. Direction of fiber—rectus
    5. Number of heads or origin—biceps
    6. Points of attachment—sterno-cleidomastoid
  B. Muscle action
    1. Antagonists—muscles working opposite to each other
    2. Synergists—muscles working together
  C. Muscles of facial expression and mastication
    1. Epicranius—surprise
    2. Orbicularis oculi—squinting
    3. Orbicularis oris—pouting
    4. Nasalis—smelling
    5. Zygomaticus—smiling
    6. Buccinator—chewing
    7. Mentalis—doubt
    8. Triangularis—sadness
    9. Platysma—sadness
    10. Masseter—chewing
    11. Temporalis—sternness
    12. Pterygoid—consternation
    13. Genioglossus and styloglossus—speaking, chewing, swallowing
  D. Muscles that move the head
    1. Sternocleidomastoid—forward
    2. Semispinalis capitis—extension, rotation
  E. Muscles that move the shoulder and arm
    1. Trapezius—lifts shoulder (shrugging)
    2. Levator scapulae—elevates scapula
    3. Rhomboid (major and minor)—elevate and retract scapula
    4. Latissimus dorsi—sweeps arm downward
    5. Pectoralis (major and minor)—draw shoulder forward, medially, and downward
    6. Serratus anterior—moves scapula forward (pushing)
    7. Coracobrachialis—moves arm forward (flexion)
    8. Deltoid—abducts arm
    9. Teres (major and minor)—rotate arm medially and laterally
  F. Muscles that move the forearm and hand
    1. Biceps—flexes forearm
    2. Brachialis—flexes forearm
    3. Triceps—extends forearm
    4. Pronator teres—pronates hand
    5. Supinator—supinates hand and forearm
    6. Flexor carpi (radialis, ulnaris)—flex, abduct, adduct hand
    7. Extensor carpi (radialis, ulnaris)—extend, abduct, adduct hand
    8. Palmaris longus—flexes hand
    9. Flexor digitorum—flexes fingers
    10. Extensor digitorum—extends fingers
    11. Adductor and abductor pollicis—adduct and abduct thumb
  G. Muscles of respiration
    1. Intercostals—enlarge thoracic cavity
    2. Diaphragm—enlarges thoracic cavity
       a. Weakness—diaphragmatic hernia
  H. Muscles of the abdominal wall
    1. External oblique, internal oblique, transversus abdominis—form the abdominal aponeurosis or girdle
       a. Weaknesses—umbilical and femoral hernias

2. Rectus abdominis—flexes vertebral column and pelvis

I. Muscles that move the thigh
1. Iliopsoas—flexes thigh
2. Gluteus (maximus, medius, minimus)—extend, abduct, rotate thigh
3. Adductor (magnus, longus, brevis)—adduct, flex thigh

J. Muscles that move the leg and foot
1. Biceps femoris, semitendinosus, and semimembranosus—flex leg, extend thigh
2. Sartorius and gracilis—flex leg and thigh
3. Quadriceps—extends leg; flexes thigh
4. Gastrocnemius and soleus—extend foot; flex leg

5. Tibialis posterior—extends foot
6. Peroneus (longus, tertius, brevis)—extend and evert foot
7. Tibialis anterior—dorsally flexes foot
   a. Foot drop

VI. Disorders affecting muscles and related connected tissue
A. Paralysis
1. Spastic
2. Flaccid
B. Fibrillation
C. Muscular dystrophy
D. Myasthenia gravis
E. Primary fibrositis
F. Tenosynovitis
G. Ganglion
H. Clubfoot

STUDY QUESTIONS AND PROBLEMS

1. Give a general description of the three types of muscle, indicating their anatomy and function and the nature of their contractions.
2. Of what value are ATP, CP, and glycogen to muscle? Describe their roles.
3. Can we create an oxygen debt? Describe it.
4. What is rigor mortis? What is its cause?
5. Sketch a single muscle twitch and define an anatomy of a muscle contraction.
6. What part do actin and myosin play in muscle contraction?
7. Can a muscle contract without calcium? Magnesium?
8. What is a threshold stimulus? What is meant by summation effect?
9. Define extensibility and elasticity. Why are they important?
10. What do the intensity of stimulus and number of stimuli have to do with a muscle's action?

11. Define tonus and the all-or-none law. Do they complement each other or contradict each other? Why?
12. Distinguish between an isotonic and an isometric contraction.
13. What is complete tetany? Why does it not occur in cardiac tissue?
14. Differentiate between fibrillation and spasm.
15. How is a nerve impulse transmitted to a muscle?
16. State at least four ways in which muscles are named. Give examples.
17. Name some inhibitors of acetylcholine and show how they achieve this effect.
18. What are the origin and the insertion of a muscle?
19. What are flexors, extensors, rotators, supinators, and depressors?
20. What is peristalsis?

8

The Voluntary Nervous System

LEARNING OBJECTIVES

- Distinguish between voluntary and involuntary control of the body

- Describe the structure of a neuron

- Classify neurons according to their structure, function, and position

- Describe myelin, and relate its importance to a peripheral nerve

- Describe the importance of myelin in nerve regeneration

- Explain resting membrane potential and action potential

- Explain the all-or-none principle as it is applied to a single neuron

- Describe the mechanism of synaptic transmission as well as abnormalities of conduction at the synapse

- Explain a simple reflex

- Distinguish between a simple versus complex reflex

- List the spinal plexuses, and identify the abnormality created by damage to the plexus' major nerve

- Name and describe the functions of the ascending and descending tracts of the spinal cord

- Differentiate between the white and gray matter of nerve tissue

- Identify the layers of the meninges and relate their function

- Distinguish between prosencephalon, mesencephalon, and rhombencephalon

- Describe the cerebrum, and identify the lobes by their location and function

- List the organs of the hind brain and relate their function to the brain and spinal cord

- Describe learning

- Differentiate between the right brain and left brain phenomenon

- Identify the twelve cranial nerves by their number, function, and distribution

IMPORTANT TERMS

afferent	convulsion	gyrus	neuron	stroke
anesthetic	cortex	meninges	palsy	sulcus
aphasia	dementia	myelin	peduncle	synapse
ataxia	dendrite	nerve fiber	plexus	
axon	efferent	neuralgia	receptor	
commissure	fissure	neuritis	seizure	
conduction	ganglion	neuroglia	somesthetic	

Once some of the body systems have been studied, it becomes fairly obvious that not one of these systems is capable of functioning alone. The systems are interdependent. All must work together as one functioning unit so that homeostasis may be maintained within the body. The mechanism that ensures that the organs and systems operate in smooth coordination is the nervous system. Conditions within and outside the body are constantly changing, and one purpose of the nervous system is to respond to these internal and external changes so that the body may adapt itself.

The nervous system is divided into two parts: the central nervous system (CNS) and the peripheral nervous system (PNS). The CNS consists of the brain inside the skull *and* the spinal cord, which runs up inside the vertebral column and expands into the brain. Communicating centers within the CNS and various nerve tracts make possible the appropriate unconscious or conscious response to sensory stimulus. The PNS is made up of a network of nerves and sense organs that gathers information from the rest of the body and feeds it into

the brain. For convenience, peripheral efferent nerve fibers distributed to smooth muscle, cardiac muscle, and glands are referred to as the autonomic nervous system (ANS).

THE NEURON

The basic unit of nervous tissue is the nerve cell, or **neuron** (Figure 8-1). Neurons are composed of a cell body containing a large nucleus with one or more nucleoli, mitochondria, Golgi apparatus, Nissl bodies (which are associated with the conduction of a nerve impulse), and numerous ribosomes. Extending from this cell body are two types of threadlike projections of cytoplasm known as **nerve fibers: dendrites,** which conduct impulses to the cell body, and **axons,** which conduct impulses away from the cell body.

Types of Neurons

Neurons are classified according to their function as follows:

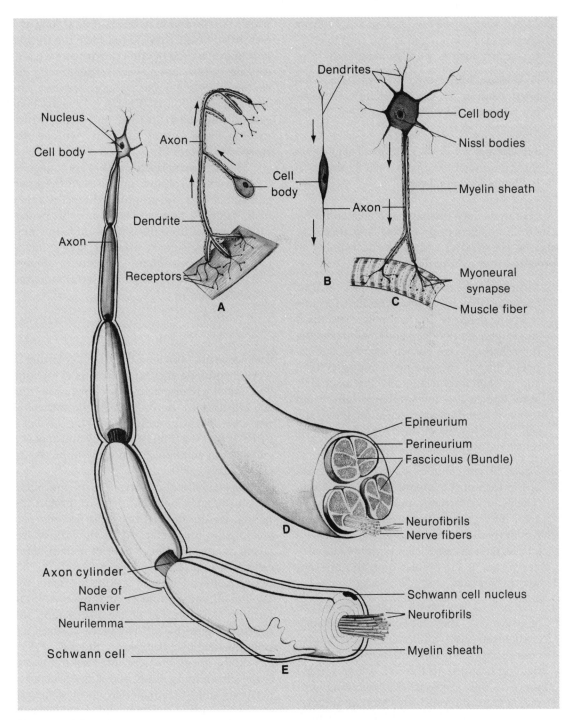

Figure 8-1 The structure of a neuron. A, Monopolar sensory neuron. B, Bipolar retinal neuron in the eye. C, Multipolar motor neuron, showing a muscle-nerve synapse. D, Structure of a typical nerve. E, An enlarged extension of an axon, showing its protective myelin sheath and neurilemma.

1. *Motor neurons.* These pass messages from centers in the brain and spinal cord to effector organs, such as muscles and glands.

2. *Association (interneuron) neurons.* Also called *internuncial* or *intercalated neurons,* these relay messages between neurons within the brain and spinal cord.

3. *Sensory neurons.* These receive messages from the environment and pass them to centers in the brain and spinal cord.

Neurons vary in shape and size and have one or more fibers or processes. Thus they may also be classified according to the number of processes they exhibit (Figure 8-1):

1. *Monopolar.* All neurons must have at least one axon; they may have none, one, or more than one dendrite. In monopolar neurons, a single process divides shortly after leaving the cell body; one branch conveys the impulse toward the central nervous system, and is, therefore, technically the axon. The other branch goes toward the periphery and conveys impulses from sense organs to the cell, acting as the dendrite. Such unipolar cells have only a sensory function.

2. *Bipolar.* These neurons have two processes, one at each of the opposite poles of the cell. One dendrite carries impulses to the cell; one axon conveys impulses from the cell to centers in the central nervous system. As an example, neurons associated with the auditory nerves are of the bipolar type and are sensory in function.

3. *Multipolar.* These are the most common type of nerve cell. They are usually star shaped (stellate) and have numerous branching processes. Short processes—dendrites—convey ingoing impulses to the cell; a large, single process—the axon—carries outgoing impulses from the cell body. Most multipolar neurons are motor or associative in function.

The dendrites of sensory neurons differ markedly from those of other neurons. They are usually single and may be short or as long as 3 feet. In any case, they do not have the treelike appearance typical of other dendrites. Each sensory nerve fiber has a special structure, the **receptor** (also called the *end-organ*), where the stimulus is initiated (Figure 8-1). Sensations such as pain, touch, hearing, and seeing, which involve these sensory neurons, will be discussed in Chapter 11.

The sensory nerve fibers that are connected with receptors conduct impulses to the brain and spinal cord. When grouped together, they form **afferent** nerves. The motor fibers that carry impulses from the centers out to the muscles and glands form **efferent** nerves. Some nerve structures contain a mixture of afferent and efferent nerve fibers, and often are referred to as *mixed nerves.*

Structure of Nerve Fibers

The nerve fiber is a long thin tube consisting of a cell membrane enclosing a core of cytoplasm. The end of a fiber may be a few feet away from the cell nucleus. As in all cells, the nucleus is necessary for the prolonged life of the cell, but if a fiber is severed from its cell body and kept in a suitable salt solution, it can carry impulses for several days.

Satellite cells, also referred to as **neuroglia,** which probably help nourish and protect the nerves, are attached to nerve cells. *Schwann cells* are satellite cells wrapped in thin concentric layers around the nerve fiber of certain nerve cells, which produce a myelin sheath or coat (white matter). On these myelinated fibers, the satellite cells are wound many times in a spiral, forming thick sleeves of fatty substance, the **myelin** (Figure 8-1). Each sleeve is separated from the next by a depression, the *node of Ranvier.* At the nodes, the nerve fiber has free access to the surrounding fluids, allowing exchange of nutrients and waste materials. Most of the peripheral nerves of the body and also the nerves found in the outer periphery of the spinal cord are myelinated. Nonmyelinated nerves, which appear gray, do not need the protection afforded by the layered myelin; these generally are found in the organs themselves, where the organ tissue provides protection and nourishment.

The axon, when it leaves the central nervous system (that is, the brain and spinal cord), usually acquires a covering of myelin. This, in turn, is encased in a neurilemmal sheath (a thin membrane) continuous with the cell body. Bundles of axons with their coverings make up the peripheral nerves. Each bundle has a connective tissue sheath, the *perineurium* (Figure 8-1). A number of bundles are encased in *epineurium*.

All accessory or supporting nerve cells of the central nervous system that neither receive nor conduct impulses are grouped within the category, neuroglia. The neuroglia are comparable to the Schwann cells of the peripheral nervous system.

PHYSIOLOGY OF NERVOUS TISSUE

Regeneration of Nerve Fibers

Although nerves do not reproduce themselves, they can regenerate new parts under special circumstances. For example, when an axon is cut, the severed end of the axon degenerates. If an injury destroys a number of axons supplying a muscle block, the muscle can no longer function and ultimately atrophies. Frequently, the damage can be repaired surgically when the injury is localized. The connection between the cut ends of the fibers is reestablished by carefully bringing the severed ends of tissue together. When this is done the nerve fibers often grow into the sheaths remaining in the severed tissue. This delicate operation is based on the discovery that, provided the sheath of myelin is intact, growth of the existing fiber can take place and new myelin can be generated.

This process, called *Wallerian degeneration*, occurs in the axon stump and in the neuronal cell body. The part of the axon distal to the damage becomes swollen and then fragments by the third day. The myelin sheath around the axon also undergoes degeneration followed by phagocytosis of the remains. During the first few days following damage, regenerating axons begin to invade the tube formed by Schwann cells that have multiplied and grown toward the proximal portion of the axon uniting the proximal and distal portions of the axon. Axons from the proximal area grow at the rate of about 2 to 3 mm/day across the area of damage, find their way into the distal portion, and unite with the axons in the distal area (Figure 8-2). Thus, sensory and motor connections are reestablished.

The most rapid to regenerate are the sympathetic fibers, and the first sign of a return of function to the affected part is usually an improvement in skin color as a result of the restoration of vasomotor activity. Sensory fibers are restored next, with a return of sensitivity first to pressure or pinching; then to pain, heat, and cold; and finally to joint movements and touch localization. Motor function is the last to be restored.

Transmission of the Impulse Along the Nerve Fiber

In all living cells there is a difference in electrical potential between the inside of a cell and the area just outside the cell wall. Consequently, if these two areas are connected by a conductor, a current will flow. The potential difference comes from the unequal distribution of ions between the inside and outside of the cell (Table 8-1).

Although the nerve impulse is electrical in nature, its transmission along the fiber is the result of chemical reactions involving potassium and sodium ions. The impulse relies on a series of changes in electrical polarity as it moves along the cell membrane. The charge is the result of the movement of potassium and sodium ions (charged atoms) across the dif-

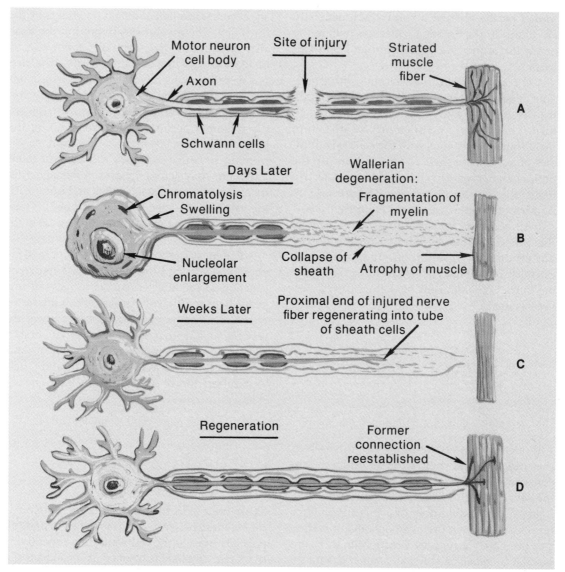

Figure 8-2 If a myelinated nerve fiber is injured **(A)**, the proximal portion of the fiber may survive, but the portion distal to the injury is likely to degenerate **(B)**. In time, the proximal portion may develop extensions that grow into the tube of sheath cells previously occupied by the fiber **(C)**, and the former connection may be reestablished **(D)**.

ferentially permeable membrane of the fiber cells (active transport theory).

The operation of nerve impulse transmission depends on the separation of the positively charged sodium and potassium ions on either side of the cell membrane. During the resting stage the differentially permeable membrane is polarized, with the inner surface negative with respect to the outer surface. This difference in electrical potential is due to the greater concentration of sodium ions outside the cell and of potassium ions inside. Membrane polarity is

Table 8-1　Average Electrolyte Composition of Body Fluids

Electrolyte	Extracellular Fluid	Intracellular Fluid
Sodium (Na$^+$)	142 mEq/L	10 mEq/L
Potassium (K$^+$)	5 mEq/L	150 mEq/L
Calcium (Ca$^+$)	5 mEq/L	less than 1 mEq/L
Magnesium (Mg^{++})	3 mEq/L	40 mEq/L
Total cations	155 mEq/L	200 mEq/L
Chloride (Cl$^-$)	103 mEq/L	0–3 mEq/L
Bicarbonate (HCO$_3^-$)	27 mEq/L	10 mEq/L
Phosphates (PO$_4^-$)	2 mEq/L	150 mEq/L
Organic acid anions	6 mEq/L	—
Sulfates (SO$_4^-$)	1 mEq/L	—
Proteins	16 mEq/L	40 mEq/L
Total anions	155 mEq/L	200 mEq/L

maintained by the neuron through energy derived from cellular respiration. Large numbers of mitochondria, the cellular bodies where energy is generated, are found in nerve fibers.

The excitation of a nerve by a stimulus produces a momentary change in the permeability of the membrane, resulting in sodium ions passing into the fiber and potassium ions moving out. The exchange of ions causes a drop in electrical activity as measured by voltage (from −70 mv to +40 mv), and that part of the fiber becomes depolarized (Figure 8-3), creating the nerve impulse (action potential or stimulus of 110 mv).

Depolarization is not sufficient in itself to cause the release of transmitter. In addition, a supply of calcium ions must be present in the extracellular environment. Depolarization seems to be the means by which an inward current of calcium ions is induced to flow through the membrane of the presynaptic terminal. In this respect too, synaptic transmission resembles other known secretory processes. In every secretory process that has been studied, secretion is triggered by an increase in the concentration of calcium inside the secretory cell. The increase is the result of the entry of calcium from the external environment or the release of calcium from internal stores. In neurons the entering calcium ions probably promote the fusion of special intracellular vesicles (synaptic vesicles) into the presynaptic membrane. The vesicles are stationed near the inner surface of the membrane at the site of transmitter release and are filled with the transmitter substance.

The depolarization lasts for only an instant; however, many more sodium ions continue to flow into the cell. As an excess of positive ions is created inside the cell, an excess of negative ions occurs outside the cell wall. The potential, therefore, has reversed.

The negative stimulated point of the membrane sets up a local current, with the positive point adjacent to it acting as a stimulus, thus

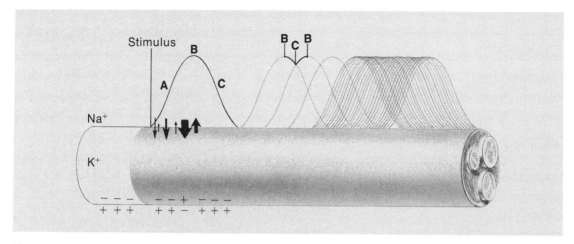

Figure 8-3 Conduction of a nerve impulse. Size of the arrows suggests the ease with which K^+ and Na^+ pass through the membrane during the transmissions of the impulse along the fiber. The + and − signs indicate the quantity of positive ions outside and inside the nerve membrane. **A,** Polarized zone with more sodiums outside the membrane than inside, and more potassiums inside than outside. **B,** De-polarized zone: after the membrane has been stimulated, the sodium ions are allowed to flow in at a greater rate than the potassium ions flow out, creating a wave of negativity on the outside of the membrane. **C,** Repolarized zone. The sodium ions are forced back out (sodium pump) and the potassium ions diffuse, restoring the greater quantity of positive ions on the outside.

creating a depolarization of that part of the membrane by reversing its polarity from positive to negative (Figure 8-3). This cycle of reversal of potential is repeated in rapid succession, and the impulse travels from point to point over the full length of the fiber, much as a ripple of water moves toward the shoreline. This movement of an impulse or wave of negativity is called **conduction.**

Repolarization results partly from the "sodium pump" of the active transport theory but is due mainly to the membrane's decreased permeability to sodium, resulting in an increased permeability to potassium. The latter diffuses outward, creating an excess of positive ions on the outside, thus restoring the resting potential of the membrane.

In the nervous system, the impulse begins at the dendrite or cell body and is then conducted throughout the length of the axon.

Myelinated and unmyelinated axons do conduct impulses differently. In an unmyelin-ated axon the impulse propagates by means of a local and temporary flow of ions, and thus of electrical current, into the depolarized region of membrane and out through adjacent membrane. These local circuits depolarize the nearby membrane in a continuous, sequential manner. Hence the signal moves down the axon like a wave. The mechanism is called *continuous conduction.* At the end of the axon the depolarization triggers the release of a neurotransmitter, a substance that can depolarize the membrane of a neuron that receives it. If the axon ends on a muscle, the neurotransmitter can cause the muscle to contract.

In a myelinated axon the process of conduction is different. Here the axonal membrane is exposed to the extracellular environment only at the nodes of Ranvier. The remainder of the axonal membrane is covered by a myelin sheath, which has a much higher electrical resistance and a much lower electric capacity than the axonal membrane itself. When the membrane is depolarized at a node of Ranvier, the local circuit generated cannot flow through the adjacent membrane. Instead it excites the membrane only at the next node. Because the excitation of the membrane jumps from node to node, the conduction of signals is faster than it is in a fiber without a myelin sheath.

Moreover, in a myelinated axon a modest flow of sodium and potassium ions will serve to depolarize the small area of axonal membrane that is exposed at a node of Ranvier. Hence a modest amount of energy is involved in pumping sodium ions back out of the axon and potassium ions back in to repolarize the membrane. In an unmyelinated fiber the entire axonal membrane must depolarize and then repolarize. That is why the signal conduction in a myelinated fiber is less demanding of energy. The form of signal conduction in a myelinated fiber is called *saltatory conduction* (from the Latin word *saltare* meaning to leap).

Velocity of Conduction

The velocity at which a neuron conducts an impulse is a function of its diameter. In humans, large motor neurons conduct impulses at a velocity of 80 to 100 m/second. Sensory nerves generally have a similar diameter but transmit more slowly. Nonmyelinated fibers conduct impulses more slowly than do myelinated ones. Their velocity is about 0.5 m/second.

All-or-None Response

Not all changes in the permeability or electrical characteristics of the membrane result in action potentials. Changes less than 10 to 12 millivolts produce local excitation, but the activity does not result in an action potential. To state this another way, the increments of change below the threshold (subthreshold stimuli) do not initiate action potentials. However, if a second subthreshold stimulus is applied to the membrane before the first one has been removed, the second stimulus will have an additive effect on the first and an action potential may be initiated. Once the threshold level has been reached,

the action potential will occur. This is an example of an all-or-none process. Either the action potential will occur completely or not at all. (This concept is also discussed in Chapter 7.)

Conduction of an Impulse Across Synapses

Nerves are interconnected at junctions called **synapses,** which means contact. These junctions, where two or more neurons are in contact, consist of fingerlike processes whose ends are sufficiently close to one another (about 250 Å apart) to allow certain chemicals to diffuse rapidly across the intervening space.

A close look at a presynaptic neuron (Figure 8-4) discloses that it contains many mitochondria and vesicles. It is now generally believed that when an impulse reaches the terminal knob, it causes the excitatory chemical, acetylcholine, to be released by the vesicles. The acetylcholine then flows across the gap and comes in contact with the membrane of the neuron, causing an increase in membrane permeability. As a result, sodium ions rush in and permit the passage of the impulse from the terminal knob to the neuron. Once the impulse has been sent across the synapse, the acetylcholine is broken down, as in muscle, into two components, acetyl and choline, by the enzyme cholinesterase.

The main feature of the point at which motor fibers enter a skeletal muscle—the neuromuscular junction or motor end-plate (Figure 8-4)—include a terminal branch of an axon and a gap similar to the neuronal synapse. Acetylcholine is the common transmitter for this gap. This type of synapse is commonly called a *myoneural junction.*

As noted in Chapter 7, the body has a means of resupplying the necessary acetylcholine for synaptic transmission. After the acetylcholine has been broken down by cholinesterase, the two free compounds, acetyl and choline, diffuse back across the gap and are reunited by a second enzyme, choline acetylase, in the terminal knob vesicles to await a new impulse.

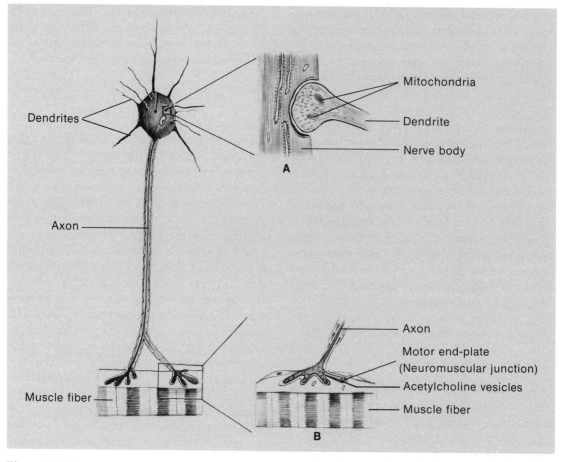

Figure 8-4 Types of synapses. **A,** Nerve-nerve body synapse. **B,** Nerve-muscle synapse (myoneural junction).

Neurotransmitters

Several different substances are thought to act as transmitters at different synapses in the central nervous system. Acetylcholine is no doubt an excitatory transmitter (although it is inhibitory at certain synapses). Norepinephrine, dopamine, and 5-hydroxytryptamine may be excitatory or inhibitory, glutamic acid may be excitatory in the brain, gamma-aminobutyric acid (GABA) is inhibitory in the brain, and glycine is inhibitory in the spinal cord (the last three are amino acids).

THE CENTRAL NERVOUS SYSTEM

The central nervous system (CNS) is composed of the brain and spinal cord. In the embryo, the central nervous system develops from a neural tube. The anterior end of the tube enlarges and greatly modified, forms the brain, which reaches its greatest complexity in humans. The original tubular structure still may be seen in the intercommunicating ventricles of the brain and in the central canal of the spinal cord. We will first consider the anatomy and function of the spinal cord and the nerves arising from it.

THE SPINAL CORD

The spinal cord occupies the entire spinal canal of the embryo and thus extends down into the tail portion of the vertebral column. However, the column of bone grows much more rapidly than the nerve tissue of the cord; therefore, the end of the cord soon is too short to extend into the lower part of the spinal canal. This disparity in growth increases during development, and in the adult the cord ends between the first and second lumbar vertebrae in the region just below the area to which the last rib attaches.

Examination of the spinal cord reveals that it has a small, irregularly shaped internal section of gray matter (nerve cell bodies) and a larger surrounding area made up of white matter (nerve fibers). A cross section of the cord shows that the gray matter is arranged in columns of cells extending up and down dorsally, one on each side. Another pair of columns is found in the ventral region; a third, less conspicuous pair is situated between the ventral and dorsal columns. These three pairs of columns of gray matter give, in cross section, an H-shaped appearance (Figure 8-5).

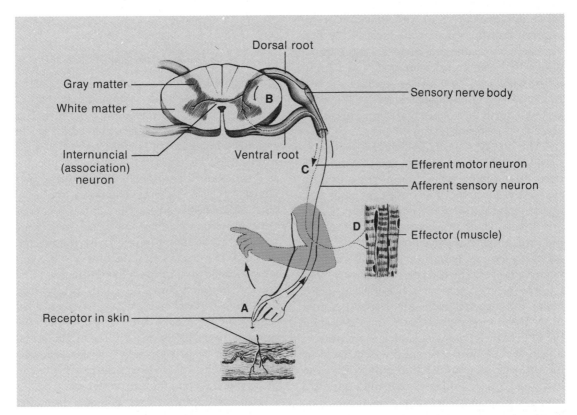

Figure 8-5 Diagram demonstrating a simple reflex arc. An impulse is initiated at the finger by a receptor **(A)** in the skin. The impulse travels over the sensory afferent neuron to the spinal cord, where it is transmitted via the internuncial neuron **(B)** to the efferent motor neuron. The impulse then travels via the efferent motor neuron **(C)** to the muscle **(D),** which effects a response.

The white matter, consisting of thousands of myelinated nerve fibers, is arranged in three columns outside the gray matter on each side. The function of these columns is to convey information up and down the cord.

Each column (or *funiculus*) is subdivided into tracts—large bundles of nerve fibers arranged in functional groups (Figure 8-6). Ascending (sensory) tracts transmit impulses to the brain, and descending (motor) tracts transmit impulses from the brain to the various levels of the spinal cord. Information can thus be conveyed up and down the cord.

At each small space between the vertebrae (intervertebral space), spinal nerves arise from each side of the cord and distribute themselves in the tissues. There are about as many pairs of spinal nerves as there are vertebrae, but at the lower end of the spine at the level of the first lumbar vertebra, the cord divides. All the spinal nerves that emerge from the vertebral column below that point continue as separate strands. Above this point, the spinal nerves arise at each vertebral level, and dorsal and ventral roots are present on each side (Figure 8-5).

The dorsal roots contain the afferent (sensory) fibers that bring information into the CNS from the receptor, where the impulse is initiated. At each of the intervertebral spaces, the fibers of the dorsal roots may or may not be myelinated. Their cell bodies are located outside the cord proper in a swelling called the *dorsal root ganglion*. (A **ganglion** is a collection of nerve cells along the course of a sensory nerve.)

The ventral roots carry the efferent motor fibers, and their cell bodies are inside the spinal cord. These motor fibers carry impulses from the CNS to the muscle or gland, and are heavily myelinated. A third neuron, the association or internuncial neuron, is found only within the brain and spinal cord, and transmits impulses from the afferent to the efferent neurons.

Spinal Reflexes

The least complicated example of how the sensory mechanism operates is the simple reflex. If you prick your finger, you will pull back your hand. The basic events that occur here are shown in the diagram of the reflex arc in Figure 8-5. The receptors at *A* are stimulated. An impulse travels over the sensory afferent nerve to the spinal cord, where it synapses at *B* with an internuncial neuron. This, in turn, synapses with the motor efferent fiber at *C*, and an impulse travels out to stimulate the muscle at *D*, which contracts and withdraws the hand.

In a very few cases (such as the patellar and pupillary reflexes), the sensory fiber synapses directly with the motor fiber. Other associative fibers may convey the impulse to other motor fibers at different levels in the spinal cord (even to the brain) to deliver the information that a stimulus has been applied.

The Spinal Nerves

Strictly speaking, the spinal nerves are part of the peripheral nervous system rather than of the CNS, but we are considering them here as part of our discussion of the spinal cord. Each of the thirty-one pairs of spinal nerves that arise from successive levels of the spinal cord begins as a series of fibers from both the dorsal and ventral roots of the cord. The dorsal root contains only sensory fibers; the ventral root contains only motor fibers and acts as a final pathway for all motor impulses leaving the spinal cord. The first cervical nerve leaves the vertebral canal by passing between the first cervical vertebra and the skull. Below this, successive cervical nerves pass between adjacent vertebrae. Below the cervical region the spinal nerves likewise leave the vertebral canal below their respective vertebrae: for example, the first thoracic nerve leaves the canal between the first and second thoracic vertebrae.

As each spinal nerve emerges from between its respective vertebrae, it divides into two small branches. The first of these, the *dorsal*, or *posterior*, *ramus*, passes posteriorly (dorsally) to supply a specific segment of the skin, bones, joints, and longitudinal muscles of the back (Figure 8-7). The other *ventral*, or *anterior, ramus*

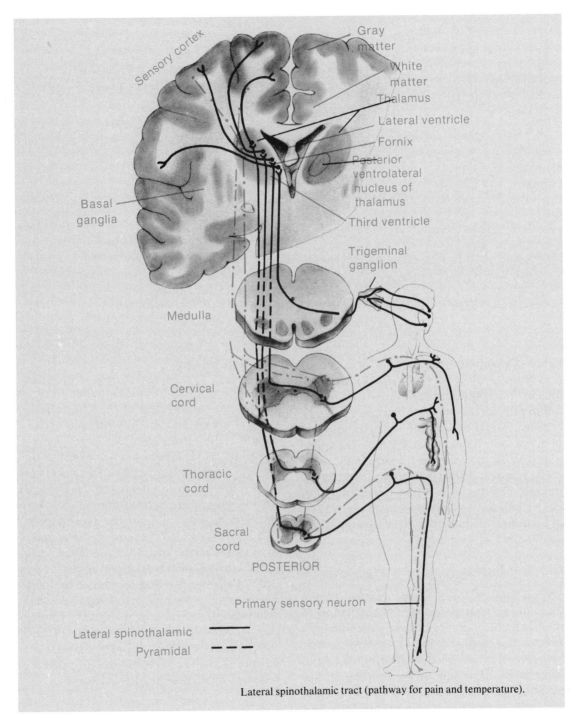

Lateral spinothalamic tract (pathway for pain and temperature).

Figure 8-6 Elements of ascending (sensory) and descending (motor) tracts of the spinal cord. The lateral spinothalamic tract is sensory for pain and temperature. The pyramidal tract conveys motor impulses primarily for voluntary muscle movements. (Modified from *Structure and Function in Man,* 3rd Ed. by S. W. Jacob and C. A. Francone. Philadelphia: W. B. Saunders Co., 1974.)

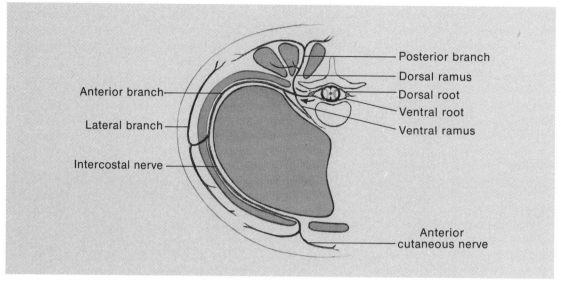

Figure 8-7 Diagram of the course and branches of a typical intercostal nerve.

supplies a portion of the anterior or lateral aspect of the trunk or the limbs. The ventral rami are larger than the dorsal rami and behave quite differently. In the cervical, lumbar, and sacral regions, ventral rami form extensive networks of nerves called **plexuses** (Table 8-2). The spinal nerves in the thoracic region course in the intercostal spaces to supply primarily the intercostal muscles and the skin overlying them. Most nerves arising from these plexuses carry fibers of neurons from more than one segment of the spinal cord (Figure 8-8).

Cervical Plexus

Ventral rami of the first four cervical nerves form the cervical plexus. Peripheral branches innervate the skin and certain muscles of the back of the head, the neck, and the shoulder. The *phrenic nerve*, which is distributed to and controls the diaphragm, arises from this plexus. Crushing damage to the phrenic nerve at its origin will result in respiratory paralysis and death. Severe reflex disorders, such as hiccups, can be arrested by pinching the phrenic nerve to paralyze the diaphragm temporarily.

Brachial Plexus

This large plexus of the neck and shoulder originates from ventral rami of the fifth to eighth cervical nerves and the first thoracic nerve. It supplies the neck, shoulder, arm, forearm, wrist, and hand. Among the nerves that supply the arm are the *axillary, median, radial,* and *ulnar nerves.* Injuries to the nerves of the upper extremities result in the loss of use of the part of the extremity that the specific nerve supplies. Damage to the radial nerve results in an inability to extend the hand at the wrist, called *wristdrop.* Median nerve damage results in an inability to flex and abduct the thumb and index finger, a critical function of the hand. Loss of the ability to oppose the thumb and any of the fingers prevents a person from picking up small objects. Damage to the ulnar nerve impairs wrist function and adduction, and the fingers cannot be spread apart, resulting in a *clawhand.*

Lumbar Plexus

Ventral rami of the second to fourth lumbar nerves divide into anterior and posterior sec-

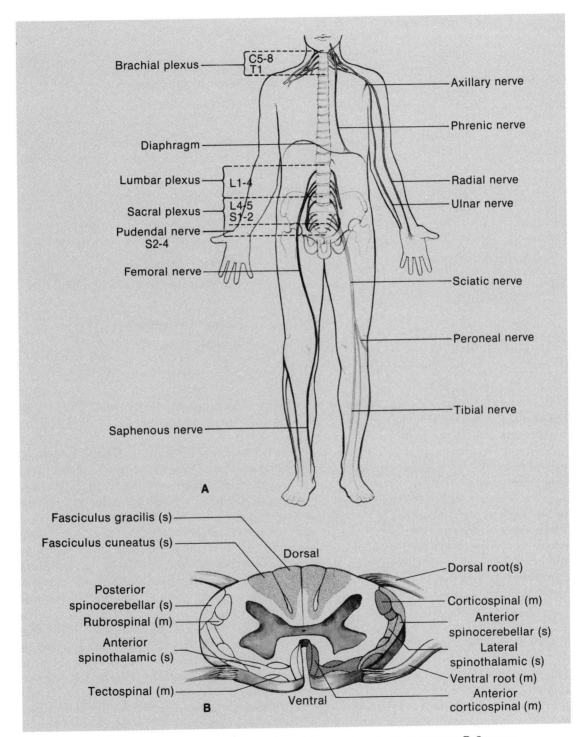

Figure 8-8 A, Spinal plexuses. Branches of the spinal cord, anterior aspect. **B,** Sensory and motor tracts from the spinal cord to the brain.

Table 8-2   Major Nerve Plexuses of the Body

Plexus	Origin	Principal Nerve(s)	Result of Damage
Cervical	C1-4	Phrenic	Respiratory paralysis; death
Brachial	C5-8; T1	Axillary	Weakened abduction and rotation of arm
		Median	Impaired flexion and abduction of hand; loss of flexion and abduction of thumb and index finger; loss of thumb apposition
		Radial	Wristdrop
		Ulnar	Clawhand
Lumbar	L2-4	Femoral	Loss of extension of leg and flexion at hip
			Lumbago—inflammation
Sacral	L4-5; S1-4	Sciatic	Loss of extension at hip and flexion at knee
			Sciatica—inflammation
		Peroneal	Footdrop

tions to form the lumbar plexus, located on the inside of the posterior abdominal wall. Its nerves contribute to the supply to the muscles and skin of the abdominal wall, buttocks, medial thigh, and front part of the leg and foot. The major nerve arising from this plexus is the *femoral nerve*, which supplies all the flexor muscles of the thigh, as well as portions of the hip and leg. Injury to the femoral nerve may result in loss of the ability to extend the leg and to flex the hips. Lumbago is characterized by an inflammation of the nerves at the base of the spine and buttocks area.

Sacral Plexus

Ventral rami of the fourth and fifth lumbar and first four sacral nerves form the sacral plexus. This group of nerves combined with the nerves of the lumbar area is commonly referred to as the *lumbosacral plexus,* since both groups supply the lower extremities. The major nerve of this plexus is the *sciatic,* the largest nerve in the body. It supplies the flexor muscles of the thigh

and divides into the *peroneal* and *tibial nerves,* which supply all the muscles of the lower leg and foot.

Dislocated hips and herniated intervertebral disks are common causes of damage to the sciatic nerve. Injury to the nerve itself results in an impairment of extension at the hip and flexion at the knee. Inflammation of this nerve is characterized by sharp, shooting pains along the course of the nerve, and, if the condition persists, the muscles supplied by the nerve may deteriorate or atrophy. Damage to the peroneal nerve results in loss of flexion of the foot and toes (footdrop).

The *pudendal nerve,* formed by the rami of the second, third, and fourth sacral nerves, is regarded by some authorities as being the major nerve of a separate plexus. However, it is generally considered part of the sacral plexus. It supplies the muscles and skin of the perineum—the floor of the pelvis. This nerve is blocked by local anesthesia for childbirth, a procedure known as *saddle block anesthesia.*

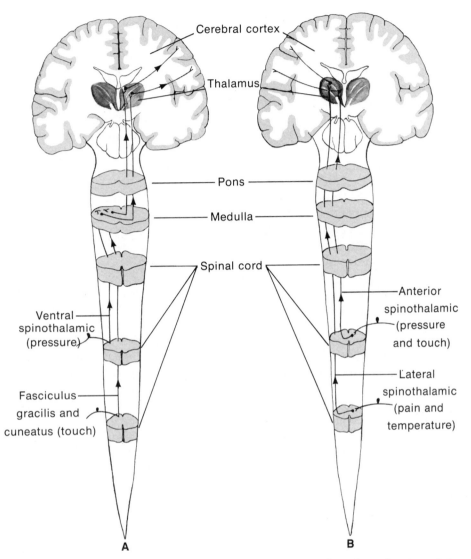

Figure 8-9 Sensory pathways to the brain. **A,** Touch and pressure pathways from the trunk and limbs via the fasciculus gracilis and cuneatus nerves.

B, Pressure and touch, and pain and temperature pathways from the trunk and limbs.

Conduction Pathways

The spinal cord, as we have seen, carries signals in both directions between the brain and the remainder of the body. This takes place along the ascending and descending tracts (Figures 8-6 and 8-9). In most cases, the name of the tract will indicate the funiculus in which it travels,

the location of its cells of origin, and the level of location of its axon termination.

Ascending Tracts

The ascending tracts carry messages along the spinal cord in the direction of the brain (see Figures 8-6 and 8-9). The *fasciculus gracilis* and *fasciculus cuneatus* are two major afferent tracts that form the first link in the pathway for the sensations of motion, movement, touch, and pressure. The cell bodies of the neurons composing these tracts lie in the dorsal ganglia of spinal sensory nerves, and their fibers extend upward to the medulla of the brain.

The *lateral* and *anterior spinothalamic tracts* lie in the white matter of the spinal cord and convey stimuli of pain, temperature, pressure, and touch. They originate in the posterior cell column and ascend to the thalamus of the brain, where they cross to the opposite side in the anterior region of the cord. Damage to the lateral spinothalamic tract results in loss of the sensations of pain and temperature on the side of the body opposite the injury.

The *anterior* and *dorsal spinocerebellar tracts* convey impulses from the muscles of the legs and trunk between the sixth cervical and second lumbar regions. They carry stimuli from the proprioceptors in muscles, tendons, and joints, and terminate in the cerebellum of the brain. These tracts transmit impulses concerned with unconscious muscle sense and play an important role in posture and muscle tone. Damage to these tracts results in uncoordinated voluntary movements due to poor skeletal muscle tone.

Descending Tracts

The descending tracts travel from the brain down the spinal cord and terminate in the gray matter of the cord.

The *rubrospinal tract,* or *extrapyramidal tract,* conveys impulses from the cerebellum to the motor cells of the anterior column. The fibers cross immediately from the thalamus and descend to various levels of the cord where they make connections with spinal motor nerve roots (see Figure 8-6). This pathway is concerned with reflexes that aid in righting the body and with the tone of muscles affecting posture.

The *corticospinal tract (pyramidal tract)* forms the great motor pathways between the cerebral cortex and spinal nerves, and carries impulses for voluntary movement, especially those concerned with skilled movements of the fingers and hands. Most of the corticospinal fibers cross in the medulla and descend as the large tract in each side of the spinal cord. Owing to the crossing of the fibers, the left half of the brain controls the skeletal muscles on the right side of the body, and the right side of the brain controls those on the left side of the body.

Injuries and Disorders of the Spinal Cord

Spinal Cord Injuries

The spinal cord may be partially injured or entirely severed (transected). Partial injury to the ascending (sensory) tracts causes lack of coordination, since spinal reflexes are disturbed. Walking becomes uncoordinated, movements are jerky, and loss of balance occurs. Partial injury can also produce paresthesias (abnormal sensations), weakness, exaggerated reflexes, and a decreased ability to sense pain and temperature.

Complete transection of the spinal cord results in paralysis of the skeletal muscles below the level of injury. Sensation is abolished, since all descending and ascending pathways are interrupted. If the injury occurs in the pyramidal tract or if the motor neurons of the cord are injured, paralysis may be complete and of the flaccid type. Spastic paralysis is caused by certain extrapyramidal tract injuries of the brain or cord and is indicated by an exaggerated tonicity and uncoordinated reflexes.

Disorders Involving the Spinal Cord

Poliomyelitis, an acute viral disease most commonly found in children, affects both the spinal cord and the brain. The virus may destroy the motor nerve cells in the spinal cord (Figure 8-10), in which case paralysis of one or more limbs results.

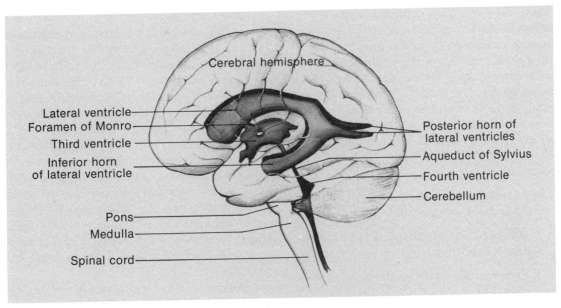

Figure 8-15 The brain ventricles, lateral aspect.

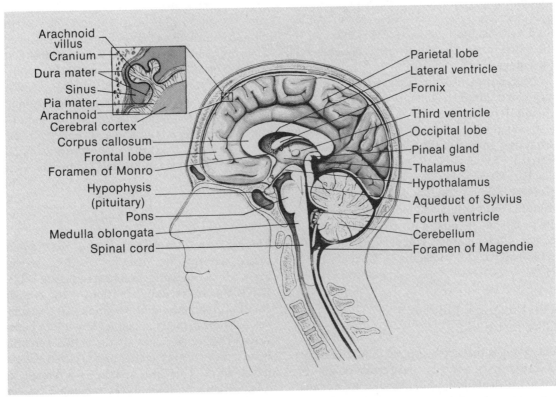

Figure 8-16 Midsagittal section of the human brain, showing circulation of cerebrospinal fluid in the brain and spinal cord. Inset: the cranial meninges.

brain, and the fluid is returned to the bloodstream through arachnoid villi, which absorb it. Because production of fluid is more rapid than absorption, flow out of the ventricles into the subarachnoid space is constant. Should the openings of the fourth ventricle become blocked, the fluid will accumulate and distend the ventricles, leading to the condition of *hydrocephalus* ("water on the brain") (see Figure 5-24).

Cerebrospinal fluid has a threefold purpose:

1. It protects the brain and the spinal cord from sudden changes in pressure.

2. It maintains a stable chemical environment for these structures.

3. It removes the waste products of cerebral metabolism.

Readily accessible and reflecting many biochemical and cell-shedding alterations in central nervous system diseases, cerebrospinal fluid analysis has long been a valuable tool in the laboratory diagnosis of many diseases and conditions.

Cerebrospinal fluid is a clear, colorless, watery blood filtrate. It contains most of the blood chemicals but not the large molecules or formed elements (blood cells). Some of its constituents are water, glucose, sodium chloride, potassium, traces of protein, and a few white cells.

Each day, the brain produces 400 to 500 mL of cerebrospinal fluid—about 0.35 mL every minute. Fluid volume at any time ranges from 10 to 60 mL in infants to 100 to 160 mL in adults. Under normal volume conditions, the cerebrospinal fluid pressure is 70 to 200 mm H_2O.

Cerebrospinal Fluid as a Diagnostic Aid

The changes that cerebrospinal fluid undergoes in disease have proved to be of major diagnostic value, particularly for infants, who are not able to describe their ailments. In acute meningitis, the vessels of the inflamed meninges allow protein and white blood cells to proliferate in the spinal fluid. The fluid is therefore purulent and turbid. Tuberculous meningitis is detected by a marked decrease in the quantity of chloride and the presence of tubercle bacilli in the fluid. The condition of the fluid in poliomyelitis is similar, except that the white blood cell count is higher, the chloride quantity is normal, and no bacilli are present. Precipitation tests, such as the colloidal gold test, assist in the detection of such diseases as aseptic and tuberculous meningitis, encephalitis, poliomyelitis, multiple sclerosis, and syphilitic paralysis. The ratio of the proteins albumin and globulin in spinal fluid change in disease. As discussed on p. 221 cerebrospinal fluid is obtained through a lumbar puncture.

Brain tumors can sometimes be detected by compressing large veins in the neck, creating a back-up of blood into the cranial sinuses and a resultant increase in cerebrospinal fluid pressure. Pressure will increase sharply and remain elevated until the venous blood is allowed to flow normally again. If the pressure does not increase or fall rapidly when released, it is possible that a tumor or lesion exists in the area. A pneumoencephalogram (ventriculogram), an x-ray picture of the skull, is obtained after some of the fluid has been removed and replaced by air by means of a lumbar puncture. Intracranial pathological states can be detected in this manner.

The Cerebrum

The cerebrum, the largest part of the human brain, accounts for up to four-fifths of the brain's total mass and weight (see Figure 8-14). It fills the entire upper portion of the cranial cavity and contains billions of neurons and synapses. These form an amazing network of pathways. The outer nerve tissue of the cerebral hemispheres is gray matter and is called the *cerebral cortex*. This gray matter is arranged in folds or convolutions (**gyri**), separated by depressions (**sulci**) or deep grooves called **fissures**. Internally, the cerebral hemispheres are largely white matter with a few areas of gray matter.

Fissures of the Cerebral Cortex

The major *longitudinal fissure* extends from the posterior to the anterior of the cerebrum, dividing it into the two cerebral hemispheres. The cerebral hemispheres with their coverings constitute the true brain. Each has a set of controls for sensory and motor activities of the body. The controls for the right side of the body are located in the left cerebral hemisphere, and vice versa. When one area of the brain is damaged, the corresponding area of the other hemisphere often can take over the functions of the damaged region.

The corpus callosum is located in the longitudinal fissure. It is made up of nerve fibers that connect the two hemispheres and is concerned with learning and memory.

The *lateral sulcus* (*fissure of Sylvius*) spreads dorsally and separates the temporal from the frontal and parietal lobes (Figure 8-14).

The *central sulcus* (*fissure of Rolando*) starts at the midpoint of the upper border of the cerebrum and extends downward toward the lateral sulcus, separating the frontal and parietal lobes. The central sulcus also separates centers for sensory and motor functions.

Localization of Cerebral Function

As a result of numerous experiments on animals and close observations of humans, physiologists have been able to localize certain areas of the brain that control motor, sensory, and other activities. In no case, however, is the control of a given function limited to a single area. All parts of the cerebrum are connected, and changes in nervous activity of any one part alter the excitability of the entire brain.

MOTOR AREA The motor area is the thickest part of the cerebral cortex and contains cells from which most motor impulses originate. It lies in a band on each hemisphere in the frontal lobe just anterior to the central sulcus and extends down to the lateral sulcus on each side (Figure 8-17). The many kinds of cells in the motor cortex include those of the pyramidal tracts. The impulses for voluntary muscular movement originate from the motor area of the cortex, and when a small electric shock is administered to this area, contractions are observed in muscles on the opposite side of the body. By varying the location of the stimuli, it has been found that the regions of the motor cortex relate to muscles at various levels of the body in a distinct but reverse order.

Knowledge of the location and function of the motor area has proved of great value in the diagnosis of brain injuries. Motor paralysis of a given part can result from injury or pressure involving the motor area of the brain. Cerebral hemorrhage, a rupture of a major vessel in the base of the cerebrum, has the same effect as injury or trauma. If a blood clot forms or a vessel is ruptured, preventing nervous impulses from reaching the muscles, a motor paralysis of the opposite side of the body occurs.

The motor area for speech (Broca's area) is located at the base of the frontal lobe and is thought to be concerned with the formation of words both in speaking and writing. As before, the motor area for speech is located on the opposite side of the brain from the side of the body actually involved in the activity; that is, the left hemisphere for a right-handed person, and vice versa.

Dysfunction of the frontal lobe, if the entire motor area is involved, results in *hemiplegia* (paralysis of the opposite half of the body). Generally, the paralysis is more severe in the upper extremities than in the lower. Irritation of the precentral gyrus creates convulsive seizures of the jacksonian type. These are seizures that begin in one part of the body and extend to another part.

SENSORY AREAS The sensory cortex is concerned with the interpretation of sensory impulses and is located mainly in the postcentral region (see Figure 8-16). The area is divided into a large number of sensory regions like those of

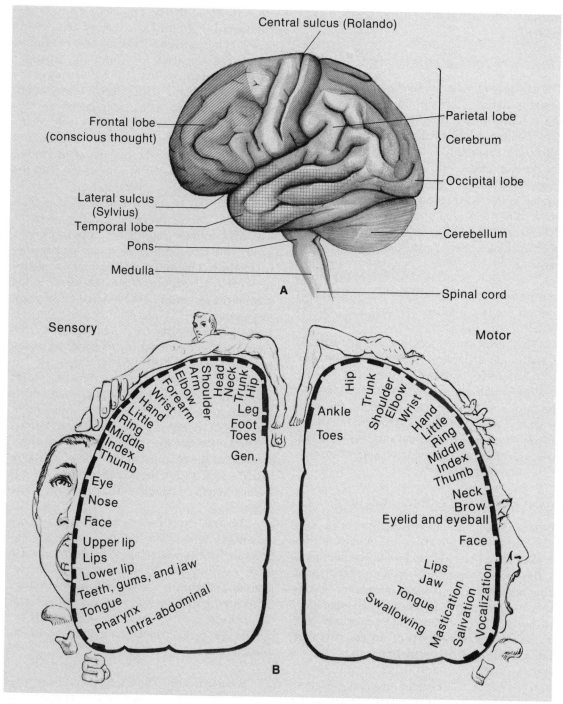

Figure 8-17 A, Centers for cerebral function, lateral aspect. **B,** Location of sensory and motor areas of the cortex showing parts of the body affected.

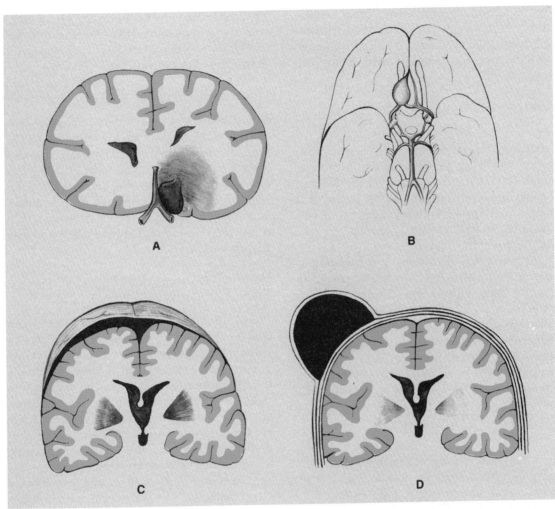

Figure 8-20 Disorders of the brain. **A,** Ruptured aneurysm intracerebrally. **B,** Aneurysm of the ante-rior cerebral artery. **C,** Hemorrhage of the pia mater. **D,** Subdural hematoma.

(*arteriosclerosis*) and generally occur after the age of 40 years. The effects of a stroke depend on the extent and location of arterial involvement. Various forms of paralysis may result, depending on the depth of the affected white matter in the lower part of the cerebrum. Categories of paralysis include:

- *Monoplegia*—paralysis of a single extremity

- *Hemiplegia*—paralysis of half the body and its extremities

- *Paraplegia*—paralysis of both lower extremities and the lower part of the body

- *Diplegia*—paralysis of both upper or both lower extremities

- *Quadriplegia*—paralysis of all four extremities

Paralysis may be spastic, resulting from damage to the extrapyramidal pathways or upper motor neurons and characterized by increased muscle tone due to the removal of the controlling influence of higher brain centers. Or it may be flaccid, with a loss of muscle tone and deep tendon reflexes, resulting from damage to the pyramids or to the lower motor neurons.

Parkinson's Disease (Paralysis Agitans)

Parkinson's disease is a disease of brain degeneration involving portions of the extrapyramidal (involuntary) motor system. It is a chronic disease that develops later in life. It is characterized by involuntary tremors, diminished motor faculties, rigidity, and a "pill-rolling" tremor of the fingers.

Degeneration of neurons in the basal ganglia (see Figure 8-19) is responsible for irregularities in body movements. Clinical findings include a deficiency of a neuronal transmitter substance, dopamine. Treatment with L-dopa, a substance that is converted to dopamine in the brain, has proven beneficial with relief of symptoms. The cause of Parkinson's disease is unknown. Heredity is offered as a possible factor.

Cerebral Palsy

Cerebral palsy is a congenital brain disorder. **Palsy** is characterized by diverse disorders of muscles that vary in degree from weakness to complete paralysis. The disease may involve the muscles of the lower extremities and speech functions as well. Continuous muscle reeducation, speech training, and other physio-occupational therapies have helped patients with cerebral palsy considerably.

Epilepsy

Epilepsy, once called "falling sickness," is a disease of the CNS characterized by recurring convulsions and loss of consciousness as an uncontrolled wave of electrical discharges sweeps through the brain (Figure 8-21). In most cases the cause of the disorder is not known. Severe seizures are termed *grand mal;* if the pa-

tient suffers no more than brief periods of dizziness or loss of memory, they are called *petit mal.* Most of those afflicted may lead normal, active lives if they use appropriate medications.

Chorea (St. Vitus' Dance)

Chorea is a convulsive nervous disease characterized by involuntary jerking movements. The type called *St. Vitus' Dance* commonly follows a disease such as rheumatic fever. *Huntington's chorea* is a hereditary mental disorder characterized by irregular movements, speech disturbances, and progressive mental deterioration. It is inherited through a dominant allele and usually appears late in life.

Reye's Syndrome (Viral Encephalopathy)

The syndrome seems to occur in the aftermath of a viral infection, particularly influenza or chicken pox. Nearly all the victims are children or teenagers. The first sign of Reye's syndrome is intractable vomiting. Next, the child shows signs of brain dysfunction, including disorientation, lethargy, and personality changes such as unprovoked shouting and use of abusive language. Finally, if the disease progresses, the child may become comatose.

Diagnostic tests include the presence of blood ammonia, elevated serum transaminase, and hypoglycemia (low glucose levels). Histologically, the outer margin of liver lobules shows striking fatty filtration and glycogen depletion. The disease remains a medical mystery, but treatment of the symptoms can reduce the mortality and prevent severe brain damage in the survivors.

Tumors

Tumors of the brain may develop at any age but are more common in the young and in middle-aged adults. The severity, growth, and destructiveness of the lesion depend on the degree to which it compresses the brain tissue. Early surgery offers hope in some cases.

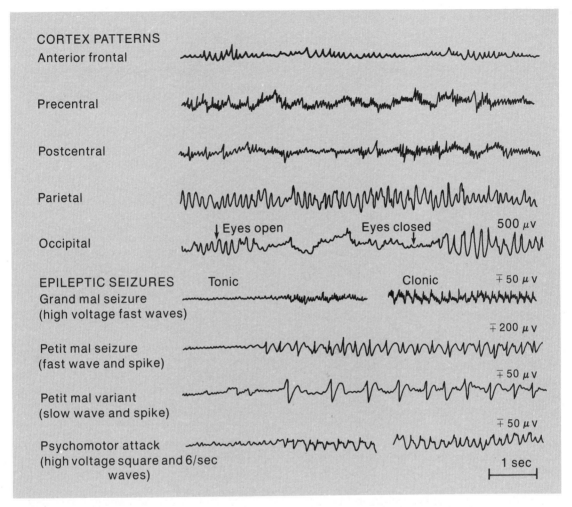

Figure 8-21 Typical EEG tracings in several types of epileptic seizures. Compare these with the normal EEG patterns in waking and sleeping.

THE CRANIAL NERVES

Twelve pairs of cranial nerves emerge from the lower surface of the brain and pass through the foramina in the base of the cranium. These nerves form part of the peripheral nervous system. They are classified as motor, sensory, and mixed nerves. The origin of the cranial nerves is comparable to that of the spinal nerves. The motor fibers of the spinal nerves arise from cell bodies in the ventral column of the cord, and the sensory fibers arise from cell bodies in the ganglia outside the cord. The motor cranial nerves arise from cell bodies within the brain, their nuclei of origin. The sensory cranial nerves arise from groups of nerve cells outside the brain. These cells may form ganglia on the trunks of the nerves or may be located in the sense organs. The central processes of the sensory nerves run into the brain and end around nerve cells that form their nuclei of termina-

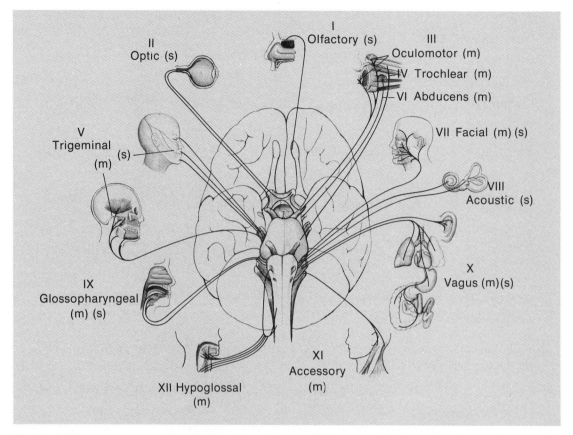

Figure 8-22 Cranial nerves. Twelve pairs of nerves arise from the undersurface of the brain to supply the head and neck and most viscera. They may be sensory (s), motor (m), or mixed in function.

tion. The nuclei of both motor and sensory nerves are connected with the cerebral cortex.

The cranial nerves are numbered according to the order in which they arise from the brain (Figure 8-22); they are also named to describe their nature, function, or distribution:

I. *Olfactory* (sensory)—carries impulses of smell to the brain.

II. *Optic* (sensory)—special nerve of the sense of sight.

III. *Oculomotor* (motor)—concerned with contraction of most of the eye muscles (the extrinsics).

IV. *Trochlear* (motor)—supplies motor fibers for movement of the superior oblique muscle of the eye.

V. *Trigeminal* (mixed)—the largest cranial nerve and chief sensory nerve of the face and head; motor fibers extend to the muscles of mastication (chewing).

VI. *Abducens* (motor)—supplies motor fibers to the lateral rectus muscle of the eye.

VII. *Facial* (mixed)—sensory fibers supply the anterior two-thirds of the tongue (taste) and the lacrimal and salivary glands; motor fibers supply the muscles of facial expression.

VIII. *Acoustic* or *vestibulocochlear* (sensory)—contains sense fibers for hearing and balance (semicircular canals of the internal ear).

IX. *Glossopharyngeal* (mixed)—sensory fibers supply the posterior third of the tongue (taste), tonsils, and pharynx; motor fibers control the swallowing muscles in the pharynx.

X. *Vagus* (mixed)—the longest cranial nerve, and the only one to leave the cranial region; motor and sensory fibers innervate most of the organs of the thoracic and abdominal cavities; the nerve acts as a cardiac inhibitor and bronchial constrictor.

XI. *Accessory* (motor)—controls the major trapezius and sternocleidomastoid muscles of the neck.

XII. *Hypoglossal* (motor)—controls movements of the tongue.

Disorders Involving the Cranial Nerves

Damage to the optic nerve (II) is an important cause of blindness. Eye fluid pressure changes (glaucoma), poisons (such as wood alcohol), or certain infections (e.g., syphilis) may cause destruction of optic nerve fibers.

Lower motor neuron lesions of vagus (X) and glossopharyngeal (IX) nerves may cause difficulty in swallowing (dysphagia) and development of nasal quality of consonant sounds (dysarthria).

Damage to the sensory fibers of the vagus nerve results in loss of taste in the pharyngeal region of the mouth. The vagus nerve also conducts sensory stimuli from the heart, bronchi, esophagus, stomach, small intestine, and ascending colon. Vagal stimulation may be responsible for the unpleasant sensation of nausea. The chief function and dysfunction resulting from abnormalities of the vagal system involve the operation of the visceral reflexes:

1. *Salivary–taste reflex.* Presence of food stimulates the salivary glands to secrete saliva.
2. *Carotid sinus reflex.* Increased pressure in the bloodstream stimulates special receptors in the wall of the carotid sinus and sends an impulse to the vasomotor center in the medulla to react accordingly.
3. *Carotid body reflex.* Special chemoreceptors in the carotid body respond to changes in carbon dioxide and oxygen content of the circulating blood. Stimulation of the respiratory center in the medulla speeds up or inhibits the rate of respiration.
4. *Cough reflex.* Irritation of the vagus nerve as well as the mucosa of the larynx, trachea, or bronchial tree stimulates the respiratory center in the medulla to bring about forced expiration.
5. *Gag reflex.* Touching the posterior wall of the pharynx or the uvula of the soft palate is followed by constriction and elevation of the pharynx.
6. *Vomiting reflex.* Forceful emptying of the stomach is brought about by relaxation of the cardiac sphincter and reversed peristalsis aided by contraction of the abdominal and thoracic muscles. A general elevation of intracranial pressure often causes vomiting, and it may also occur if there is localized pressure on the medulla.

Complete unilateral destruction of the abducens nerve (VI) makes it impossible to turn the eye outward. The unopposed pull of the medial rectus muscle causes the eye to turn inward, producing an internal strabismus, or internal squint. The result is diplopia (double vision) since the images do not fall on corresponding points of the left and right retinae. With bilateral abducens nerve paralysis neither eye can be moved in a lateral direction past the mid-position.

Isolated lesions of the trochlear nerve (IV) result in impairment of ability to turn the affected eye downward and outward.

An aneurysm (blood-filled sac in the wall of an artery), tumor, or hemorrhage may exert pressure on the oculomotor nerve (III), resulting in loss of ability to turn the eye inward, drooping of the eyelid (ptosis), and dilation of the pupil (mydriosis).

An interruption of the facial nerve (VII) causes total paralysis of the muscles of facial expression. The muscles of one side of the face sag, and the normal lines around the lips, nose, and forehead seem to disappear. The patient loses control of salivation, facial expression, and the ability to close his eye, leading to irritation and predisposition to infection. This condition is called *Bell's palsy* and, fortunately, it lasts for only 1 or 2 months.

Peripheral lesions of the motor division of the trigeminal nerve (V) cause atrophy and weakness of the muscles of the jaw. *Tic douloureux,* or *trigeminal neuralgia,* is a disorder characterized by attacks of unbearably severe pain over the distribution of one or more branches of the trigeminal nerve. No cause for the disease has been discovered.

Neuralgia is severe pain along the course of a nerve. Neuritis, on the other hand, is an inflammation or degeneration of a nerve. Both are painful, but neuralgia does not result in any organic damage to the nerve.

OUTLINE

I. Neuron
  A. Anatomy of the neuron
  B. Classification of neurons
    1. According to function
      a. Motor neurons
      b. Association (internuncial) neurons
      c. Sensory neurons
    2. According to number of processes
      a. Monopolar
      b. Bipolar
      c. Multipolar
  C. Structure of nerve fibers
    1. Schwann cells
    2. Myelin
    3. Nodes of Ranvier
II. Physiology of nervous tissue
  A. Regeneration of nerve fiber
  B. Transmission of nerve impulse
    1. Conduction theory
      a. Wave of negativity
      b. Saltatory conduction
    2. Velocity of conduction
    3. All-or-none response
    4. Summation of action potentials
    5. Conduction of an impulse across synapses
      a. Acetylcholine cycle
III. Central nervous system
  A. Spinal cord
    1. Spinal reflexes
      a. Simple
      b. Complex
    2. Spinal nerves
      a. Thirty-one pairs
      b. Dorsal and ventral rami
      c. Plexuses
        (1) Cervical
        (2) Brachial
        (3) Lumbar
        (4) Sacral
    3. Conduction pathways—tracts
      a. Ascending (sensory)
      b. Descending (motor)
    4. Injuries and disorders of the spinal cord and spinal nerves
      a. Flaccid and spastic paralysis
      b. Poliomyelitis
      c. Tabes dorsalis
      d. Multiple sclerosis
      e. Congenital defects
      f. Neuritis
      g. Shingles
    5. Spinal puncture and anesthesia
  B. Coverings of the brain and spinal cord
    1. Meninges
      a. Dura mater
      b. Arachnoid
      c. Pia mater

C. Brain
  1. Divisions of the brain
  2. Ventricles and cerebrospinal fluid
     a. Four ventricles and foramina
     b. Cerebrospinal fluid production, composition, and function
     c. Cerebrospinal fluid as a diagnostic aid
  3. Cerebrum
     a. Fissures of cerebral cortex
     b. Localization of cerebral function
        (1) Motor area
        (2) Sensory areas
        (3) Association areas
        (4) Right brain and left brain
  4. Thalamus and hypothalamus
  5. Cerebellum
     a. Peduncles
     b. Muscle coordination
  6. Brain stem
     a. Pons
     b. Medulla
  7. Disorders of the brain
     a. Stroke—paralysis
     b. Parkinson's disease
     c. Cerebral palsy
     d. Epilepsy
     e. Tumors
     f. Chorea
D. Cranial nerves
  1. Motor, sensory, or mixed
  2. Location in reference to brain
  3. Listing according to number, type, and function
  4. Disorders involving cranial nerves

STUDY QUESTIONS AND PROBLEMS

1. Name the three types of neurons according to their structure, function, direction, and myelinization.
2. Is a neuron capable of regeneration? Explain.
3. Explain an action potential with reference to the electrical theory of conduction.
4. What is acetylcholine? What is its main function in a synapse?
5. Define a reflex arc. Compare it with a complex reflex.
6. Trace the reception and interpretation of pain from the toe to the brain.
7. John Jones is a watchmaker who has lost the ability to use his hands as precisely as before. Relate the loss of manual dexterity to the brain and conduction pathways.
8. Mary Smith fell on her back and felt a twinge run down her left leg. Can you trace the possible impairment of nerve function from the nerve plexus down to her muscles?
9. Differentiate between the symptoms caused by a crush injury at the level of the fourth lumbar vertebra and at the level of the tenth thoracic vertebra.
10. What are the meninges? Describe their anatomical and pathological significance.
11. What is cerebrospinal fluid? Where is it produced and for what functions?
12. How would you obtain a sample of cerebrospinal fluid? Once the sample is obtained, what significant procedures can be followed?
13. List the three major sulci of the cerebrum and describe their anatomical and physiological significance.
14. What part of the brain is responsible for all motor activities of a conscious nature?
15. What do we mean by "localization of function" of the cerebrum?
16. Compare the pons and medulla, as lobes of the brain stem, with a railroad terminal.
17. Relate the function of the thalamus to that of the parietal association area of the cerebrum.
18. Why is the medulla called the "vital reflex center" of the body?

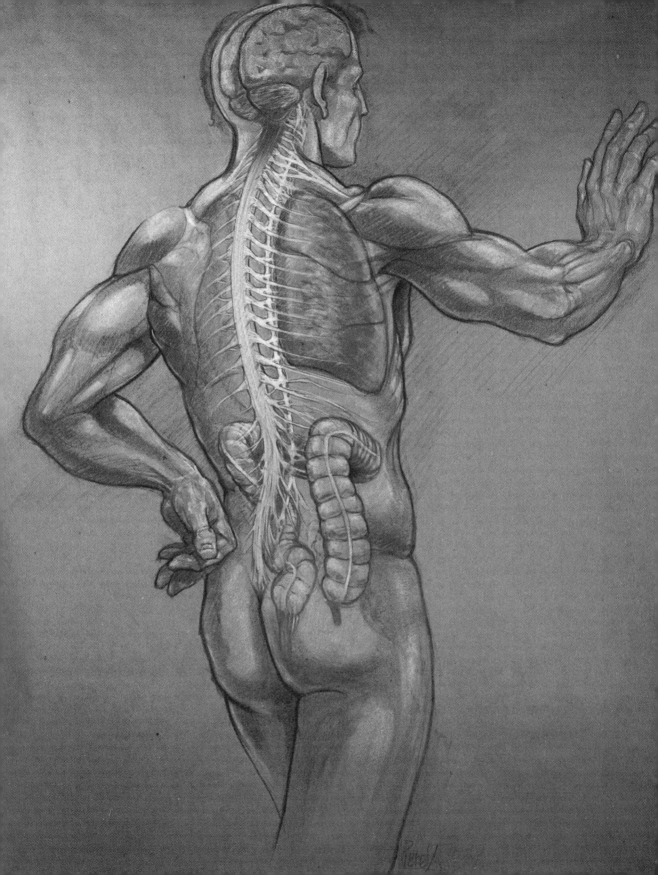

9

The Involuntary Nervous System

LEARNING OBJECTIVES

- Describe the anatomy of the autonomic nervous system

- Differentiate between a white and gray ramus and their contents

- Integrate the autonomic nervous system with the central nervous system

- Describe the sympathetic trunk and its preganglionic and postganglionic fibers

- Explain the chemical action of both the sympathetic and parasympathetic divisions

- Identify adrenergic and cholinergic fibers and their receptors

- Explain how a visceral reflex operates

- Define antagonism

- Detail the functions of the hypothalamus

- Describe the possible reason why we eat too much, or not enough

- Cite a few common disorders of the autonomic nervous system

IMPORTANT TERMS

adrenergic
autonomic
 nervous system
celiac

cholinergic
enteric
hypogastric
involuntary

neurotransmitters
obesity
plexus
vasoconstriction

vasodilation
vasomotor

The organism becomes aware of changes in its external environment through external receptors such as the eyes and ears, and it uses its skeletal muscles to adapt to these changes. In like manner, internal receptors signal information about the state of the internal environment to the vital centers of the brain.

Although the internal organs, such as the heart, lungs, and stomach, contain sensory nerve endings and nerve fibers for conducting sensory messages to the brain and cord, most of these impulses do not reach the individual's consciousness. Sensory neurons from the organs are grouped with those that come from the skin and voluntary muscles. On the other hand, the efferent motor neurons that supply the glands and the involuntary, or smooth, muscles are arranged very differently from those that supply the voluntary muscles. This variation in location and arrangement of the visceral efferent neurons has led to their being classified as part of a separate division—the **autonomic nervous system.**

STRUCTURE OF THE AUTONOMIC NERVOUS SYSTEM

The autonomic, or involuntary, nervous system (ANS) has many special parts, particularly *ganglia*, collections of nerve fibers, that serve as relay stations. Instead of a neuron alone transmitting an impulse from the spinal cord to a muscle, an autonomic pathway has a relay ganglion between the cord and the muscle. A visceral efferent pathway has two types of components: *preganglionic fibers* and *postganglionic fibers* (Figure 9-1). Many fibers pass to and from each ganglion, and each of the structures is able to discriminate in its relaying. In the case of voluntary muscle cells, each nerve fiber extends all the way from the spinal cord to the muscle, with no intervening relay station. The principal parts of the autonomic nervous system are the sympathetic and parasympathetic divisions and the autonomic plexuses.

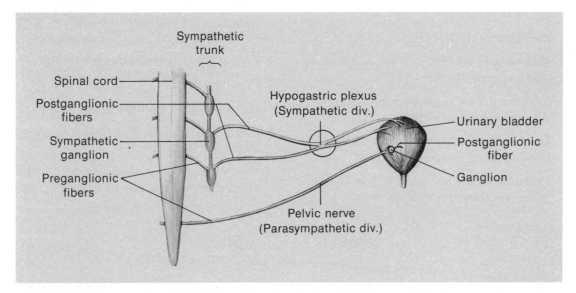

Figure 9-1 Diagram of the lumbosacral segment of the autonomic nervous system, showing the arrangement of ganglia, preganglionic fibers, and postganglionic fibers of both the sympathetic and parasympathetic divisions to the spinal cord and visceral organs.

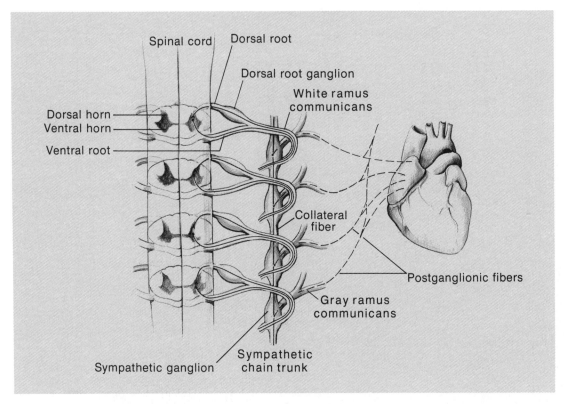

Figure 9-2 Pathways for distribution of sympathetic fibers from the spinal cord to the viscera.

The Sympathetic Division

The pathway of the sympathetic division of the autonomic nervous system begins in the thoracolumbar area of the spinal cord. Preganglionic fibers emerge from the cord and extend to spinal nerves by way of ventral roots (Figure 9-2). They then leave the spinal nerves and branch off into white communicating branches (*rami communicantes*) that lead to the adjacent sympathetic (or lateral) ganglia. The beadlike lateral ganglia form two chains, the *sympathetic trunks,* one on each side of the spinal cord. The lateral ganglia contain the cell bodies of the second set of neurons (postganglionic), whose fibers emerge as gray rami communicantes and rejoin

the spinal nerves for distribution to glands, blood vessels of the limbs, and involuntary muscle tissues (Figure 9-1).

The arrangement of the lateral ganglia in the sympathetic trunk is related to the segmentation of the spinal column. However, the ganglia of the cervical region (called the *superior, middle cervical,* and *stellate* ganglia) are fused to make three larger ganglia, the *ciliary, sphenopalatine,* and *otic* (Figure 9-3). Each lateral ganglion is joined to those above and below it, and there are many more postganglionic fibers than preganglionic. Therefore a visceral efferent impulse arising from the brain has a more diffuse effect than a comparable somatic motor impulse.

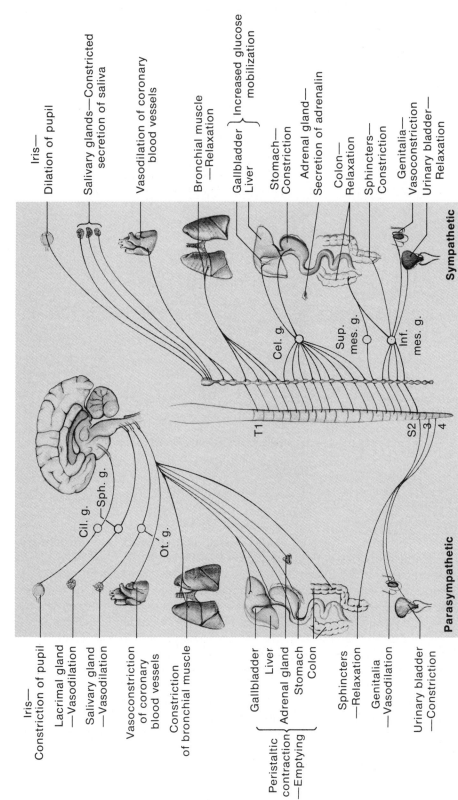

Iris—
Dilation of pupil

Salivary glands—Constricted
secretion of saliva

Vasodilation of coronary
blood vessels

Bronchial muscle
—Relaxation

Gallbladder ⎫ Increased glucose
Liver ⎭ mobilization

Stomach—
Constriction

Adrenal gland—
Secretion of adrenalin

Colon—
Relaxation

Sphincters—
Constriction

Genitalia—
Vasoconstriction

Urinary bladder—
Relaxation

Sympathetic

Cel. g.

Sup.
mes. g.

Inf.
mes. g.

T1

S2
3.
4

Parasympathetic

Cil. g.

Sph. g.

Ot. g.

Iris—
Constriction of pupil

Lacrimal gland
—Vasodilation

Salivary gland
—Vasodilation

Vasoconstriction
of coronary
blood vessels

Constriction
of bronchial muscle

Gallbladder ⎫
Liver ⎪ Peristaltic
Adrenal gland ⎬ contraction
Stomach ⎪ —Emptying
Colon ⎭

Sphincters
—Relaxation

Genitalia
—Vasodilation

Urinary bladder
—Constriction

Figure 9-3 Diagram of the autonomic nervous system, showing the chain of parasympathetic ganglia on the left and the sympathetic ganglia and their nerves on the right. Abbreviations: Cil. g., ciliary ganglion; Sph. g., sphenopalatine ganglion; Ot. g., otic ganglion; Cel. g., celiac ganglion; Sup. mes. g., Superior mesenteric ganglion; Inf. mes. g., inferior mesenteric ganglion. The outer columns indicate functions of the sympathetic and parasympathetic divisions.

244

time, sympathetic nerves to the digestive system are relatively inactive. Blood vessels supplying your digestive tract are wide open, but those supplying limb muscles and kidney tissue are constricted. Suppose now that you decide to go swimming before giving your system enough time to digest the food. The strong stimulus triggers chemical messages that race through your body, inhibit parasympathetic nerves, and excite sympathetic nerves. Digestion stops; your skeletal muscles quiver with the excitement of fresh forced blood, and you race into the water. As you exert your skeletal muscles while swimming, your smooth muscles lining the digestive tract remain inactive, allowing the partially digested food to remain stationary causing cramps. Only when you stop swimming and begin resting again do the antagonistic pathways shift priorities, so that digestion resumes and the cramp goes away.

The ANS is under the influence of the CNS at all times, especially certain higher centers of the cerebral cortex, thalamus, hypothalamus, medulla, and spinal cord. Afferent impulses keep these centers informed as to the physiological state of the body and its parts, and efferent impulses are sent from them to the autonomic system to make adjustments as required to maintain the homeostatic state.

SYMPATHETIC SYNDROMES

Certain syndromes are characteristic of diseases of the sympathetic nerve trunks. Dilation of the pupil of the eye on the same side as a penetrating wound of the neck is evidence of a disturbance in the cervical sympathetic cord; temporary paralysis of the bowel following fracture of a lumbar vertebra can indicate loss of peristalsis and distention of the intestine by gas; compression fractures of the upper thoracic vertebrae can produce marked variations in pulse rate and rhythm. Raynaud's disease, a condition in which attacks of pallor or cyanosis of the extremities, brought on by cold or emotion occur intermittently, is commonly treated by sympathectomy, the cutting of the sympathetic nerve trunks to the affected limb. With the sympathetic innervation interrupted, the blood vessels that were constricting too much dilate passively and provide a better blood supply, preventing disability and the potential development of gangrene.

OUTLINE

I. Structure of the autonomic nervous system
   A. Visceral efferent fibers
      1. Preganglionic
      2. Postganglionic
   B. Ganglia
      1. Sympathetic
      2. Parasympathetic
   C. Divisions
      1. Sympathetic
         a. Begins in thoracolumbar area of spinal cord
         b. Rami communicantes—communicating branches
         c. Sympathetic trunks—lateral ganglia
      2. Parasympathetic
         a. Begins in craniosacral region
         b. Ganglia usually located within organ innervated
   D. Autonomic plexuses
      1. Cardiac—heart
      2. Celiac—stomach
      3. Hypogastric—pelvis
      4. Enteric—digestive tract
         a. Contains both sympathetic and parasympathetic fibers
      5. Reflex activity
II. Central control of autonomic nervous system

   A. Medulla oblongata—heart, vasomotor tone, respiration
   B. Thalamus—sensory receiving center
   C. Hypothalamus
      1. Regulates viscera
      2. Coordinating center
         a. Body temperature
         b. Appetite
III. Chemical transmitters
   A. Adrenergic fibers
      1. Sympathetic fibers
      2. Norepinephrine
      3. General stimulation
   B. Cholinergic fibers
      1. Parasympathetic fibers
      2. Acetylcholine
      3. General inhibition
IV. Functions of ANS
   A. Sympathetic and parasympathetic nerves produce opposite effects—antagonists
      1. Sympathetic—response to stress
      2. Parasympathetic—conservative; resting conditions
   B. Maintains homeostasis
V. Sympathetic syndromes
   A. Pupil dilation
   B. Bowel paralysis
   C. Pulse and heart rate irregularities
   D. Raynaud's disease

STUDY QUESTIONS AND PROBLEMS

1. What is the main function of the autonomic nervous system?
2. Name the two divisions of the ANS, and show how they work during and following a moment of extreme fear.
3. Explain how stimulating the vagus nerve can cause the heart rate to slow or stop altogether.
4. Discuss the role of the ANS in the control of body temperature.
5. Define a plexus. What is a ganglion?
6. Distinguish between preganglionic and postganglionic fibers.
7. Discuss the antagonism of the ANS in reference to chemical mediation and anatomical variation.
8. How does the ANS play a part in the physiological phenomenon of homeostasis?

10

The Endocrine System

LEARNING OBJECTIVES

- Explain the different mechanisms of hormone action

- Explain the relationship between the central nervous system and endocrine glands

- Explain the action of a hormone on a cell

- Identify the prostaglandins and explain their role in cellular control

- Describe the feedback mechanism

- Identify each hormone of the hypophysis by its name, target gland, function, and clinical significance

- Explain the anatomy and histology of the thyroid and parathyroid glands

- Describe the action of thyroxine on the basal metabolic rate

- Relate the levels of calcium and phosphorus to tetany, stone formation, and bone abnormalities

- Relate the different glandular products of the adrenal gland to their sites of secretion

- List the effects of adrenal steroid production

- Relate adrenal medulla production with stress

- Explain the source of insulin and glucagon

- Explain the different stages of diabetes mellitus

- Differentiate between hyperglycemia and hypoglycemia in terms of insulin and glucose level in the blood

- Identify the sources of both male and female hormones

- Describe the functions of the estrogens, progesterone, and testosterone

- Explain the relationship of the pineal gland to ovarian hormone production

IMPORTANT TERMS

acromegaly
adenohypophysis
adenoma
adrenal glands
adrenocorticotropic
  hormone (ACTH)
aldosterone
allergy
androgens
antidiuretic
  hormone (ADH)
autoimmune
  disease
calcitonin
catecholamine
coma
cortisone
endocrine glands

epinephrine
estrogens
follicle-stimulating
  hormone (FSH)
glucagon
glucocorticoids
gluconeogenesis
glycogenesis
goiter
growth hormone
  (GH)
hormone
hydrocortisone
hyperactive
hyperemia
hyperglycemia
hypoactive
hypoglycemia

hypophysis
insulin
interstitial cell-
  stimulating
  hormone (ICSH)
lactogenic hormone
  (ITH)
luteinizing
  hormone (LH)
melanocyte-
  stimulating
  hormone (MSH)
mineralocorticoids
neurohypophysis
norepinephrine
oxytocin
pancreas
parathormone (PTH)

parathyroid glands
pineal gland
pituitary gland
polyuria
progestin
prolactin
prostaglandin
releasing factors
steroid
testosterone
thymus gland
thyroid gland
thyroid-stimulating
  hormone (TSH)
thyrotropin
vasopressin

The activities of the body are controlled and integrated by three means: the central and autonomic nervous systems, discussed in Chapters 8 and 9, and the endocrine system. The endocrine glands are considered a system only for convenience. Actually, the endocrine glands have various embryonic origins, widely different functions, and locations in different regions of the body (Figure 10-1). The endocrines differ from other (exocrine) glands in that they have no ducts. They secrete their secretions, known as **hormones,** directly into the bloodstream. For this reason, the endocrines are referred to as *ductless glands.*

Just as the nervous system has important functions beyond its role in homeostasis, some endocrine functions are associated with activities beyond those necessary for the immediate regulation of organs. For example, the nervous system controls the skeletal muscles, which allow an animal to respond behaviorally to its external environment. Similarly, some endocrine glands exert a slower, longer-lasting control: the hormones secreted by the reproductive glands, for instance, determine patterns of sexual and maternal behavior.

HORMONES

Chemical Nature of Hormones

Hormones are organic compounds of varying structural complexity that are carried by the blood to other parts of the body where they exert their specific effects. They are produced in one gland or part of a gland and are effective in another gland or in another region of that same gland. In simple terms, hormones are chemical messengers that pass via the bloodstream to the target organ or process. They may either stimulate or inhibit a function, but in general they do not initiate a process. Hormones are either proteins, glycoproteins (combination of a protein and a sugar), or steroids (substance made of carbon). The one thing that all hormones have in

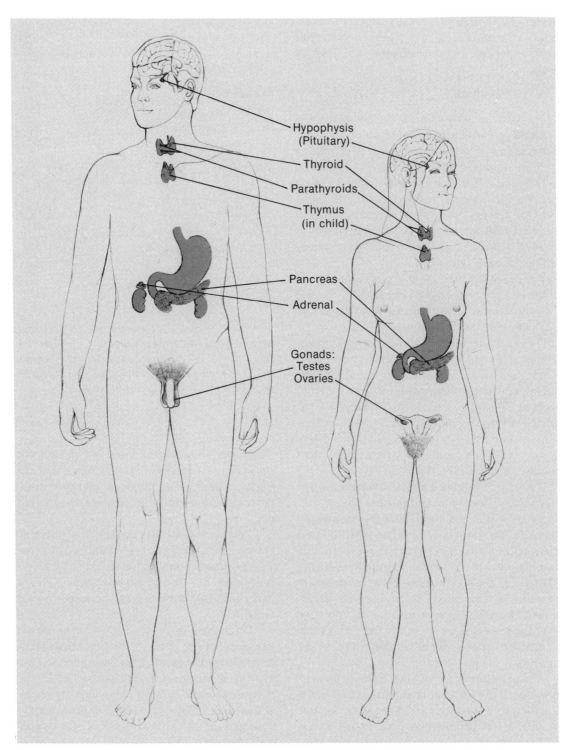

Figure 10-1 General location of the major endocrine glands of the body.

common, whether protein, glycoprotein, or steroid, is the function of maintaining homeostasis by modifying the physiological activities of cells.

Actions of Hormones

Hormones are effective in remarkably small quantities. For example, the injection of a few micrograms of epinephrine into a dog causes a definite increase in the rate of heart beat. There is little doubt that each hormone has some effect on the fundamental metabolism of its target cells or tissues. Hormones, therefore, have a marked influence on such basic life processes as growth, development, reproduction, energy utilization, and cell permeability.

Much is known about the way hormones work on their target tissues or cells. Protein and **catecholamine** (steroid stimulating) hormones act by first interacting with receptor sites on the cell membrane (Figure 10-2). The cell membrane contains the adenyl cyclase system. The hormone binds to specific receptors in the cell membrane and subsequently activates adenyl cyclase. This enzyme converts adenosine triphosphate (ATP) into 3,5-cyclic AMP, which acts as the secondary messenger (Figure 10-3). Cyclic AMP, the secondary messenger, then moves to other structures.

Although it is not known how the group of hormones called **prostaglandins** directly act upon cells, it is postulated that they somehow regulate the formation of cyclic AMP (Figure 10-4) and may be involved in the responses of target tissues to hormonal stimulation. Prostaglandin A (PGA) causes a decrease in blood pressure accompanied by an increase in regional blood flow to areas such as the coronary and renal systems. Its effect involves relaxation of smooth muscle fibers in the walls of the arteries.

Prostaglandin E_1 (PGE$_1$) relaxes smooth muscles in the airways to the lungs and has a long-lasting mucosal constriction effect. Because of this, it has proved beneficial to asthmatics and other problems involving nasal congestion.

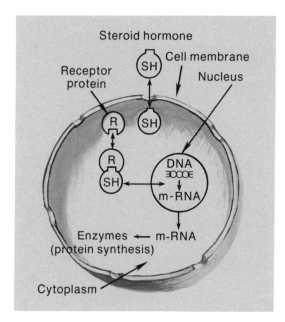

Figure 10-2 Hormone effect on mRNA. Hormones bind to intracellular receptor proteins, which carry them to the cell nucleus. Here, steroid hormones modify the formation of mRNA and protein synthesis, ultimately affecting gene transcription.

PGE$_1$ also inhibits breakdown of fats (lipolysis) normally produced by catecholamines and glucagon as well as produces an increased urine flow and excretion of sodium ions in the kidney.

Prostaglandin E_2 (PGE$_2$) inhibits gastric secretion, increases vasopermeability, increases venous muscle tone, and is found in high concentrations during antigen reactions. PGE$_1$ and PGE$_2$ are responsible for an increased level of cyclic AMP.

Prostaglandin $F_{2\alpha}$ (PGF$_{2\alpha}$) is found in lung parenchyma and causes bronchoconstriction by constricting the arteries and veins of the airways to the lung. PGF$_{2\alpha}$ is also associated with allergic reactions, since it stimulates the release of histamine. PGF$_{2\alpha}$ stimulates uterine contractions and can be used to induce labor and accelerate delivery. PGF$_{2\alpha}$ has also been found to

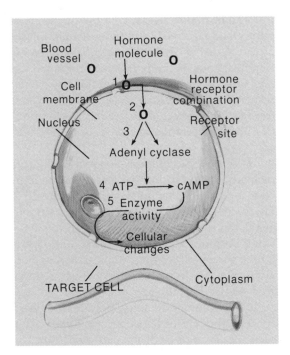

Figure 10-3 Transformation of ATP into cyclic AMP. Hormone molecules (1) reach target cells by means of body fluids and (2) combine with receptor sites on the cell membrane. As a result, molecules of adenyl cyclase (3) diffuse into the cytoplasm and (4) cause the change of ATP into cyclic AMP, which in turn brings about various cellular changes.

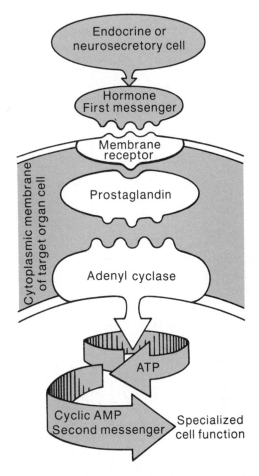

Figure 10-4 Mechanism of hormone action. Hormones act as messengers delivering their message to a membrane receptor in the target cell. Prostoglandins appear to regulate hormonal activity by influencing adenyl cyclase and cyclic AMP activity within the cell. Cyclic AMP serves as the second messenger for specialized cell functions.

induce menstruation and, therefore, is a possible contraceptive device. PGF compounds also affect intestinal motility and are required for normal peristalsis.

Feedback Control Systems

Hormones are secreted continually, and their rate of secretion is regulated by the demands of the body needs. The nervous system controls the endocrine system either directly or indirectly. The direct influence is minimal and best illustrated by the effects of the sympathetic nervous system on the secretion of the adrenal medulla. Indirect control is most common and is centered around the role of the hypothalamus. The hypothalamus secretes certain hormones

and transfers these hormones to the posterior pituitary where they are stored. Also, the hypothalamus secretes chemical substances, known as **releasing factors,** that are released into the vascular bed between the hypothalamus and the anterior pituitary. These releasing factors are specific and regulate the release of the anterior pituitary hormones (Figure 10-5).

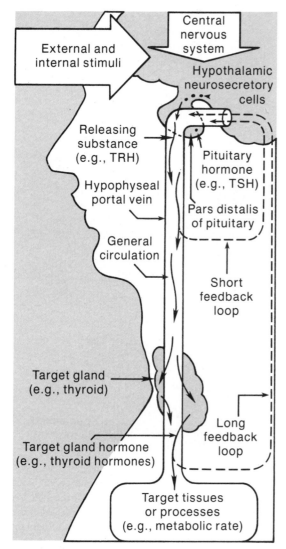

Figure 10-5 Feedback control of endocrine gland activity. External or internal stimuli influence the release of releasing factors from the hypothalamus, which in turn trigger the release of hormones from glands that control important vital functions of the body.

The hormone released by the target endocrine gland may feed back and influence release of the releasing or inhibiting substance of the hypothalamus or the pituitary hormone. If the effect of the feedback is to inhibit the overall response of the system, it is termed *negative feedback.* If the effect of the feedback is to stimulate further hormone release, it is *positive feedback.*

In many cases the secretion of a hormone is regulated by both positive and negative feedback mechanisms as illustrated in Figure 10-5. Thyroxine, which is released by the thyroid gland, can regulate the release of thyroid-stimulating hormone in two ways. First, high levels inhibit the release of thyrotropin-releasing hormone by the hypothalamus, which reduces TSH release by the anterior pituitary. Second, thyroxine inhibits the release of TSH by direct effect on the cells of the anterior pituitary. This particular type of feedback pathway (from target gland to pituitary to brain) is called a *long feedback loop.* Additionally, the pituitary hormones may feed back to the brain to influence the releasing or inhibiting substances; this pathway is called a *short feedback loop.*

ENDOCRINE GLANDS

Either overactivity or underactivity of a gland is the malfunction that causes endocrine diseases. If a gland secretes an excessive amount of its hormone, it is **hyperactive** or overactive. This condition is sometimes caused by an enlarged gland (one that has hypertrophied) or by a glandular tumor.

A gland that fails to secrete its hormone or secretes an inadequate amount of it is **hypoactive,** or under active. The gland may be diseased or tumorous, or it may have been adversely affected by surgery or radiation. A gland that has atrophied has decreased in size and consequently is secreting inadequately (Table 10-1).

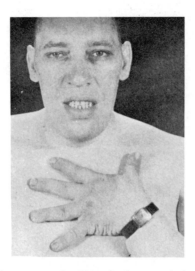 

Figure 10-8 Acromegaly. Note the large and elongated head and the large hand, nose, ears, and lips. Projection of the jaw and increased interdental spaces are also noticeable. (From R.H. Williams, *Textbook of Endocrinology*, 6th Ed., Philadelphia: W.B. Saunders, 1981.)

endocrine organ vital in early pregnancy. It does not produce these effects if injected into a nonpregnant woman because it requires the levels of estrogen and progesterone secretion characteristic of pregnancy.

MELANOCYTE-STIMULATING HORMONE (MSH), OR INTERMEDIN **MSH** stimulates the production of melanin in the skin, providing its pigmentation. It is produced by the anterior pituitary. In the absence of the hormone, the skin may be pallid. An excess of MSH may cause darkening of the skin. High estrogen and progesterone levels in pregnancy often cause darkening of the face, as well as the nipples and genitalia. As mentioned earlier, ACTH also produces increased pigmentation of the skin. Structurally, several of the same amino acids occur in the same sequence in both MSH and ACTH.

Anterior Pituitary Releasing Factors

As mentioned early in this chapter, the hypothalamus produces releasing factors, or hormones, that are secreted into the vascular bed (pituitary portal system) between the hypothalamus and the anterior pituitary (Figure 10-9). The names and functions of these releasing factors are as follows:

- Growth-releasing hormone (GRH)—stimulates the adenohypophysis to secrete GH

- Growth-inhibiting hormone or somatostatin (GH-GIH)—inhibit GH release

- Thyrotropin-releasing hormone (TRH)—stimulates the adenohypophysis to secrete TSH

- Corticotropin-releasing factor (CRF)—stimulates the adenohypophysis to secrete ACTH

- Follicle-stimulating hormone releasing hormone (FSH-RH)—stimulates the adenohypophysis to secrete FSH

- Luteinizing hormone releasing hormone (LH-RH)—stimulates the adenohypophysis to secrete LH

- Prolactin releasing factor (PRF) and prolactin inhibiting factor (PIF)—act in a dual control system to regulate prolactin blood levels

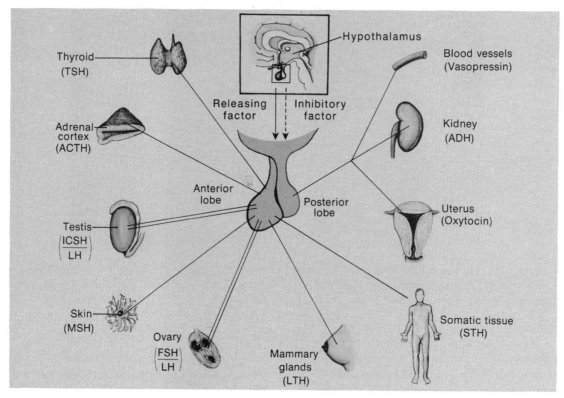

Figure 10-9 Control of the endocrine system by the nervous system. Note the hypothalamus controls the pituitary through releasing and inhibiting factors. The anterior lobe of the pituitary gland then releases tropic hormones that act on target glands (thyroid, adrenals, and gonads).

Posterior Pituitary Gland

The neurohypophysis releases at least two known and distinct hormones, **oxytocin** and **vasopressin** (also known as **antidiuretic hormone** or **ADH**), which exert certain significant physiological actions. Oxytocin and vasopressin are not actually produced in the neurohypophysis. Instead, they are manufactured in the neurons of the hypothalamic region (supraoptic and paraventricular nuclei) of the brain and migrate through the axons to the posterior pituitary, where they are stored (see Figure 10-6).

OXYTOCIN This hormone promotes a generalized contraction of the smooth muscles of the body. While almost all such muscles are affected, the greatest response is from the uterine muscle, especially in late pregnancy. The hormone is frequently injected after delivery to promote the contraction of the empty uterus, thereby slowing bleeding from any minor lacerations that may have occurred during childbirth.

In the lactating woman, the infant's sucking at the nipple sends impulses to the hypothalamus, leading to oxytocin release. Because of the increase of the level of the hormone in the blood, the smooth muscle of the breast contracts and ejects the milk.

VASOPRESSIN (PITRESSIN) OR ANTIDIURETIC HORMONE (ADH) Changes in the osmotic concentration of the blood and body fluids stimulate particularly sensitive neurons in the hypothalamus, resulting in the release of

ADH from the neurohypophysis. It travels by way of the blood to the distal and convoluted tubules of the kidney and promotes resorption of water into the bloodstream by making the tubules more permeable to water. Lower concentrations of ADH in the blood cause a less efficient resorption of water, resulting in increased urine output or diabetes insipidus.

In the absence of ADH, water is not absorbed and is lost in the urine. **Polyuria,** or excessive urination, results. No sugar is present in the urine, and the water loss leads to dehydration. A physical sign indicating this problem is an insatiable thirst (polydipsia).

A second, less important, function of ADH is its role as a **vasoconstrictor.** By narrowing a blood vessel it can produce a mild increase in blood pressure, and for this reason is called vasopressin. In humans, the vasoconstrictor effect occurs mainly in the coronary vessels, where it tends to lower the cardiac output; therefore, no dramatic increase of systemic blood pressure is experienced.

Posterior Pituitary Control

Both ADH and oxytocin are synthesized in the neurosecretory cells of the hypothalamus and move down the axons to the posterior pituitary from which they are released into the blood circulation. It is thought that the specialized neurosecretory cells also conduct nerve impulses, which may somehow be involved in controlling the release of the neurohypophyseal hormones.

Effects of dehydration on ADH levels has proved to be a factor of control. An increase in the osmotic pressure of extracellular fluid stimulates ADH secretion. A decreased urine output tends to increase the volume of extracellular fluid, which in turn decreases the osmotic pressure back to normal. The opposite is true if the osmotic pressure of extracellular fluid decreases.

As mentioned before, stimulation of the areolar region of the nipples by an infant's nursing affects oxytocin production. Suckling initiates sensory impulses that travel to the paraventricular nucleus of the hypothalamus, stimulating it to synthesize oxytocin. The posterior lobe of the pituitary in turn increases the release of oxytocin, which causes the ejection of milk into the ducts of the breast.

Pituitary Disorders

Hypopituitarism can be partial (mild) or total (severe); total hypopituitarism is termed *panhypopituitarism.* A basal skull fracture, tumor, ischemia (lack of blood), or congenital defect can cause pituitary necrosis. Marked destruction results in a progressive loss of hormonal function. The thyroid gland will cease to function without TSH, resulting in metabolic deficiencies and mental lethargy. Lack of ACTH causes the adrenal cortex to atrophy. Inadequate cortical hormones alter the salt balance, creating improper metabolism of nutrients and sexual malfunctions. Gonadal hypofunction results in amenorrhea (absence of menstruation) or aspermia (no formation of sperm) and loss of pubic and axillary hair. All or any of these conditions must be treated by giving those with hormonal deficiencies the hormones they lack throughout their lives. Administration of thyroxine, cortisone, growth hormone, and sex hormones can compensate for the dysfunction of the pituitary gland.

Hyperpituitarism is most often associated with an **adenoma,** a functioning benign tumor. The most dramatic example is an acidophilic adenoma, which produces excessive skeletal as well as tissue growth. If the adenoma develops during the period of skeletal growth, gigantism results. If the adenoma develops after growth, acromegaly results. Patients with pituitary tumors have symptoms indicative either of hormonal dysfunction or a space-occupying lesion. Endocrine symptoms are those of insufficiency, since the tumor compresses and destroys the surrounding normal tissue. This compression phenomenon can often be seen in hyperfunctioning tumors. Symptoms associated with a space-occupying lesion may be those of a cerebral tumor. If the tumor remains small, there will be no symptoms, but symptoms caused by pressure on neighboring structures may develop and indicate the correct diagnosis. Tumors pressing on surrounding bone, causing enlargement of the sella turcica, can be easily diagnosed by x-ray examination.

THYROID GLAND

The thyroid gland is located in the anterior aspect of the neck, just below the larynx (voice box). It consists of right and left lateral lobes connected by an isthmus that lies upon the upper part of the trachea (Figure 10-10). Internally, the gland contains numerous secretory cells (follicular and parafollicular-C) responsible for the production of thyroid hormones (Table 10-1). The thyroid gland's cells secrete two types of hormones: (1) thyroxin (T_4) and triiodothyronine (T_3), which are modified amino acids and are involved with iodine metabolism and regulation of the basal metabolic rate (BMR), and (2) calcitonin (thyrocalcitonin), a polypeptide involved in calcium metabolism.

Most of the body's iodine is obtained from the diet and appears in the blood as inorganic iodide. The vast majority of iodide is taken up by the follicles of the thyroid gland and is oxidized to iodine and combined with an amino acid tyrosine to form **thyroxin,** or tetroiodothyronine (T_4). A variation of this multiple combination is **triiodothyronine** (T_3). The tyrosines are not available in a free state but are part of a glycoprotein molecule named *thyroglobulin* that is stored in the follicles. When the thyroid is actively secreting, thyroglobulin molecules are broken down, releasing T_3 and T_4 into the bloodstream as the thyroid hormones. They are then bound to plasma proteins such as thyroxine-binding globulin (TBG). In the tissues, T_3 and T_4 are freed from the binding proteins and leave the circulation. Normally, the level of T_4 is considerably higher (approximately fifty times) than T_3.

The thyroid hormones regulate metabolism of the tissues by mobilizing proteins in adults and increasing protein synthesis, which favors growth in children. They also lower serum cholesterol levels and influence the action of epinephrine with regard to the breakdown of glycogen (**glycogenolysis**) and the subsequent elevation of blood glucose (**hyperglycemia**). Thyroid hormones are also instrumental in lipid metabolism including synthesis, mobilization, and desaturation of fats. In their absence, tissue oxygen consumption falls to about half its

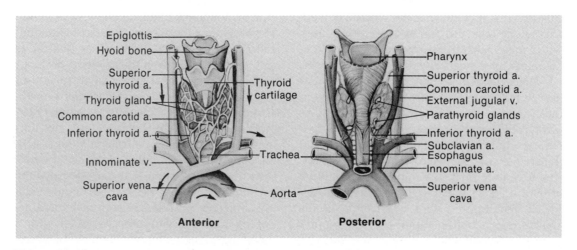

Figure 10-10 Thyroid and parathyroid glands and their relationships to the blood supply of the trachea   (a = artery; v = vein).

Parathyroid Control

One of the controlling factors of PTH is TCT. Secreted by the thyroid gland, it has an opposite effect to PTH and acts to maintain a moderate calcium level of the blood.

PTH is not controlled by the pituitary gland but by a negative feedback control system involving calcium levels (Figure 10-16). When the calcium ion level of the blood falls, more PTH is released. Conversely, when the calcium ion level of the blood rises, less PTH (and more TCT) is secreted.

Parathyroid Disorders

Hypercalcemia due to hyperparathyroidism leads to greater urinary secretion of calcium but also causes decreased PTH secretion. This de-

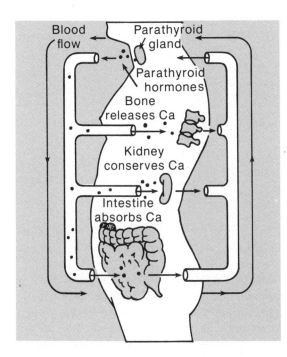

Figure 10-16 Effect of the parathyroid hormone on calcium. Parathyroid hormone stimulates the release of calcium from bone, the conservation of calcium by the kidney, and the absorption of calcium by the intestines. The resulting increase in blood calcium level inhibits the secretion of the hormone.

crease diminishes the excretion of phosphorus (hypophosphatemia) and raises the solubility product, causing a net deposition of bone. Kidney stones, in particular, frequently occur.

Treatment of hyperparathyroidism involves excision of the parathyroid tumor or partial removal of all four parathyroid glands. Treatment of hypoparathyroidism involves the administration of calcium salts, PTH, and vitamin D to encourage intestinal absorption of calcium.

ADRENAL GLANDS

The two adrenal (or suprarenal) glands lie, one on each side, just above the kidneys and usually are covered by perirenal fat. Each appears to be a compact unit but actually comprises two endocrine glands (Figure 10-17). The outer part, the *cortex*, secretes several **steroid** hormones synthesized from cholesterol; the core of the gland, the *medulla*, has an entirely separate function, the secretion of epinephrine and norepinephrine (Table 10-1).

Adrenal Cortex

The adrenal cortex secretes so many hormones that the classical experiment of removing the glands from animals led to a bewildering series of results. In general, however, the animals showed poor performance and a low tolerance to infection, temperature changes, or any form of stress. They also suffered profound water and electrolyte disturbances. These animals usually died after a few weeks, but their life could be extended, for a short time, by giving them unlimited access to salt and water.

Histologically, the cortex is subdivided into three zones (Figure 10-17). The outermost zone, or *zona glomerulosa*, is responsible for the secretion of **mineralocorticoids**, principally aldosterone. The middle zone, or *zona fasciculata*, is the widest of the three zones and secretes the **glu-**

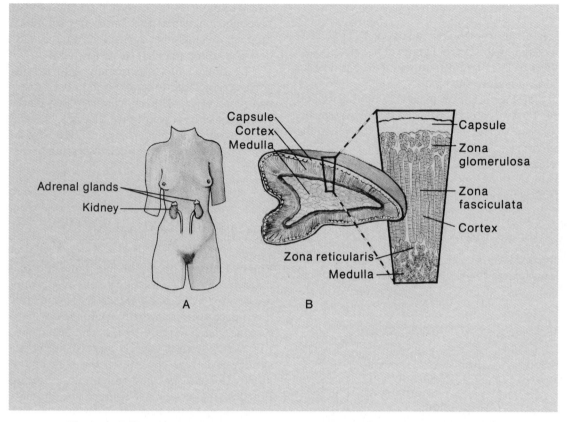

Figure 10-17 Adrenal glands. **A,** Location of the adrenal glands. **B,** Cross section, showing the cortex and medulla.

cocorticoids. The inner zone, or *zona reticularis,* synthesizes mostly sex hormones.

MINERALOCORTICOIDS These steroids regulate electrolyte metabolism. **Aldosterone** is the principal mineralocorticoid and has a regulatory effect on the relative concentrations of the mineral ions (especially sodium and potassium) in the body fluids and therefore on the water content of the tissues. An insufficiency of the mineralocorticoid results in an increased excretion of sodium and chloride ions and water into the urine. This is accompanied by a fall in sodium, chloride, and bicarbonate concentra-

tions in the blood, resulting in a lowered pH (acidosis).

GLUCOCORTICOIDS These steroids have a principal role in carbohydrate metabolism. They promote the deposition of liver glycogen from proteins (**glyconeogenesis**) and inhibit the utilization of glucose by the cells, thus increasing the blood sugar level.

Cortisone and **hydrocortisone** (cortisol) are the primary glucocorticoids, and a deficiency in either results in depression of the gluconeogenic potential. It may be noted that they diminish allergic response, especially the more serious inflammatory types, such as rheumatoid arthritis and rheumatic fever.

Glucocorticoids also aid the body in stressful situations. A sudden increase in available glucose by way of gluconeogenesis from amino acids makes the body more alert. It gives the body energy for combating extreme temperatures, fright, bleeding, infection, and other stresses.

Glucocorticoids decrease vasodilation and tissue edema associated with inflammation and are therefore called anti-inflammatory substances. In this regard, they inhibit **hyperemia** (swelling due to increased blood supply), cellular migration, and cellular permeability. Adversely, they can inhibit the immune response in that they impair humoral antibody production and inhibit proliferation of germinal centers of spleen and lymphoid tissue in the primary response to antigen.

SEX HORMONES The **androgens** are the male hormones and **estrogens** are female hormones. Their effects are similar to those of the hormones produced by the testes and ovaries. However, the adrenal cortex secretes both male and female hormones regardless of the sex of the individual. The androgens secreted by the cortex do not have strong masculinizing properties, except for testosterone, but this is in trace quantities. Small amounts of androgens secreted by the female cortex have been known to support sexual behavior, and in extreme quantities, it may cause growth of a beard. The adrenal cortical hormones as well as the gonadal hormones are steroids that are synthesized from cholesterol.

Adrenal Cortex Control

Release of the mineralocorticoids (aldosterone) from the adrenal cortex is influenced by a kidney secretion, *renin*. Renin acts on a precursor substance called *angiotensinogen* that is produced by the liver and found in its inactive state in the blood. Renin converts angiotensinogen to angiotensin I, which is in turn converted in the

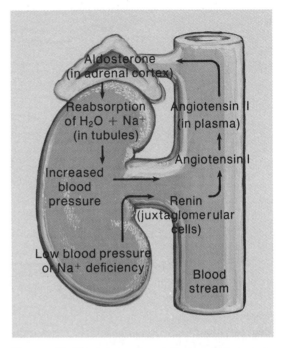

Figure 10-18 Effect of falling blood volume on the kidneys. When the blood volume falls because of dehydration, lack of salt, or massive bleeding, a complex process takes place. The kidneys sense the loss and secrete into the blood a hormonelike substance called renin. Renin causes the formation of angiotensin, which in turn stimulates the adrenal cortex to secrete aldosterone. The circuit is complete when aldosterone acts on the kidneys to prevent them from excreting salt in the urine, thereby helping them to retain salt in the body. The blood volume increases and tends to return to normal because most of the blood is simply a weak solution of salt water.

lungs to angiotensin IIA (Figure 10-18). Angiotensin II stimulates aldosterone release from the adrenal cortex. When blood pressure in the afferent arterioles of the kidney decreases below a certain level, it stimulates the renin-angiotensin mechanism to secrete aldosterone and increase water retention in the kidney, thereby raising the blood pressure to normal. The blood potassium concentration also helps regulate aldosterone secretion. A high concentration of potassium in the blood stimulates aldosterone secretion; a low blood potassium inhibits it.

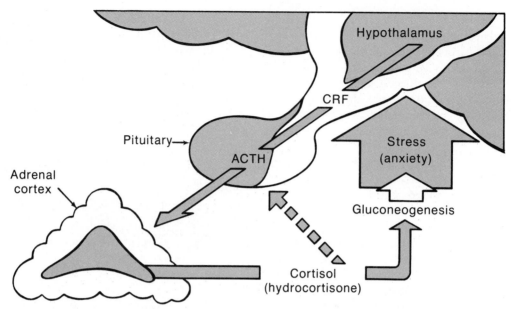

Figure 10-19 The effect of hydrocortisone on the body. Hydrocortisone from the adrenal cortex influences many body functions, including sugar metabolism, heart contractions, and blood pressure. It also acts to control its own secretion through negative feedback. The pituitary hormone ACTH stimulates secretion of cortisol (hydrocortisone), then negative feedback partially shuts off ACTH to keep the system in balance. The brain hormone CRF regulates both hormone levels. In times of stress or anxiety, increased secretion of CRF steps up the amounts of hydrocortisone, with effects on sugar, heart, blood pressure, and other functions to help handle the stress.

Glucocorticoid release is under the control of ACTH from the anterior pituitary, which in turn is influenced by CRF from the hypothalamus. A number of stressful situations may cause a release of ACTH and result in an increased glucocorticoid secretion (Figure 10-19). Once released, cortisol is bound to a plasma glycoprotein (transcortin) for transport in the blood.

Adrenal Cortex Disorders

Hypofunction of the adrenal cortex in humans can be either acute or chronic. Acute adrenocorticosteroid insufficiency can be caused by hemorrhagic destruction of the glands infected by N. meningitides. This destruction is characterized by a sudden decrease in blood pressure (hypotension) and extensive hemorrhage into the skin. Acute adrenocorticoid deficiency can also be caused by a rapid withdrawal of corticosteroid (cortisone) medication from patients who have been treated with the corticosteroid for a period of time.

Chronic adrenal insufficiency, also known as *Addison's disease*, is a pathological process involving all zones of the cortex and their respective hormonal secretions. It is characterized by hyperpigmented (bronzing) skin (see Figure 10-20A), muscular weakness, loss of appetite, weight loss, and hypotension. Addison's disease can be caused by tuberculosis, metastasis to the adrenal gland from a neoplasm located elsewhere, and as a manifestation of an autoimmune process. Adrenal antibodies have been found in high filters. These antibodies react with antigens in the adrenocortical tissue and cause an inflammatory reaction that leads to tissue destruction. Endocrine diseases like histoplasmosis (infectious), Hashimoto's thyroiditis,

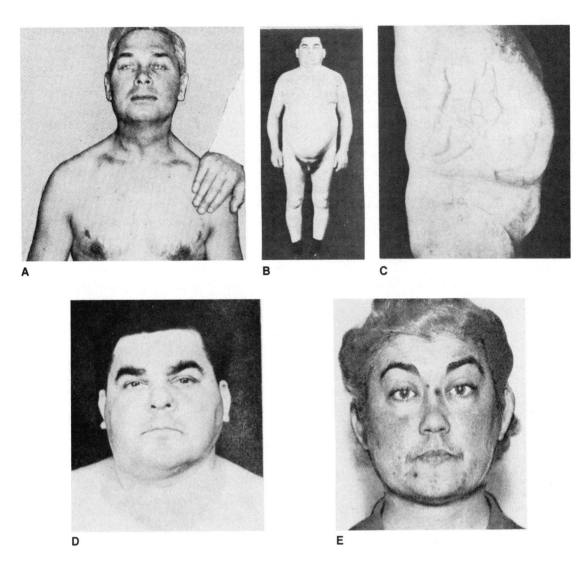

A   B   C

D   E

Figure 10-20 A, Addison's disease secondary to tuberculosis of the adrenals. Note diffuse brown pigmentation of variable intensity. The scars above and lateral to each breast are pigmented. **B,** Cushing's syndrome showing truncal obesity with prominent abdomen and relatively thin extremities. **C,** Cushing's syndrome showing purplish red striae. **D,** Cushing's syndrome showing round fiery-red face, short thick neck—typical moon face. **E,** Red moon faces with hirsutism. (From R. H. Williams, *Textbook of Endocrinology*, 6th Ed., Philadelphia: W. B. Saunders, 1981.)

diabetes mellitus, and hypoparathyroidism are known to occur concurrently with Addison's disease. Hormonal (cortisol) administration may alleviate the previously described symptoms. Diagnosis is based partly on the clinical symptoms and detection of low serum cortisol levels.

Hyperfunction of the adrenal cortex may result from an excessive production of cortisol or from long-term administration of glucocorticoids. Hypercortisolism may produce *Cushing's syndrome*, which is characterized by accumulation of adipose tissue ("moon face"), weight gain, osteoporosis, weakness, hypertension, diabetes, and hairiness (see Figure 10-20*B*). In 10 percent of cases this syndrome occurs from a pituitary adenoma of basophil cells that secrete ACTH or from an adrenal cortex adenoma. Surgical removal of the primary adrenal tumor usually restores normal adrenal function. Physically induced (iatrogenic) Cushing's syndrome is the most common cause of the disease. It is seen in patients with conditions such as asthma, rheumatoid arthritis, lymphoma, and generalized skin disorders who receive glucocorticoids or anti-inflammatory agents. Gradual cessation of the drugs (cortisol) usually returns the patient to normal.

Hyperaldosteronism, or *Conn's syndrome*, is characterized by potassium depletion and expansion of the extracellular fluid area of the tissues, resulting in edema and hypertension.

Oversecretion of the androgenic and estrogenic hormones may produce an adrenogenital syndrome, resulting in either a virilizing or feminizing effect. Excess androgens in a newborn female infant promotes masculinization: the child's clitoris becomes enlarged and pseudohermaphroditism develops. Excessive androgens in the male child hastens his male organ development. Treatment of the adrenogenital syndrome involves administration of cortisol to depress ACTH production. Masculinization in women is usually caused by an adrenal tumor that can be removed surgically.

The Adrenal Medulla

The adrenal medulla makes up the pulpy center of the gland (see Figure 10-17), in which groups of irregular cells are located amidst veins that collect blood from the sinusoids. This gland produces two very similar hormones, **epinephrine** (80 percent) and **norepinephrine** (20 percent), also known as adrenaline and noradrenaline, respectively (Table 10-1). Their molecular structures are relatively simple: both are amines, and

they are sometimes referred to as *catecholamines*. Both hormones elevate the blood pressure: epinephrine accelerates the heart rate, and norepinephrine acts as a vasoconstrictor. They are directly innervated by preganglionic cells of the sympathetic and parasympathetic divisions of the autonomic nervous system. In all visceral effectors, preganglionic fibers first synapse with postganglionic neurons before innervating the effector. In the adrenal medulla, however, the preganglionic fibers pass directly into the cells of the gland; therefore the secretion of hormones is directly controlled by the autonomic nervous system. The two hormones together promote glycogenolysis (glucogenesis), the breakdown of liver glycogen to glucose, causing an increase in blood sugar concentration. Of the two, epinephrine functions most closely as the emergency hormone for the body. Its functions of increasing the blood glucose level (by glycogenolysis) and the blood flow to the skeletal muscles (vasodilation of arterioles), heart, and viscera serve to shunt or direct the circulation of blood where necessary during exertion or increased stress.

The adrenal medulla, unlike the thyroid and other endocrine glands, is not essential for life. It can be removed surgically without causing untreatable damage.

Adrenal Hormones and Stress

When an individual is under severe emotional stress, such as rage or fear, a marked excess of epinephrine is poured into the blood, thereby enabling the person to perform feats of muscular exertion of which he would not ordinarily be capable (Figure 10-21). It is not certain just how this is accomplished, but since a major response to overwhelming stress is vascular collapse, some investigators have suggested that the glucocorticoids may enhance the vasoconstriction of norepinephrine. At the same time, the glucocorticoids may prevent an excessive vasoconstriction that can lead to tissue ischemia (blood deficiency) by acting directly on vascular

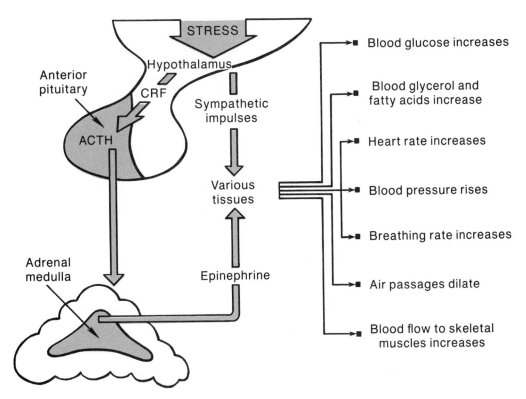

Figure 10-21 Hypothalamus. At times of stress, the hypothalamus helps prepare the body for "flight or fight" by triggering sympathetic impulses to various organs.

smooth muscle and stimulating heart muscle. This has been called the emergency functions of the adrenals, or the *fight or flight response.*

PANCREAS

The pancreas lies close to the posterior abdominal wall behind the stomach (Figure 10-22*A*). It functions as both an exocrine and endocrine gland. Its exocrine functions involve the secretion of digestive enzymes into the small intestine; this will be discussed in the chapter on metabolism and digestion. The endocrine function of the pancreas is carried on by groups of special tissue scattered throughout the pancreatic tissue called the *islets of Langerhans* (Figure 10-22*B*). The number of cells in these islets vary from several hundred thousand to a few mil-

lion. Each islet consists of several types of cells but those designated as *alpha* and *beta* predominate. The hormone **insulin** is secreted by the beta cells and another hormone, **glucagon,** by the alpha cells. Both hormones are polypeptide in nature and can be synthesized commercially (Table 10-1).

The earliest studies on the effect of insulin on carbohydrate metabolism showed four basic effects of the hormone:

1. *Increases rate of glucose metabolism.* Glucose cannot pass through the cell pores but, instead, must enter by some transport mechanism through the membrane matrix. Glucose combines with a carrier substance and is then transported to the inside of the cell. This process can occur in either direction. In the absence of insulin, the rate of glucose

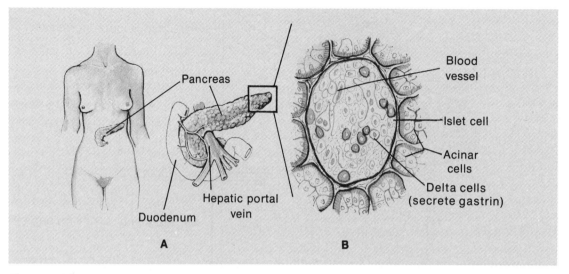

Figure 10-22 Pancreas. **A,** Location of the pancreas. Its close proximity to the duodenum facilitates transfer of digestive enzymes from the pancreas, where they are produced, to the duodenum, where they are utilized. **B,** Microscopic section, showing the arrangement of the islet cells of Langerhans.

transport into the cell is about one-fourth the normal value. Conversely, an excess of insulin multiplies the normal rate; therefore insulin increases the rate of glucose transport through the membranes into most cells in the body.

2. *Regulates blood glucose concentration.* In the absence of insulin, glucose absorbed from the gastrointestinal tract cannot be transported into the tissue cells; consequently, the normal blood glucose level of 90 mg/100 mL may rise to as high as 1000 mg/100 mL. In the presence of insulin, glucose is transported into the tissue cells so rapidly that the glucose level may drop as low as 20 mg/100 mL. Therefore the rate of insulin secretion by the pancreas must be regulated to facilitate a constant and normal glucose value in the blood.

3. *Increases glycogen storage in the tissues.* Glucose is preferentially metabolized by the cells to supply all the energy needed by the body. If glucose is inadequate, the major share of energy is then supplied by fats. Large deposits of fat are broken down (lipolysis), with resultant release of free fatty acids into the blood. Acetoacetic acid (a ketone body), a product of fatty acid metabolism, is released by the liver into the blood, causing the condition called *ketosis.*

4. *Promotes a positive nitrogen balance.* Insulin facilitates the movement of amino acids into cells and therefore aids the synthesis of proteins and promotes a positive nitrogen balance.

The activities of glucagon are generally the opposite to those of insulin. Glucagon decreases glucose oxidation and promotes excess quantities of glucose (hyperglycemia) in the blood. It does this by stimulating the breakdown of liver glycogen (glycogenolysis), which increases the blood glucose concentration from sources other than carbohydrates. Glucagon is also responsible for formation of ketone bodies (ketogenesis) through its action of lipid breakdown in liver and adipose tissue. In general, glucagon is the principal hormone preventing hypoglycemia.

Current research findings indicate the presence of another polypeptide hormone secreted by the pancreas. A definite distribution of cells associated with the glucagon-secreting beta cells are now known to secrete a hormone that is elevated in diabetes and is influenced by blood levels of protein and lipids. The hormone increases both gastric secretions (digestion) and the production of glucagon (endocrine). The hormone is called *pancreatic polypeptide* (PP), or *gastrin* produced by the delta cells of the islets of Langerhans.

Hyperglycemia

In insulin deficiency, or *hypoinsulinism*, the glucose uptake by the cells from the blood is reduced, although absorption from the intestines and breakdown of liver glycogen continue. The blood glucose level rises, resulting in **hyperglycemia,** and exceeds the renal threshold (approximately 150 mg/100 mL), causing glucose to be lost in the urine **(glycosuria).** This disease, *diabetes mellitus,* is caused by hyposecretion of insulin by the pancreas. The result of the cell's inability to utilize glucose is the accumulation of ketone bodies (fatty acid and amino acid breakdown), leading to acidosis. Acidosis, along with an electrolyte derangement, may result in paralysis, coma, and death.

The major pathological change in diabetes is an early onset of arteriosclerosis, especially in the eyes, kidneys, and heart. Gangrene, blindness, heart disease, and uremia are generally found to be the end results of diabetes mellitus.

Diabetes may occur before the age of 20 years (juvenile diabetes) or after the age of 40 years (maturity onset diabetes). It is generally known to follow this sequence:

1. *Prediabetic stage.* No biochemical abnormality of glucose metabolism is detectable. Prediabetes is suspected in children born of two diabetic parents, in women who give birth to babies weighing over 9 pounds, and in patients with renal glycosuria (glucose in the urine at normal blood glucose levels) in the nonpregnant state.

2. *Chemical (latent) stage.* A high glucose level in the blood is detected in the blood glucose tolerance test (GTT). This stage is manifested only during times of stress, pregnancy, obesity, or infection. Latent diabetics do not usually complain of symptoms of diabetes.

3. *Overt stage.* Patients develop symptoms of abnormal glucose tolerance. They are unable to maintain normal fasting blood glucose levels, and hyperglycemia is accompanied by glucosuria. At this stage, patients develop an increased urine output (polyuria), increased thirst (polydipsia), weight loss, weakness, fatigue, and sleepiness.

Patients with juvenile-onset diabetes generally have a rapid onset of polydipsia, polyuria, weight loss, polyphagia (excessive hunger), and sleepiness. They may become ill, develop ketoacidosis (low pH due to excessive keto acids in the blood), and die if treatment is not carried out within a short time. Maturity-onset diabetes patients may never know they are diabetics because the clinical symptoms are not demonstrated. An elevated fasting blood glucose level followed by an abnormal GTT is indicative of the disease.

The cause of diabetes is still unknown. It is postulated that one or more of the following factors may lead to diabetes mellitus:

- Gene interaction
- Environmental agents altering beta cell function
- Viruses
- Autoimmunity
- Inheritance
- Inflammatory disease
- Metabolic disorders
- Lesions of the pancreas leading to insulin insufficiency
- Abnormal glucagon secretion

Complications of Diabetes Mellitus

Complications of diabetes mellitus can be grouped into two categories: acute metabolic complications and long-term vascular complications.

The most serious acute metabolic complication is diabetic ketoacidosis. With severe insulin insufficiency, patients develop severe hyperglycemia and glycosuria, along with ketoacidosis resulting from lipolysis. The production of keto acids (acetone) is so great in the blood that it can be smelled (sweet) on the breath of the patient. Continued production of ketones in plasma causes ketosis and eventually a decreased hydrogen ion concentration and metabolic acidosis. Marked glycosuria leads to polyuria, loss of electrolytes, and loss of consciousness called *diabetic coma*. During this time, the blood glucose level increases (over 350 mg/ mL), and unless treated, the patient may die, which occurs in 10 percent of cases. The administration of insulin and intravenous sodium bicarbonate generally alleviates diabetic ketoacidosis.

The long-term vascular complications involve thickening of the capillary basement membranes of small vessels of the retina and large vessels (atherosclerosis/arteriosclerosis). Since all the chemical components of the basement membrane can be derived from glucose, hyperglycemia promotes an increased formation of vessel cells, creating a thickening of the walls and decreased diameter of the vessel. Another frequent complication involving hyperglycemia is a disturbance in the production of fructose from glucose (glucosesorbitolfructose) in the lens of the eye. An increase in glucose concentration also increases the content of sorbitol in the lens, leading to formation of cataracts and blindness. Increased sorbitol and fructose in nerve tissue results in neuropathies (axonal loss, paresthesia, decreased proprioception, loss of tendon reflexes, muscle weakness, and atrophy).

Diabetes mellitus may also promote spontaneous abortions in pregnancies. The rate of intrauterine fetal death and premature infants with respiratory distress has markedly decreased in diabetic mothers who have demonstrated more rigorous diabetic control during pregnancy.

Hypoglycemia

Increased insulin production (hyperinsulinism) occurs occasionally, generally as the result of a pancreatic tumor. As previously noted, the central nervous system acquires essentially all its energy from glucose. If the blood glucose level drops below normal (hypoglycemia), CNS metabolism becomes depressed, and the person falls into an insulin shock syndrome.

Insulin shock, also called *hypoglycemic shock*, results from too much insulin, not enough food, or excessive exercise. The patient feels lightheaded and faint, trembles, and begins to perspire. Taking sugar in some form, as in candy or orange juice, may be adequate treatment at this stage. If the glucose level is not raised, the condition becomes more serious. The patient's speech becomes thick and walk, unsteady. The low level of glucose (30 to 50 mg/ mL) affects the brain causing double vision possibly followed by a loss of consciousness. Intravenous injections of glucose and epinephrine generally alleviate the problem.

In addition to hypoglycemic shock in the diabetic resulting from too much insulin, insufficient food, and strenuous exercise, other factors can cause hypoglycemia. A tumor of the beta cells of the pancreas results in hypersecretion of insulin, which lowers blood glucose. A patient with Addison's disease (hypoactivity of the adrenal cortex) secretes an inadequate amount of glucocorticoids to raise the level of blood glucose.

Treatment of hypoglycemia depends on its cause. For the diabetic, precautionary measures can usually prevent its development: exact insulin dosage, careful observance of diet and times for meals, and exercise within the prescribed range. A tumor causing excessive insulin secretion should be removed; the patient with Addison's disease requires hormonal supplements.

Control of Pancreatic Hormones

The regulation of insulin secretion, like that of glucagon secretion, is directly determined by the level of glucose in the blood and is based on a negative feedback system (Figure 10-23). Other hormones, however, have a marked influence on insulin production. GH raises the blood glucose level, which triggers insulin secretion. ACTH, by stimulating the secretion of glucocorticoids, causes hyperglycemia and resulting stimulation of insulin release. Somatostatin (GIF) inhibits the secretion of insulin.

Effective diagnosis of blood glucose levels is carried out by the GTT. This 3-to-4-hour test can assess the body's handling of an increased quantity of glucose (Figure 10-24). The patient drinks a prescribed quantity of glucose. Blood samples are taken before the test (fasting) and every hour for 3 or 4 hours after the drink, and the glucose contents of these samples are determined. If the blood glucose does not return to

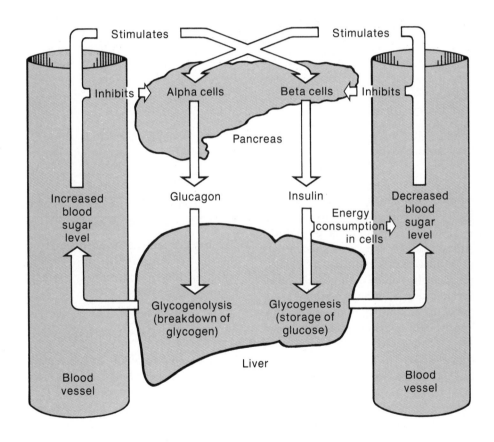

Figure 10-23 Regulation of the secretion of glucagon and insulin. When the glucose level rises, insulin is released; when the glucose level falls, insulin is no longer secreted. Glucagon and insulin act together to maintain a constant blood sugar level.

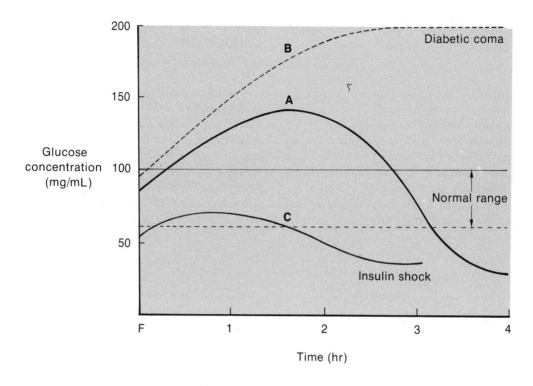

Figure 10-24 Glucose tolerance curves. **A,** Normal person. **B,** Potential diabetic. **C,** Person with hypoglycemia.

OVARIES AND TESTES

normal within 2 hours, insulin deficiency is indicated. The glucose concentration in the blood sometimes attains a very high or a very low level. At either of these extremes, a coma may result, and it must be determined whether the patient is in diabetic coma (high-glucose level) or hypoglycemic shock (low-glucose level).

Because of the accumulation of keto acids in the blood of patients in diabetic acidosis, an approximation of the concentration of acetone and/or acetoacetic acid in the blood sometimes is of clinical interest. A serum acetone test can be performed by a long complicated procedure or by using commercially available Acetest Tabs or Ketostix strips, which indicate the quantity of serum acetone in the blood and urine. Both employ the nitroprusside reaction that indicates the quantity of acetone present by a color change (lavender).

The ovaries and testes are primary sources of the sex hormones that control the maturation and function of the reproductive system (see Figure 10-1 and Table 10-1).

Female Sex Hormones

The principal function of the female sex hormones is the maintenance of the reproductive tract for the development of the fertilized egg into a new individual. The ovary secretes two specific types of hormones, estrogens and progestins, both of which are steroids.

Estrogen is a collective name for three different but closely related steroids. The most common of these is *estradiol;* the others are *es-*

STUDY QUESTIONS AND PROBLEMS

1. Differentiate between the terms exocrine and endocrine.
2. What is a hormone?
3. Explain the anatomic relationship between the hypothalamus and the neurohypophysis.
4. Giving examples, explain the feedback mechanism for control of the secretion of a hormone.
5. Which gland is called the "master gland"? Why?
6. What is the relationship between suckling and oxytocin?
7. Why is thyroxine sometimes prescribed for obese patients?
8. Describe the problem leading to cretinism. Differentiate between cretinism and dwarfism.
9. What is tetany? Is it the same as muscle tetany?
10. What is the relationship between the parathyroid gland and kidney stones?
11. How do glucocorticoids elevate the blood sugar?
12. What is the physiological response to stress?
13. What hormone enhances and prolongs sympathetic effects?
14. Compare diabetes mellitus with diabetes insipidus in terms of pathology and character of urine output.
15. Explain the relationship between hyperthyroidism and the BMR and PBI.
16. Explain the "self-regulation" of the adrenal cortex by the interaction of ACTH and the glucocorticoids.

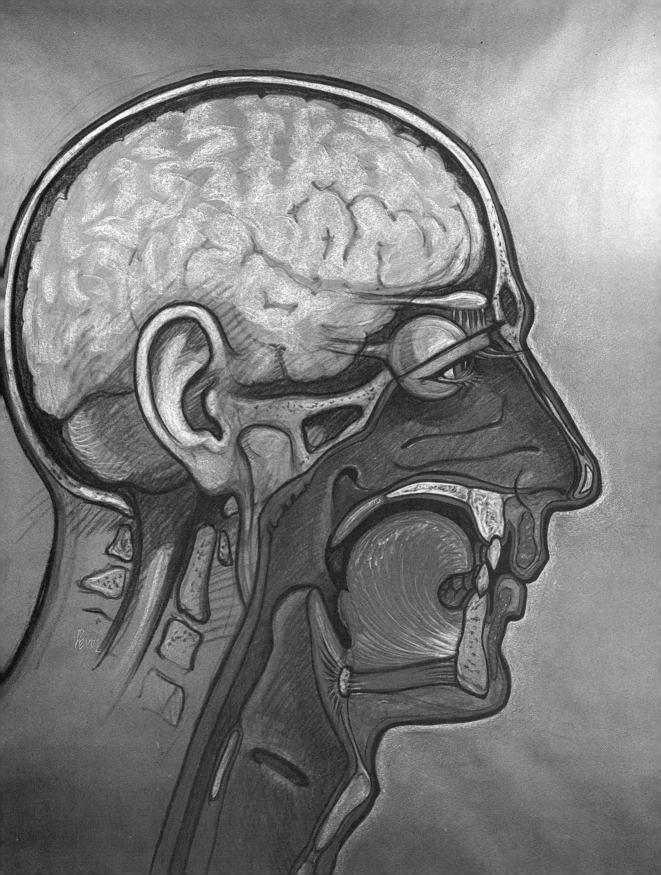

11

The Special Senses

LEARNING OBJECTIVES

- Describe the sensory mechanism

- Differentiate among interoceptors, proprioceptors, and exteroceptors, and give examples of each

- Define chemoreceptor, pressoreceptor, and photoreceptor, and relate their significance

- Explain the law of projection and apply it to phantom limb pain

- Describe where the cutaneous senses are located in the skin

- Explain why thickness of skin is relative to a sharp or dull sensitization

- Describe referred pain

- Define enkephalin and endorphin, and describe their action

- Explain why a blind man has such an acute awareness of his body position

- Define the phenomena of hunger and thirst, and relate them to the law of projection

- Explain the mechanism of taste discrimination, and describe its neural pathway

- Identify the primary taste sensations

- Give an explanation for olfactory and gustatory correlation

- Describe the anatomy of the eyeball and its protective structures

- Explain the physical phenomenon of refraction and how it operates in focusing

- Explain how rods and cones operate in dim and bright light

- Describe the physiology of color vision and color blindness

- Explain depth perception and relate binocular vision with diplopia and hemiopia

- Describe the anatomy of the ear

- Follow the transmission of sound from tympanic membrane to the baselor membrane

- Explain the Place theory of hearing

- Differentiate between static and dynamic equilibrium
- Define cataract, glaucoma, myopia, and hyperopia with reference to anatomical structure and function
- Explain nystagmus and vertigo
- Identify three kinds of deafness

IMPORTANT TERMS

accommodation	concave	glaucoma	otoliths	sense organ
adaptation	cones	gustation	photoreceptor	strabismus
ampulla	convex	hyperopia	placebo	taste buds
anorexia	deafness	interoceptors	presbyopia	teleceptors
astigmatism	diplopia	kinesthetic	pressoreceptor	tinnitus
audiometer	endorphin	monochromatic	proprioceptors	trichromatic
binocular	enkephalin	myopia	refraction	vertigo
cataract	equilibrium	nystagmus	rhodopsin	
chemoreceptor	exteroceptors	ophthalmoscope	rods	

SENSORY MECHANISMS

A **sense organ,** or receptor, is a specialized nervous tissue situated at the peripheral endings of the dendrites of afferent neurons. The receptor's primary function is to provide the body with information, both conscious and unconscious, about degrees of change in the organism's external and internal environments.

Three important features of sense organs should be emphasized:

1. Specific receptors are particularly sensitive to specific stimuli. However, they can also respond to other stimuli of sufficient intensity. For example, pressure on the eyeball causes a sensation of light.

2. Specific sensitivity to certain stimuli is due to the structure and composition of the receptor. For example, light-absorbing pigments are found in the photoreceptors of the eye.

3. The type of sensation elicited by a receptor depends on which nerve pathways are activated, not on how they are activated. All nerve impulses are essentially alike, regard-

less of the stimulus that initiates them. The impulse is transmitted to the CNS where it is interpreted. For example, the region of the cerebrum receiving impulses from an olfactory receptor decodes and interprets them as a specific odor or aroma. If an olfactory receptor is artificially stimulated, the same nerve pathways to the brain will be activated, and the brain will interpret the arriving impulses as a specific odor.

Classification of Sense Organs

Numerous attempts have been made to classify the senses into groups, but none of the suggested systems has been entirely successful. Traditionally, the special senses are smell, vision, hearing, equilibrium, and taste; the cutaneous senses are those associated with receptors in the skin; and the visceral senses are concerned with perception of the body's internal environment.

Different sensations are evoked by stimulation of various specialized sensory receptors.

Some of these may be grouped according to the nature of the stimulus:

- *Chemoreceptors*—receptors sensitive to chemical stimuli (necessary for taste and smell)

- *Pressoreceptors*—receptors sensitive to mechanical stimuli (necessary for touch, pain, temperature, pressure, sound, and motion)

- *Photoreceptors*—receptors sensitive to visible light (necessary for sight)

Another classification of the various receptors categorizes them according to their somatic or visceral locations.

- *Teleceptors*—nerve endings that detect environmental changes occurring some distance from the body (for example, light rays, sound waves, and odors are detected by receptors of the eye, ear, and nose, respectively)

- *Exteroceptors*—nerve endings that detect environmental changes that directly affect the skin (temperature, touch, pressure, or pain)

- *Proprioceptors*—receptors that provide the body with information about its position in space (muscles, tendons, and joints)

- *Interoceptors*—receptors within organs equipped with motor innervation from the autonomic system and concerned with maintenance of the internal environment (smooth muscle contraction, peristalsis, visceral contractions)

Projection

No matter where a particular sensory pathway is stimulated along its course to the cerebral cortex (Figure 11-1), the conscious sensation produced by such stimulation is referred back to the location of the receptor. This principle is called the *law of projection*. For example, when the cortical receiving area for impulses from the left hand is stimulated, sensations are felt in the left hand, not in the head. Should you hit yourself on the left hand with a hammer, causing considerable pain, the cerebral cortex will interpret the sensation of pain but project the feeling back to the hand. It is not your brain that experiences the pain—it is your hand.

A particularly dramatic instance of projection occurs in amputees, who may complain of pain and proprioceptive sensations in the absent limb. For example, a patient has experienced a midthigh amputation of one leg and is now conscious after the operation. The nurse has placed bed covers over the patient to keep him warm and to restrict his movements. The pressure of the covers on the stump of his leg stimulates the nerve tract that once had come from the receptor in his foot; consequently, the pressure is interpreted as coming from the foot. The patient complains of discomfort caused by the "too tight" blanket on his foot, even though the foot is no longer there. This phenomenon is called *phantom limb pain* (see Figure 11-1). A reorientation of the cerebrum to new nerve stimuli is essential in all amputees following surgery. Patients must, in a sense, retrain their brain to realize that impulses coming from the amputated area are now originating along the tract of the remaining sensory mechanism.

CUTANEOUS SENSES

Touch, pressure, heat, cold, and pain are the cutaneous senses.

Touch (Tactile) and Pressure Sensations

At least four kinds of receptors—free nerve endings, Meissner's tactile corpuscles, Pacinian corpuscles, and nerve fibers at the base of each hair—give rise to touch sensations (Figure 11-2). The hair nerve ending is stimulated by a change of position of the hair; Meissner's and Pacinian corpuscles sense a quick touch or deformation of tissue, but not a sustained touch. Free nerve endings adapt slowly to continuous stimulation. For example, when a particle is removed

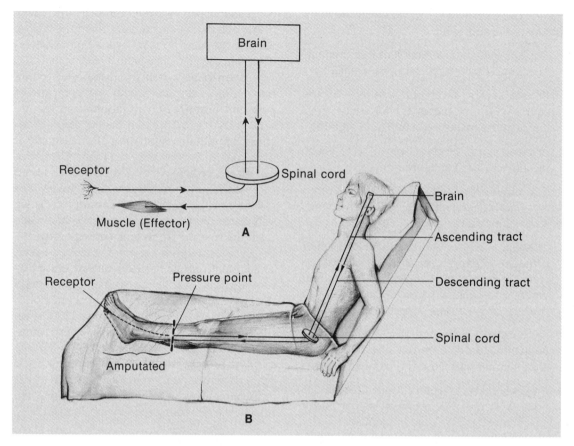

Figure 11-1 "Phantom limb" pain. **A,** The sensory mechanism, beginning with the receptor and ending at the muscle. **B,** Pressure exerted on the lower leg by the bed clothes, after surgical removal of the foot and the end of the lower leg, stimulates the nerve tract that once originated in the foot. The patient falsely interprets this impulse as coming from the foot that is no longer there.

from the eye, the free nerve endings in the cornea continue to "feel" it for a short time thereafter. Merkel's corpuscles or disks, located deep in the epidermal layer of the skin, are another example of free nerve endings concerned with touch sensation.

Pressure sensations differ from touch sensations in that they last longer. Light pressure can be discriminated from hard pressure by the depth of the receptor recognizing the stimulus. Free nerve endings in the superficial epidermis and Meissner's corpuscles are receptors of light pressure, whereas Pacinian corpuscles, found deep in the dermis, recognize deep or firm pressure.

The brain's sensory area for touch is situated in the general sensory area of the parietal lobe of the cerebral cortex, just posterior to the central sulcus (see Figure 8-14). The adaptation to touch is very rapid and exceedingly important in acquiring information from the external environment. New techniques in learning still use the age-old principle of "take it in your hands and feel it and you'll never forget it."

Temperature Sense

Humans perceive temperatures ranging from 12° to 50°C. Some regions of the body, such as the fingertips and face, are equipped with spe-

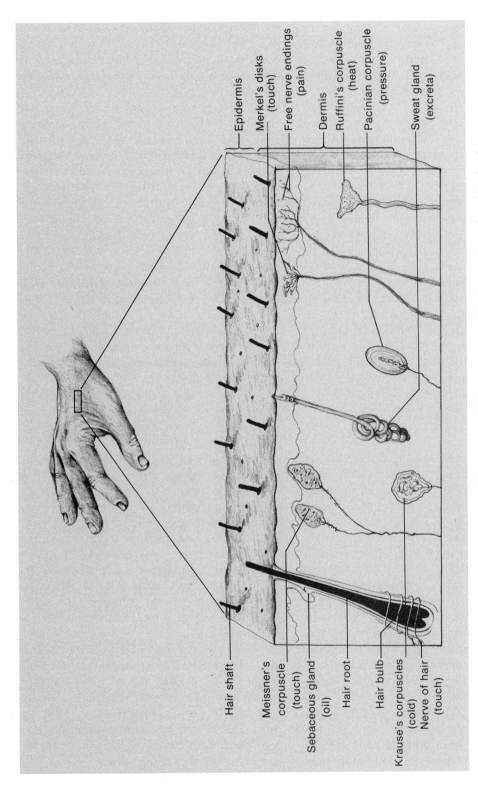

Figure 11-2 Cutaneous senses. Various types of exteroceptors and their sensations found in the skin.

cific structures for particular temperature ranges (Figure 11-2). The bulboid Krause's corpuscles, which lie in the dermis of the skin, are receptors for cold (12° to 35°C). The receptors for heat (25° to 50°C), Ruffini's end-organs or corpuscles, are also found in the dermis. In skin areas lacking specific receptors, temperature sensation is less acute and is subserved by free nerve endings. Hot and burning sensations result from the combined sensory inputs from "cold," "warm," and free nerve endings.

The sensory area for temperature, like that for touch, is in the general sensory area of the parietal lobe of the cerebral cortex, just posterior to the central fissure. Crude sensations of temperature are experienced in the thalamus. Afterimages can persist for a long time in the sensation of temperature. A cold object placed on the skin and quickly removed can still be "felt" for a few minutes. The skin of the face and hands is less sensitive to temperature changes than are the parts of the body usually covered by clothing, having adapted to frequent temperature changes over the years.

Somatic Pain

The receptors for pain are free nerve endings found in the epidermis (see Figure 12-3), muscles, joints, and tendons. Another stimulus for pain is excessive stimulation of any of the tactile senses. For example, an ice cube placed in the hand feels cold at first, but then begins to hurt. Overstimulation of the cold receptors resulted in pain.

Impulses for pain are transmitted to the thalamus by way of the lateral spinothalamic tract (Figure 11-3A). Awareness of pain takes place in the thalamus, but localization and recognition of the kind and intensity of pain occur in the parietal lobe of the cortex, just posterior to the central sulcus (Figure 11-3A; see also Figure 8-14).

Adaptation to pain does not exist. This is an extremely important defense—a warning signal. If we became accustomed to pain, damage to the body would follow our ignoring this sig-

nal. Everyone has about the same threshold of pain, but reactions to pain vary widely among individuals, depending on factors such as ethnic background, childhood experiences, and emotional status.

The term *referred pain* is used when pain that seems to be in one part of the body, particularly in an outer part such as the skin, actually originates in a different body organ or structure. For example, liver and gallbladder disease often cause referred pain in the skin over the right shoulder. Spasm of the coronary arteries that supply the heart may cause pain in the left shoulder and the left arm (Figure 11-3B). One reason for this phenomenon is that certain neurons have the twofold duty of conducting inpulses both from visceral (organ) pain receptors and pain receptors in neighboring areas of the skin. The brain cannot differentiate between these two possible sources; however, since most pain sensations originate in the skin, the brain automatically assigns the pain to this more likely site of origin.

In recent years, another group of chemical messengers in the brain has been identified. These are *peptides* and consist of chains of amino acids that occur naturally in the brain. In 1975 the first peptides, referred to as **enkephalins** (within the head), were discovered. These chemicals are similar in structure to morphine, the pain killer derived from the opium poppy. Enkephalins are concentrated in the thalamus, in parts of the limbic system, and in those spinal cord pathways that relay impulses for pain. It has been suggested that enkephalins are the body's natural pain killers. They do this by inhibiting impulses in the pain pathway and by binding to the same receptors in the brain as morphine.

Enkephalins concentrate in nerve endings located near nerves containing opiate receptors. When a pain impulse enters the spinal cord, special neurons release the enkephalins. Fitting into a nearby nerve's opiate receptors, they suppress the release of neurotransmitters that would ordinarily pass along the pain signal and thereby lessen the feeling of pain.

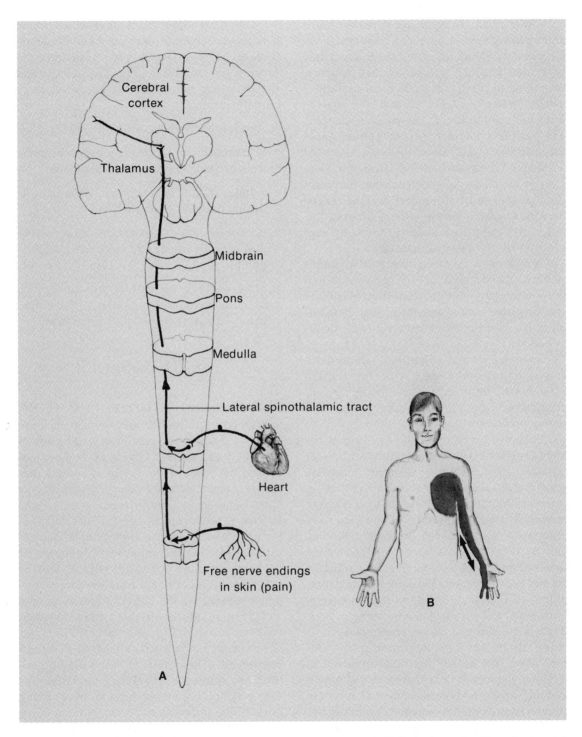

Figure 11-3 Referred pain. **A,** Nerve network that transmits pain sensation from the skin and visceral organs (heart). **B,** Pain generating from the heart is commonly interpreted as coming from the left arm.

Other naturally occurring peptides, called **endorphins,** were subsequently isolated from the pituitary gland. Like the enkephalins, they have morphinelike properties that suppress pain. One of the better studied endorphins is beta(β)-endorphin. A peptide that is believed to work with endorphins is known as substance P. The peptide is found in sensory nerves, spinal cord pathways, and parts of the brain associated with pain. When substance P is released by neurons, it conducts pain-related nerve impulses from peripheral pain receptors into the central nervous system. It is now suspected that endorphins may exert their analgesic effects by suppressing the release of substance P.

A brain peptide called *dynorphin* (dynamis = power), discovered in 1979, is two hundred times more powerful in action than morphine and fifty times more powerful than β-endorphin. Its exact functions have yet to be determined.

Enkephalins and endorphins are also involved in the **placebo** effect. About one-third of patients who think they are receiving a potent drug but actually receive an inactive sugar pill report a reduction of pain.

Enkephalin research has also helped explain the chemistry of narcotic addiction. Under normal circumstances, scientists believe, enkephalins occupy a certain number of opiate receptors. Morphine relieves pain by occupying receptors that are left unfilled. Too much morphine may cause a halt of enkephalin production, leaving receptors open. The body then craves more morphine to fill the unoccupied receptors and to cut down the pain. If denied morphine, all the opiate receptors remain empty, causing painful withdrawal symptoms.

Enkephalins may also regulate mood. These chemicals are heavily concentrated in the limbic system, the area of the brain involved in regulating emotion. Enkephalins in this region may act as the body's own "natural high" to counteract disappointment and prevent depression. The euphoria produced by morphine and other opiates lends support to this theory, but the actual process by which enkephalins influence mood is not yet known.

Pain sense differs from other senses in that continued stimulation does not result in adaptation, as has been mentioned. Sometimes the cause of pain cannot be remedied quickly, and the following measures may be necessary to help the patient:

- Application of cold to dull the senses

- Compression of the area to prevent the nerve from carrying the impulse to the brain

- Administration of analgesics, mild pain-relieving drugs (such as aspirin)

- Administration of narcotics, strong drugs that produce stupor and sleep (such as morphine)

- Inducing local or general anesthesia with chemical agents that produce lack of sensation in a specific area or total unconsciousness

POSITION SENSE

Receptors located in muscles, tendons, and joints relay impulses that help us judge the position and changes in location of body parts in relation to one another (Figure 11-4). They also inform the brain of the extent of muscle contraction and tendon tension. These widely spread end-organs—proprioceptors—are aided in their function by the semicircular canals of the inner ear. This sense of position, also called the **kinesthetic sense,** enables us to know the positions of various body parts without actually seeing them. Information from the proprioceptors is needed for the coordination of muscles and is important in everyday activities, such as walking and running, and in more complex skilled movements, such as playing a musical instrument. These muscle sense end-organs (Ruffini's end-organ and Golgi's tendon organ) also play an important part in maintaining muscle tone and posture.

The pathway for muscle sense impulses is provided by large myelinated fibers in the cranial and spinal nerves. Impulses dealing with conscious activities travel to the medulla by

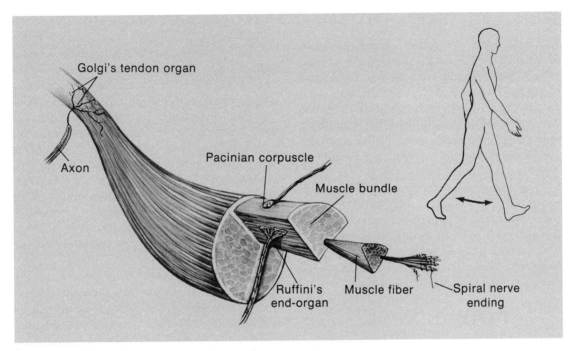

Figure 11-4 Proprioceptors in skeletal muscle.

way of posterior columns, where they are then relayed to the thalamus and cerebral cortex. Impulses concerned with reflex activities travel to the cerebellum by way of the spinocerebellar tracts. The sensory area for conscious muscle sense lies posterior to the central fissure in the sensory area of the parietal lobe of the cortex.

Syphilis and other diseases may involve the posterior spinal cord, causing its degeneration and a loss of position sense—a disorder called *tabes dorsalis*. The lower part of the body is affected first, resulting in a loss of both position sense and muscular coordination. As the disease progresses, the patient may lose the ability to walk. Once the nerve tissue has been damaged, no cure is possible; therefore, early treatment is important to prevent permanent tissue destruction.

HUNGER AND THIRST

Hunger and thirst are organic sensations that follow the law of projection. The feeling of hun-

ger is projected to the stomach; the feeling of thirst is projected primarily to the throat (pharynx).

Hunger

The feeling commonly called hunger occurs normally at a certain time before meals and presumably is due to contractions of the empty stomach that stimulate receptors distributed throughout the organ's mucous membrane. If food is not taken, hunger increases in intensity for a time and is likely to cause headaches, nausea, and overall weakness. The cliche, "never make a decision on an empty stomach," is based on the fact that a hungry individual can become quite excitable.

Appetite differs from hunger. Although it is basically a desire for food, it often has no relationship to a need for food. Appetite is also a pleasant sensation; conversely, hunger is disagreeable. A loss of appetite is known as **anorexia** and may be due to a great variety of physical and mental disorders.

Thirst

Although thirst may reflect a generalized lack of water in the tissues, the sense of thirst seems to be localized largely in the mouth, tongue, and pharynx. It is a very unpleasant sensation and is continuous until relieved or until death ensues. An excessive excretion of water, as in diabetes, may provoke excessive thirst, called *polydipsia.* The physical condition in which the tissues lack water is dehydration.

Visceral afferent impulses concerned with the organic sensations of hunger and thirst are conducted to the hypothalamus, where appetite is regulated. The location of the nerve receptors that transmit hunger and thirst impulses is still unknown. They are probably located in the muscles lining each respective organ.

TASTE AND SMELL

Taste and smell are considered to be chemical sensations because the taste buds and olfactory receptors respond to substances dissolved in fluids secreted by the oral and nasal epithelial tissues.

Taste

The sense of taste (**gustation**) involves receptors in the tongue and three different nerves that carry taste impulses to the brain. The taste receptors, called *taste buds,* are located along the edges of small depressed fissures separating the papillae of the tongue (Figure 11-5). These papillae vary in shape: pointed at the front of the tongue and fat at its base. The papillae give the tongue its characteristic rough appearance. All papillae contain taste buds.

There are essentially four kinds of taste (Figure 11-6):

- *Sweet*—acutely experienced at the tip of the tongue

- *Sour*—detected at the sides of the tongue

- *Salty*—felt at the tip of the tongue and along its sides

- *Bitter*—detected at the back of the tongue

To be tasted, substances must be in solution. At the top of each bud an opening called the taste pore admits solutions (materials dissolved in saliva), which stimulate the receptors.

The sensory area for taste is located in the lateral fissure of the cerebral cortex near the postcentral gyrus. The sensory nerve fibers from the anterior two-thirds of the tongue travel to the brain via the facial (seventh cranial) nerve and from the posterior third of the tongue by the glossopharyngeal (ninth cranial) nerve. The vagus (tenth cranial) nerve conducts taste sensations from the pharyngeal region and from the base of the tongue.

Although the taste buds are structurally alike, they are physiologically different in sensitivity. Bitterness can be detected in a dilution of 1:2,000,000; sweetness, 1:250; sourness, 1:135,000; and saltiness, 1:500. Sourness is a response to acids and is the most specific of the four taste sensations. Taste sensations other than the basic four are a result of a blending of the fundamental sensations with one another and with the sense of smell.

Smell

The olfactory receptors are located in the mucous membrane lining the upper part of the nasal cavity. Their afferent pathway is the olfactory (first cranial) nerve, which leads to the olfactory center in the temporal lobe. A layer of mucus secreted by the surrounding supporting cells covers the epithelium and dissolves and absorbs gaseous particles in inspired air (Figure 11-6). Only volatile substances in solution can stimulate the receptors of smell. The amount of air reaching the olfactory region is greatly increased by sniffing, a semireflex response that usually occurs when a new odor attracts attention.

Naked endings of the trigeminal (fifth cranial) nerve are found in the olfactory mucous membrane. These are stimulated by irritating substances (such as chlorine, peppermint, men-

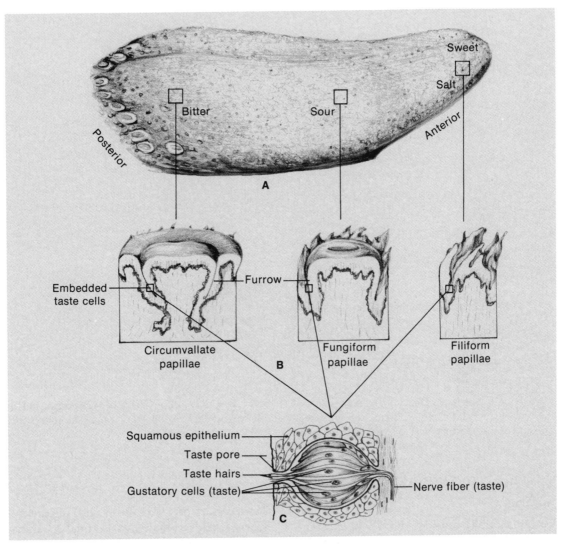

Figure 11-5 Taste. **A,** Regions of the tongue that are most sensitive to the various tastes. **B** and **C,** The tongue appears rough owing to the three types of papillae (**B**) that contain the taste (gustatory) cells or buds (**C**) embedded deep within each furrow.

thol, and others), producing sneezing, choking, and short breaths. This response can be advantageous at times, such as in using ammonia to arouse an individual from a fainting spell.

Humans' sense of taste depends a great deal on their sense of smell. If you block your nose or have a severe cold, your food is tasteless. This phenomenon is due to the gustatory nerve, which blends sensations of smell and taste. It also creates the strange feeling that you are able to taste something you actually are smelling. For example, if you enter a room that has been perfumed, you can taste sweetness from breathing in the odor. The smell of cooked cabbage can almost be tasted. A principal characteristic of smell is that only a minute quantity of the stimulating agent in the air is required to evoke the sensation.

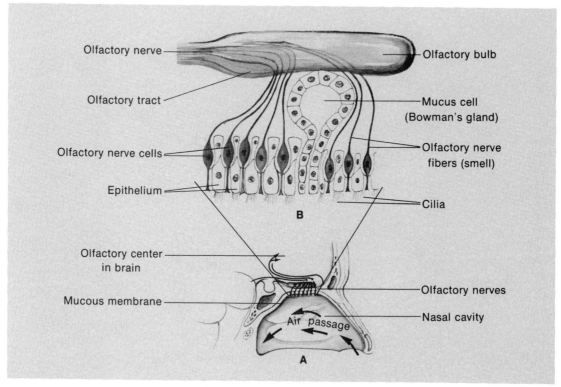

Figure 11-6 Smell. **A,** Olfactory receptive area in the roof of the nasal cavity, showing the gross relationship of the olfactory nerves to the mucous membrane. **B,** A microscopic sketch of a supporting mucus cell surrounded by olfactory cells that collectively form the olfactory nerve.

VISION

Vision is usually considered to be the most important of all the senses. Blind people can and do learn to depend on their other senses to a remarkable degree, but nothing approaches complete compensation for loss of vision. Fundamentally, the sense of vision is similar to all the other senses in that there are photoreceptors (sensitive to light), an afferent pathway (optic nerve) that conveys the impulses to the cerebral cortex, and an area (occipital lobe) of the cerebrum necessary for interpretation of these impulses. But vision is far more complex than such senses as hunger and thirst, and therefore it is necessary to consider the physics, anatomy, physiology, and biochemistry of sight.

Accessory Organs of the Eye

The accessory organs of the eye include the eyebrows, eyelids, conjunctiva, lacrimal apparatus, and muscles of the eyeball.

The *eyebrows* are thickened ridges of skin, covered with short hairs. They are positioned on the upper and lower borders of the orbit of the eye and protect the eye from strong light, foreign materials, perspiration, and so forth.

The *eyelids* are two movable folds of skin that are placed in front of the eyeball and cover it. They are lined internally with a mucous membrane, the *conjunctiva*, which begins at the lower edge of the upper lid, progresses down over the eyeball, and ends at the upper edge of the lower lid (see Figure 11-9). A small muscle is attached to the upper lid and elevates it. The orbicularis oculi muscle is arranged as a sphincter around both lids and opens and closes them (see Figure 7-16). Eyelids protect the eye and spread its lubricating secretions (tears) over the surface of the eyeball.

The conjunctiva, which lines the eyelids and is reflected onto the surface of the eyeball, is kept moist by tears flowing across the front of the eye. Inflammation of this membrane is called *conjunctivitis* and may be acute or chronic. A variety of irritants and pathogens can produce the condition. *Pink eye* is a highly contagious, acute conjunctivitis caused by a bacterium. *Trachoma*, or granular conjunctivitis, is caused by a virus and is contagious in its early stages. This disease is characterized by granulation of the eyelids, which may result in an irritation of the cornea, leading to blindness.

An eye infection of the newborn, *ophthalmia neonatorum*, is caused by venereal bacteria. Prevention of blindness in this instance, and in other possible congenital disorders of the eye due to bacterial invasion from the vaginal tract, is carried out by instilling a mild antiseptic (silver nitrate) into the infant's eyes at the time of delivery.

The eyelids bear rows of short, thick hairs, the eyelashes. The follicles of these hairs receive a lubricating fluid from sebaceous glands that open into them. The glandular secretions lubricate the edges of the eyelid and prevent adhesion of the two lids. If the glands become infected, a *sty* results. Distention of a sebaceous gland is called a *chalazion*.

The *lacrimal apparatus* consists of the lacrimal gland, lacrimal sac, and nasolacrimal duct (Figure 11-7). The lacrimal gland produces tears and is located above the eye toward one side (that is, superior and lateral to the eyeball). Tiny ducts carry the tears to the front surface of the eyeball, where they constantly wash the conjunctiva that separates the front part of the eyeball from the larger back portion (see Figure 11-9). Six to twelve minute ducts lead from the gland to the surface of the conjunctiva of the upper lid. Tears pass from the gland, through the ducts, over the surface of the eyeball, and

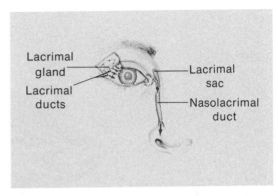

Figure 11-7 Lacrimal apparatus in relation to the eye and nose. Note that the direction of flow is from the outside in, to flush the eye continually and drain into the nose.

into two tiny lacrimal ducts at the inner angle of the eye. The lacrimal sac is the expanded upper end of the nasolacrimal duct, a small canal that opens into the nose.

Tears are secretions made up of salts, water, and mucin. These secretions normally are carried away by the nasolacrimal duct as quickly as they are formed, but under certain circumstances (conjunctivitis, intense emotion, hay fever), the secretion exceeds the drainage capacity, and the fluid overflows onto the cheeks. Because an infant's lacrimal glands do not develop sufficiently until about 4 months of age, the infant's eyes must be protected from bright light and dust. Occasionally, a newborn's lacrimal ducts become blocked and cause a great deal of "caking" of the eyelids in the early morning. The blockage may be alleviated by applying warm water to the eyelids; if the situation is serious enough, however, medical intervention may be necessary, in which case a blunt metal probe is used to unblock the ducts.

The eye is moved within the orbit by six ocular muscles innervated by the oculomotor (III), trochlear (IV), and abducent (VI) cranial nerves. These muscles and the directions in which they move the eyeball are presented in Table 11-1 and shown in Figure 11-8. The six muscles connected with each eye are ribbonlike

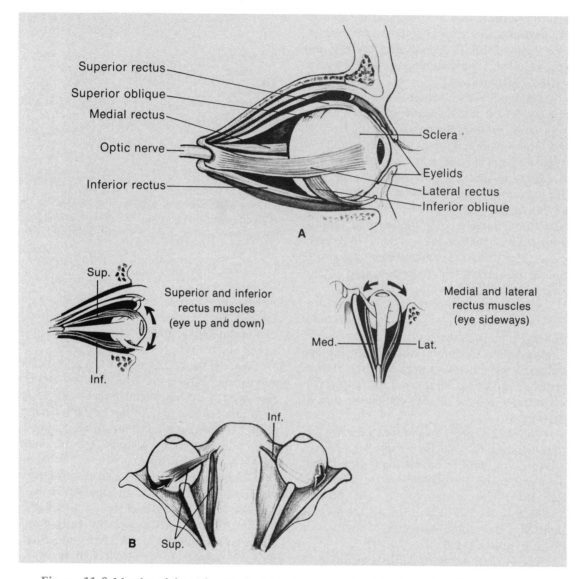

Figure 11-8 Muscles of the right eye. **A,** Extrinsic eye muscles, lateral aspect. **B,** Eye movements.

and extend forward from the apex of the orbit behind the eyeball. One end of each muscle is attached to a bone of the skull; the other end is attached to the white portion (sclera) of the eye. These muscles pull on the eyeball in a coordinated fashion that causes the two eyes to move together to center on one visual field.

Anatomy of the Eyeball

The eyeball has three coats, or layers, of tissue (Figure 11-9): sclera, choroid, and retina. The *sclera,* the opaque white outer layer of tough, fibrous tissue, gives shape to the eyeball and protects its delicate inner layers. The most ante-

Table 11-1   Muscles of the Eye

Name	Origin	Insertion	Function
Superior rectus	Apex of orbital cavity	Upper side of sclera	Rolls eyeball upward and medially
Inferior rectus	"	Under side of sclera	Rolls eyeball downward and medially
Lateral rectus	"	Lateral side of sclera	Rolls eyeball outward
Medial rectus	"	Medial side of sclera	Rolls eyeball inward
Superior oblique	Lesser wing of sphenoid above orbital cavity	Sclera	Rolls eyeball on its axis, downward and laterally
Inferior oblique	Orbital plate of maxilla	Sclera	Rolls eyeball on its axis, upward and laterally

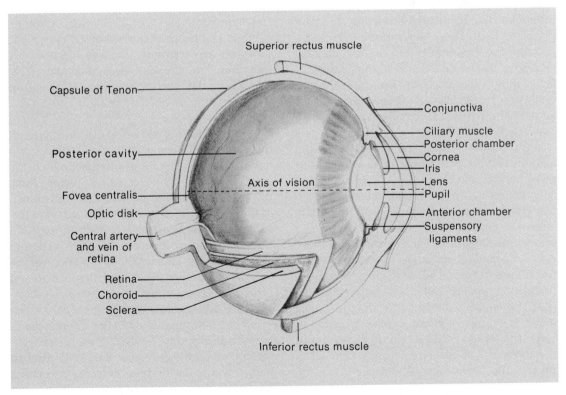

Figure 11-9 Structure of the eye, transverse section.

rior or forward continuation of the sclera is the transparent *cornea*. Both the sclera and cornea are fibrous connective tissue and lack blood vessels and nerves—a fact essential to know when transplants are undertaken.

The transparent and colorless cornea is referred to as the "window" of the eye. It bulges forward slightly and is the eye's most important refracting structure. Injuries caused by foreign objects or infection may result in scar formation and a slight opacity through which light rays cannot pass. Because the cornea is free of blood and nerves, it is one of the parts of the eyeball that can be transplanted (keratoplasty). Eye banks arrange for corneal transplants from donors to recipients who are blind because of a corneal defect. The entire transplant procedure should be performed within 72 hours after the death of the donor. The operation has been remarkably successful.

The middle layer of the eye comprises primarily the *choroid*, which is highly vascular and resembles skin. Its principal function is to supply blood to the other layers of the eyeball, especially the innermost retina. The forward extension of the choroid is a thin muscular layer, the *iris* (Figure 11-9). The iris is the colored part of the eyeball; an opening at its center allows light to pass. This circular opening is the *pupil.*

The function of the iris is to regulate the amount of light entering the eye and thus to assist in obtaining clear images. This regulation is accomplished by its smooth muscles, which contract or dilate, causing the pupil to become larger or smaller. The larger the pupil the greater the amount of light reaching the inner eye. This mechanism is an involuntary reflex that allows the eye to adjust to changes in light intensity, a property called **adaptation** (light reflex). When in a dimly lit room, the pupils dilate to permit as much light as possible to reach the retina. When a light is turned on in the room, the pupils automatically contract to protect the retina against damage from too much light.

Homatropine (a drug used in eye examina-

tions), epinephrine, atropine, and cocaine all make the pupil larger. Such drugs are called *mydriatics.* Pilocarpine and morphine cause the pupil to contract; such agents are *miotics.*

The *ciliary muscles,* which support and modify the shape of the lens, are another anterior modification of the choroid layer. The *lens,* a flexible, transparent, colorless body, consists of a more or less semisolid substance enveloped by a thin capsule. It is biconvex (thicker in the middle), and this property serves to converge light rays and focus them on the retina. During youth the lens is elastic and plays an important role in the eye's ability to focus on near objects. With aging, the lens loses it elasticity and therefore its ability to adjust by thickening, resulting in a condition known as **presbyopia.**

The third and innermost layer of the eyeball is the all-important *retina.* The word retina means a net, and that accurately describes the network of highly specialized nerve cells, called **rods** and **cones** and their processes. Each eye has approximately 110,000,000 such photoreceptors. The image is focused on this layer, the receptors are activated, and the impulses are transmitted via the optic (second cranial) nerve to the occipital lobe of the cortex. More will be said about the physiology of sight later in this chapter.

Figure 11-9 shows two spaces, one between the cornea and lens and another between the lens and retina. The more anterior cavity is the *anterior chamber,* which contains a watery fluid, the *aqueous humor.* Aqueous humor fills much of the eyeball in front of the lens and is continually produced by the ciliary body in the anterior and posterior chambers behind the iris and absorbed by the choroid blood capillaries in the anterior chamber through the canals of Schlemm. This is a matter of physiological importance because blockage of these canals can result in a buildup of fluid in the anterior chamber, producing an increased pressure on both the cornea and lens (Figure 11-10). Continued high pressure on the lens can cause premature hardening and opacity of the lens, a disorder called *glaucoma.* This is said to be the most pain-

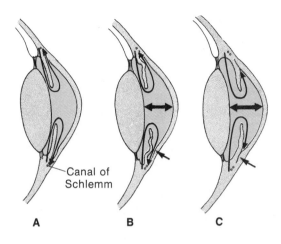

Figure 11-10 Glaucoma. **A,** Normal aqueous humor drainage through the canal of Schlemm. **B,** Open-angle glaucoma, showing a chronic condition treated by drugs. **C,** Closed-angle glaucoma requiring surgery (iridectomy) for correction.

ful of all diseases, but it can be detected in its early stages by measuring the pressure on the anterior eyeball with a drumlike instrument—a procedure called *tonometry.* Iridectomy, the surgical removal of a portion of the iris, or drugs affecting the eye blood vessels (vasodilators) have been successful in relieving glaucoma.

The eye's posterior cavity contains a proteinaceous fluid similar to that of the anterior chamber. However, the aqueous humor is water, whereas the vitreous humor of the posterior cavity is jellylike and fills the entire space behind the lens. This substance is necessary to keep the eyeball in its spherical shape.

Visual Pathways

The cell bodies that give rise to the primary fibers of the visual pathways reside in the retina. Processes from these cell bodies sweep across the internal surface of the retina to be gathered at a depressed point on the retina, the *optic disk* (see Figure 11-9). Fibers emerge from the eye-

ball at the optic nerve; neurons that innervate the medial half of each retina cross at the *optic chiasma* (Figure 11-11), whereas those that supply the lateral half do not cross but continue back on the same side.

Having two eyes rather than one makes possible true **binocular,** or stereoscopic, vision. Two optical images are received from slightly different angles, achieving the impression of distance and depth, thereby adding a third dimension to the visual field. In binocular vision it is necessary to turn the eyes inward so that two images of an object will lie on what are called *corresponding points* of the two retinas. Excitation of two corresponding points causes only one sensation, a perfect superimposition, in the occipital lobe of the cortex.

Convergence of the eyes is brought about by activation of the extrinsic eye muscles. If one or both eyes are deviated from their proper direction, convergence will be defective, and a form of **strabismus** (squinting) will result. A strabismus in which one sees two views or pictures, resulting in **diplopia** or double vision, is called *exotropia. Esotropia* (crossed eyes), another form of strabismus, may be due to a muscle defect or a weakness in coordination and results in an inability of the two eyes to work together. The affected eyeball is pulled inward (medially). Esotropia is fairly common early in life, and "lazy eye" clinics in elementary schools have played an extremely important role in detecting the condition and treating it immediately. In some patients, glasses and exercise may correct the defect; in others, surgery may be required. If correction is not accomplished early, the affected eye may become blind through disuse.

It is beyond the scope of this book to describe the many alterations in visual fields resulting from disease states. Suffice it to say that the exact disturbance depends almost exclusively on the site of interruption of the conducting pathway and on the extent of that interruption (Figure 11-11). Characteristic types of defects are associated with lesions of these pathways.

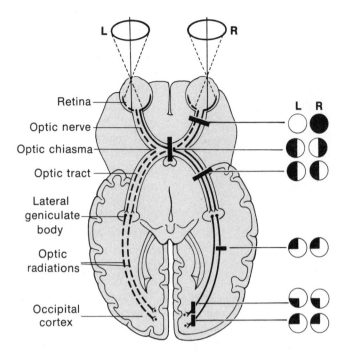

Figure 11-11 Visual pathway. On the right are diagrams of the visual fields, with areas of blindness darkened to show the effects of injuries in various locations.

Physiology of Vision

Light rays, while passing through a series of transparent, colorless eye parts including the cornea, aqueous humor, and lens, undergo a process of bending known as **refraction.** Refraction of light rays makes it possible for light from a large area to be focused (pinpointed) on a very small surface, the retina, where the photoreceptors are located (Figures 11-11 and 11-12). Light rays bend as they pass from one medium into another of a different density. Parallel light rays striking a convex lens are refracted toward a focal point behind the lens (Figure 11-12A). This focal point is on a line of axis of light passing through the center of the curvature of the lens. Conversely, a concave lens causes light rays to diverge or spread away from the axis of light (Figure 11-12B).

The greater the curvature of a lens the greater its refractive power. When the ciliary muscle is relaxed, parallel light rays striking the optically normal (emmetropic) eye are brought into focus on the retina. The problem of bringing diverging rays from objects closer than 20 feet to a focus on the retina can be solved by increasing the curvature or refractive power of the lens, a process called **accommodation.** At rest, the lens is held under tension by the suspensory ligaments (see Figure 11-9). When the gaze is directed at a distant object, the ciliary muscle contracts, producing a narrow, thin lens. This decrease in curvature facilitates focus on a distant object. Relaxation of the ciliary muscles produces a greater curvature in the lens for refraction of light rays from a near object (Figure 11-12C).

In some individuals, the eyeball may be shorter than normal or the refractive index of the lens too weak, bringing the light rays to a focus behind the retina (Figure 11-12E). This abnormality of accommodation is called **hyperopia,** or farsightedness. Sustained accommodation may compensate for this defect, but the prolonged muscular effort is tiring and may cause headaches and blurred vision. A biconvex

and exert undue pressure on the eardrum, causing earaches. The hairs of the canal protect the ear from foreign substances.

Middle Ear

The *tympanic membrane (eardrum)* separates the auditory canal from the *tympanic cavity* of the middle ear (see Figure 11-14). The eardrum is a thin layer of fibrous tissue covered externally with skin and internally with mucous membrane. The eardrum is attached so that it can vibrate freely when sound waves enter the auditory canal.

The tympanic cavity is a small, irregular bony cavity in the temporal bone. It is separated from the auditory canal by the eardrum and from the inner ear by two small openings, the *oval window* and the *round window*. In the round window there is an opening into the mastoid process; through this, an infection of the middle ear may extend into the mastoid cells, causing *mastoiditis.*

The middle ear cavity is filled with air that reaches it from the upper part of the throat (nasopharynx) by way of the eustachian tube. The walls of the cavity are lined with a mucous membrane continuous with that of the nasopharynx. Thus it is possible for infection to spread from the nose or throat to the middle ear (*otitis media*). Accumulations of pus in the middle ear can cause increased pressure on the eardrum. In this case, the membrane should be lanced (myringotomy) to alleviate this situation and to prevent further problems.

The eustachian tube acts to equalize the air pressure of the middle ear with atmospheric pressure. In inflammatory conditions the tube may become blocked, thus preventing this equalization; hearing is impaired, and permanent damage may result if the inflammation is not corrected. The pharyngeal opening of the tube is closed except during swallowing, yawning, or sneezing. It tends to remain closed when the external pressure is greater. Sudden changes in pressure exerted on the tympanic membrane

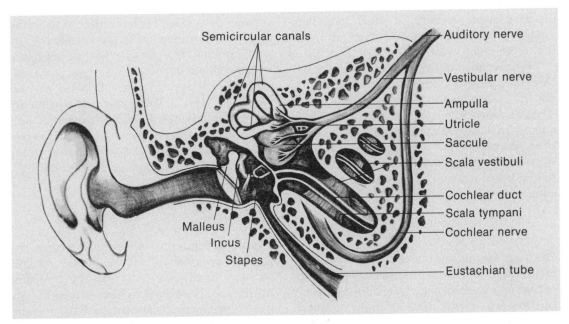

Figure 11-15 Cross section through the cochlea showing the relationship between the membranous and bony labyrinths and the structures concerned with balance.

by loud explosions or rapid falling, as in an airplane's descent, may cause the membrane to rupture. Such an occurrence often may be prevented by swallowing or yawning, which will open the eustachian tube, thereby balancing the pressure. Divers must be examined regularly to prevent ear damage that may result from prolonged intervals spent deep under the water; a diver with a blocked eustachian tube due to a cold, for example, may easily suffer ruptured eardrums.

Stretching across the cavity of the middle ear from the tympanic membrane to the oval window are three tiny, movable bones or ossicles named according to their shapes—the *malleus* (hammer); the *incus* (anvil); and the *stapes* (stirrup) (see Figure 11-15). The handle of the malleus is attached to the tympanic membrane; its head is attached to the incus. The long process of the incus is attached to the stapes, and the footpiece of the stapes fits into the oval window. These bones transmit sound waves from the outer ear to the inner ear. They are attached freely to one another by ligaments and are set in motion by any movement of the tympanic membrane. The bones of the middle ear form a series of levers, the effect of which magnifies the force of sound vibrations to about ten times that at the oval window.

Inflammation of the joints between the ossicles may prevent normal vibration and amplification of sound waves. *Otosclerosis* is a type of deafness found in adults and occurs more frequently in women. Bone changes (fusion of the joints) prevent normal vibration of the third ossicle (stapes). An operation to release the stapes so that it will move again has had considerable success.

Inner Ear

The inner ear contains receptors for hearing and for position sense. It consists of a bony labyrinth enclosing a membranous labyrinth. The *bony labyrinth* is a series of channels in the temporal bone filled with a fluid called *perilymph*. The inner *membranous labyrinth* more or less duplicates the shape of the bony channels and is filled with a fluid called endolymph. There is no communication between the spaces filled with perilymph and endolymph.

The *cochlea*, a portion of the labyrinth about 1.5 inches long, is a coiled tube resembling a snail shell. Internally, the cochlea is divided into three spiral membranous channels that run the full length of the organ. The organization and arrangement of these channels can be visualized by hypothetically unwinding the spirals of the cochlea, as shown in Figures 11-16 and 11-17. The uppermost channel, the vestibular channel (*scala vestibuli*), is attached to the oval window. At the tip (apex) of the cochlea, the vestibular channel communicates with the lowermost channel, the tympanic channel (*scala tympani*). The base end of this channel terminates at the round window, leading back to the middle ear. Both the tympanic and vestibular channels are filled with perilymph. The third and smallest channel, the *cochlear duct*, rests between the other two and is filled with endolymph. It is separated from the overlying vestibular channel by the vestibular (Reissner's) membrane, and from the underlying tympanic channel by a ledgelike projection of the bony cochlear wall and by the basilar membrane.

The actual receptors for hearing are found in the cochlear duct as several rows of specialized *hair cells*, approximately 20,000 in number (see Figure 11-17). These bear cilia that project into the endolymph from the free end of each cell. The hair cells, supporting cells, and nerves constitute the organ of Corti, which rests on the basilar membrane within the cochlear duct.

Physiology of Hearing

Hearing depends on a succession of events. The first is entry of sound waves into the external auditory canal, which causes the tympanic membrane to vibrate. These vibrations are transmitted mechanically by the malleus, incus, and stapes of the middle ear to the oval window at the entrance to the inner ear. The vibrations

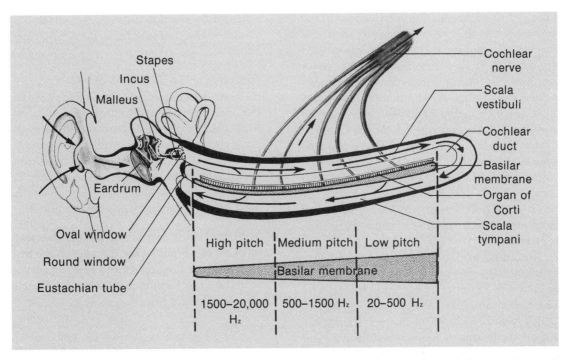

Figure 11-16 Diagram of an "unrolled" cochlea, showing the transmission of sound through the canals.

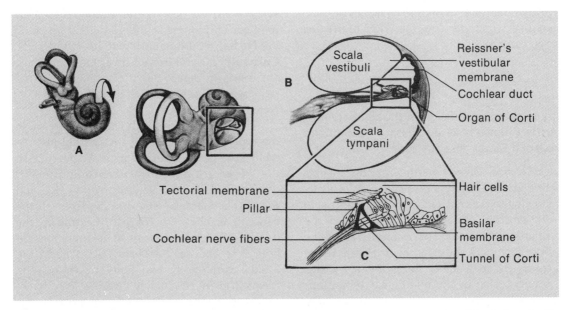

Figure 11-17 Diagrams of the inner ear. **A,** External appearance of the cochlea and semicircular canals. **B,** Internal structure of the cochlea. **C,** A microscopic sketch of the organ of Corti, showing the hair cells.

of the oval window set up pressure waves in the perilymph of the vestibular and tympanic channels. These pressure waves are transmitted through the basilar membrane to the endolymph of the cochlear duct, where the hair cells of the organ of Corti are stimulated. The resulting stimulation of the hair cells arouses nerve impulses that are transmitted by the cochlear nerve to the hearing center in the temporal lobe of the cerebral cortex. This process suggests that a certain portion of the basilar membrane and certain hair cells are affected by sound waves of one frequency. Waves of a different frequency set up vibrations in another portion of the basilar membrane.

According to the just-described place theory of hearing, age and other factors determine the range of hearing among individuals. The human ear can detect sounds ranging in frequency from 20 to 20,000 cycles/sec; sounds below or above this range are not heard. The squeaking of a mouse is higher than 20,000 cycles/sec. We do not hear it, but the cat does. A dog whistle can be made to produce sounds higher than our hearing capacity. The dog's hearing range goes as high as 24,000 cycles/sec; therefore, when we blow the whistle, the dog hears it and we don't.

Deafness

Any impairment or defect in the ability to hear sounds is known as **deafness.** The common types of deafness are as follows:

1. *Conductive.* Defects in the sound-transmitting mechanism may result from an obstruction in the external auditory canal, damage to the eardrum, or damage to or stiffness of the middle ear bones. For example, excessive wax may block the external canal; scar tissue or injury may affect the tympanic membrane; fusion of the middle ear bones (ankylosis) or stiffness of the ear bones may occur. Conduction through the air is prevented in each situation, but hearing aids devised for conduction through bone from the outside to the inner ear fluids may help these patients.

2. *Perceptive.* This kind of deafness is due to cochlear damage, usually to the hair cells in the organ of Corti and to the cochlear nerve. Individuals exposed to loud sounds continually such as pilots and artillery soldiers often are deaf to particular sounds following destruction of a specific group of hair cells in the organ of Corti. The cochlear nerve is most often impaired by growth of tumors.

3. *Central.* Central deafness is usually due to a physical or psychological disorder in the auditory sensory region on the cerebral cortex or in the tracts leading to it. This type of deafness is relatively rare.

Hearing Acuity

The instrument used to detect or measure a person's ability to hear is the **audiometer.** New methods employing sound vibrations are being used to help children born with hearing defects. Currently, all children of a certain school age are tested to detect hearing problems so that such weaknesses may be identified and corrected immediately.

Hearing tests not only help to determine the type of hearing defect present but also establish the potential of the patient's hearing. In the Weber test an inexpensive tuning fork is used to differentiate between conductive and perceptive deafness. The fork is placed on the patient's forehead to compare hearing in the two ears. In normal hearing or deafness equal in both ears, the person hears vibrations in the middle of the head. Variations suggest hearing inequality. In conductive hearing loss, bone-conducted sounds shift to the poorer ear; in perceptive hearing loss, sounds are heard louder in the better ear.

Equilibrium

In addition to the cochlea, the inner ear contains three organs—the utricle, saccule, and semicircular canals—that play a major role in the sense of equilibrium or balance (see Figures 11-15 and 11-18).

The *vestibule* is a chamber containing the first two membranous sacs, the *utricle* and the *saccule,* each of which is filled with endolymph and is concerned with static equilibrium. Within the walls of these two sacs are special thickenings, or maculae, containing receptor hair cells and nerve endings that project into the overlying endolymph. Enmeshed in the hairs are tiny particles of calcium carbonate, called **otoliths** or ear stones (Figure 11-18). Any changes in position of the head or body that cause the otoliths to bend in the direction of the pull of gravity initiate nerve impulses that travel to the brain via the vestibular nerve (Figures 11-14 and 11-18). The vestibular nerve joins the acoustic nerve to become collectively the eighth cranial nerve.

The three semicircular canals are hollow tubes containing endolymph. These are the or-

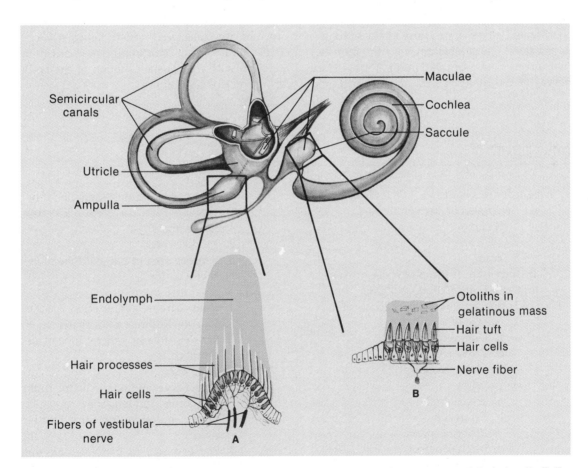

Figure 11-18 Equilibrium. Diagram of the organization of the maculae and cupula of the inner ear. Both contain receptors covered by a gelatinous mass. **A,** Otoliths move in the maculae, mechanically stimulating the hair processes of the hair cells. **B,** Endolymph movement stimulates the hair cells in the cupula. Each stimulation is carried to the brain, which interprets the movement.

gans of dynamic equilibrium. They are arranged at right angles to one another so that all three planes of space are represented (Figures 11-15 and 11-18). One end of each canal is enlarged into a swelling called the **ampulla**. Within each ampulla is a small cluster of hair cells and nerve endings that project into the endolymph. Movement of the hair cells sets up impulses in the vestibular nerve; whenever movement begins or ends, goes faster or slower, or changes direction, impulses are initiated.

A change in position of the head causes impulses from receptors in the semicircular canals and vestibule to initiate the righting reflex. Balance against gravity is thus maintained by the coordinated effective response of the antigravity muscles. The cerebellum links the impulses that arise from stimulation of the sensory nerves of the canals, joints, and so forth, and sends nerve impulses to the motor centers of the cerebrum and spinal cord.

Disorders of Equilibrium

Tinnitus is a noise in the ear, usually a buzzing or ringing sound. It may be caused by an accumulation of wax in the outer canal, by otitis media, and possibly by a degeneration of the acoustic nerve itself.

Vertigo is a sense of movement, either of self or surroundings, due to stimulation of the hair cells in the semicircular canals by movements of the endolymph. **Nystagmus,** an oscillating eye movement, is closely associated with vertigo. Both produce a sensation of nausea and are collectively called *motion sickness.* The inability of the equilibrating mechanism to keep up with the constant motion of the body brings on this condition.

OUTLINE

I. Sensory mechanisms
  A. Awareness of environment
    1. Receptor
    2. Neuron
    3. Brain
  B. Important features of sense organs
  C. Classification of sense organs
    1. Special, cutaneous, and visceral
    2. Nature of stimulus
      a. Chemoreceptors
      b. Pressoreceptors
      c. Photoreceptors
    3. Location
      a. Teleceptors
      b. Exteroceptors
      c. Proprioceptors
      d. Interoceptors
  D. Law of projection
    1. Phantom limb pain
II. Cutaneous senses
  A. Touch and pressure sensations

    1. Free nerve endings, Meissner's corpuscles, Pacinian corpuscles, hair nerve fibers
    2. Merkel's disks
    3. Sensory area in parietal lobe of cerebral cortex
  B. Temperature sense
    1. Cold—Krause's corpuscles
    2. Warm—Ruffini's end-organs
    3. Sensory area in parietal lobe of cerebral cortex
  C. Somatic pain
    1. Free nerve endings in skin, muscles, joints, and tendons
    2. Overstimulation of any tactile senses
    3. Sensory area in parietal lobe and thalamus
    4. Referred pain
    5. Enkephalins and endorphins
    6. Pain remedies

III. Position sense (kinesthesia)
  A. Muscles, tendons, and joints
    1. Ruffini's end-organ
    2. Golgi's tendon organ
    3. Proprioceptors
    4. Sensory areas in thalamus and cerebral cortex (conscious activities), and in cerebellum (reflex activities)
IV. Hunger and thirst
  A. Organic sensations
    1. Hunger—projected from body cells to stomach
      a. Appetite—pleasant sensation
    2. Thirst—projected from body cells primarily to pharynx
      a. Dehydration—tissues lack water
      b. Polydipsia—excessive thirst
    3. Sensory area for organic sensation in the hypothalamus
V. Taste and smell
  A. Chemical sensations
    1. Taste—gustatory taste buds found on tongue and oropharynx
      a. Sweet, sour, bitter, and salty
      b. Sensory area—lateral fissure of cerebral cortex
    2. Smell—receptors in nasal mucosa
      a. Volatile substance detection
      b. Sensory area in olfactory center of temporal lobe
    3. Gustatory nerve—blends sensations of taste and smell
VI. Vision
  A. Accessory organs
    1. Eyebrows
    2. Eyelids
    3. Eyelashes
    4. Conjunctiva—disease barrier
    5. Lacrimal apparatus—produce tears
    6. Six muscles (extrinsic)
  B. Anatomy of eyeball
    1. Layers
      a. Sclera—outer white layer of eye
        (1) Cornea
      b. Choroid—middle vascular layer
        (1) Iris, pupil, ciliary muscles
        (2) Lens
        (3) Adaptation
      c. Retina—inner layer of nerve cells
        (1) Rods and cones—photoreceptors
    2. Anterior chamber
      a. Aqueous humor
      b. Canal of Schlemm
        (1) Glaucoma
    3. Posterior cavity
      a. Vitreous humor
      b. Shape of eyeball
  C. Visual pathways
    1. Optic disk—convergence of fibers of optic nerve
    2. Binocular vision
      a. Strabismus
        (1) Extropia
        (2) Esotropia
  D. Physiology of vision
    1. Refraction
      a. Bending of light rays
      b. Two media of differing densities
    2. Convex and concave lenses
    3. Accommodation
    4. Hyperopia and myopia
    5. Astigmatism
    6. Presbyopia—old age
    7. Cataracts
  E. Perception of light and color
    1. Light—electromagnetic radiation
      a. Visible light-spectrum colors
      b. Rods—night vision
        (1) Rhodopsin—visual purple
      c. Cones—light vision, color vision
        (1) Macula lutea—fovea centralis
    2. Visual acuity—20/20 normal vision
    3. Color
      a. Young–Helmholtz theory—trichromatic vision

(1) Red, green, and blue cones
(2) Dichromatic and monochromatic color blindness
b. Ishihara color plate test
4. Ophthalmoscope
a. Eye examination
b. Diagnostic tool

VII. Hearing and equilibrium
A. Anatomy of the ear
1. External ear
a. Pinna (auricle)
b. Auditory canal—cerumen glands
2. Middle ear
a. Tympanic membrane (eardrum)
b. Tympanic cavity
(1) Eustachian tube—maintains equal air pressure
(2) Otitis media, myringotomy
(3) Ossicles (ear bones)—malleus, incus, and stapes
(4) Conduction of sound vibration
(5) Otosclerosis, ankylosis
3. Inner ear—labyrinths
a. Cochlea-coiled, snaillike
(1) Channels filled with perilymph and endolymph
(2) Organ of Corti—sensory hair cells for hearing
(3) Cochlear nerve

B. Physiology of hearing
1. Sound waves
2. Place theory
a. Vibrations conducted from external to inner ear
b. Hair cells of organ of Corti stimulated
c. Impulses carried to temporal lobe of cerebral cortex by cochlear nerve
d. Range of hearing—20 to 20,000 cycles/sec
3. Deafness
a. Conductive
b. Perceptive
c. Central
4. Audiometer—measures person's ability to hear
C. Equilibrium
1. Anatomy
a. Vestibule (saccule and utricle) and semicircular canals
b. Hair cells in utricle, saccule, and ampulla of semicircular canals
c. Otoliths enmeshed in hair cells
2. Righting reflex
3. Disorders
a. Tinnitus
b. Nystagmus
c. Vertigo
d. Motion sickness

STUDY QUESTIONS AND PROBLEMS

1. An individual receives a third-degree burn on the arm. The deep structures of the skin are destroyed. What sensations are lost? Which receptors are involved? Will there be much pain? Explain your answers.

2. Explain how receptors function as a protective mechanism.

3. How does one know what is happening outside one's body? Inside one's body?

4. What does the term "referred pain" mean? Give an example.

5. What is the difference between hunger and appetite?

6. Colds are hunger antagonists. Explain.

7. If you have a cold, why is it not a good practice to blow your nose hard? Describe the resultant middle ear involvement.

8. John Hones has received an injury to his right occipital lobe. Will he experience partial loss of sight in his right, left, or both eyes? Explain.

9. If my father cannot distinguish blue from black or purple, how can I explain this situation to him?

10. Describe the phenomenon of accommodation, using as examples your eyes, the windshield of your car, and a pedestrian walking across the street 50 feet in front of your moving car.

11. What is the law of refraction? Using it as a basis, explain the following:
    a. Myopia and hyperopia
    b. Astigmatism
    c. Presbyopia

12. What does "20/20" mean? If you have 20/10 vision, what is the condition of your eyes?

13. What are mydriatics and miotics? Describe their function and give an example of each.

14. How is equilibrium related to the ear? How is vertigo related to dizziness?

15. Differentiate among the types of deafness. Describe bone versus air conduction of sound.

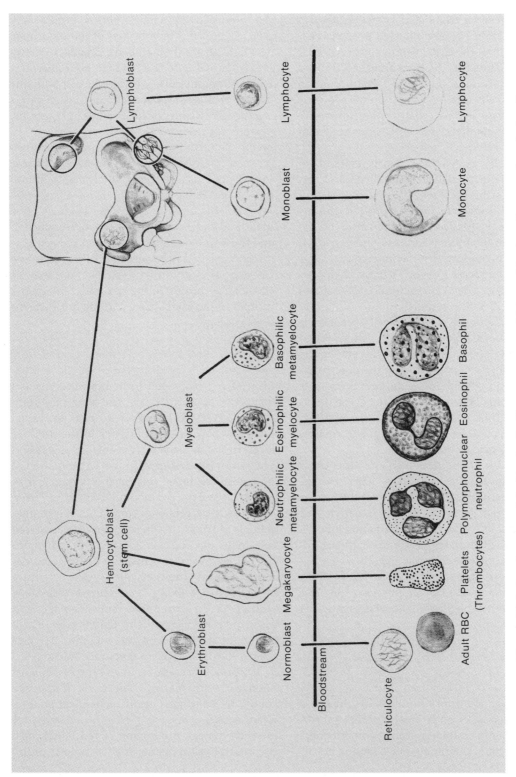

Figure 12-3 Development of the blood cells. Those cell forms above the black line are found in the marrow or lymphatic organ and are considered imma- ture. Mature blood cells are found in the bloodstream. RBC, red blood cell.

the *intrinsic factor* (IF), which promotes absorption of vitamin B_{12} (also referred to as *extrinsic factor* or *EF*) in the distal ileum. Absorbed vitamin B_{12} is stored in the liver and then utilized enzymatically to complete the maturation in red blood cells (Figure 12-4). Several proteins, folic acid, basic metals (cobalt, copper, and iron), and other vitamins (vitamin B_6) are also essential for red blood cell formation.

Normally a relatively small amount of erythropoietin circulates in the blood and acts to maintain the basal level of erythrocyte production. A decreased oxygen supply to the tissues triggers an increased production of the hormone in the kidney and resultant increased erythrocyte production (erythrocytosis). Conversely, an increased oxygen supply to the tissues results in a decrease of hormonal level and a decrease in erythropoiesis.

An index of the rate of red cell formation is the reticulocyte count. Normally, there is between 0.5 to 1.5 percent reticulocytes among the total red blood cell population. If the percentage of these young cells increases, the marrow is usually overactive or hyperplastic. If the delivery of the cells from the marrow is impaired, marrow hypoplasia is detected by a normal to low reticulocyte count.

An index of the rate of cell destruction is the haptoglobin determination. This test involves **electrophoresis** (separation of differently charged particles). It is a sensitive and reliable test of hemolysis. When hemoglobin is released, it will bind with a group of alpha globulins of plasma called *haptoglobin*. The more hemoglobin in the plasma the higher the haptoglobin level. Depending on the test method, normal haptoglobin level is as much as 100 mg/100 mL.

Red Blood Cell Disorders

Anemia

Anemia is an abnormal condition in which the blood either lacks its normal number of red blood cells or its normal concentration of hemoglobin. Certain symptoms are common to all the anemias. The anemic person is generally pale; the mucous membranes of the mouth are also light as are the nail beds. This lack of normal color is due to the hemoglobin deficiency. Fatigue and muscular weakness accompany the disease because of the inadequate oxygen-carrying capacity to the cells and tissues. The anemic person experiences *dyspnea*, a shortness of breath. To meet the need for more oxygen, respiration rate is quickened. A rapid heart rate is experienced as the heart attempts to pump more blood to the tissues.

The distinction between anemias due mainly to increased red cell loss or destruction and those due mainly to decreased red cell production is based on evaluation of the quantitative aspects of erythropoiesis. Anemias may be classified according to the size and hemoglobin content of the erythrocyte and the mechanisms of its etiology.

PRIMARY ANEMIAS Primary anemias are characterized by a disturbance in the mechanism of red blood cell formation, leading to a deficiency of erythrocytes. The most important primary type is *pernicious anemia*. Its initial cause is a permanent deficiency of the intrinsic factor in the stomach responsible for absorption of vitamin B_{12}, resulting in the accumulation of immature red blood cells in the bone marrow. In this disease the red blood cell count may drop as low as 20 percent of normal; a blood examination also reveals the presence of immature red cells in the bloodstream (Figure 12-5). When untreated, pernicious anemia can bring about deterioration of the nervous system, causing difficulty in walking, generalized weakness, stiffness of the extremities, brain damage, and permanent damage to the spinal cord. Early treatment by injection of vitamin B_{12} and proper diet may assure a favorable outcome.

Aplastic anemia is another type of primary anemia. Here, marrow failure is frequently caused by poisonous agents. Chemical poisons (lead, benzene, arsenic, nitrogen mustard, and gold) discourage production of red blood cells, whereas physical agents (x-rays, atomic radia-

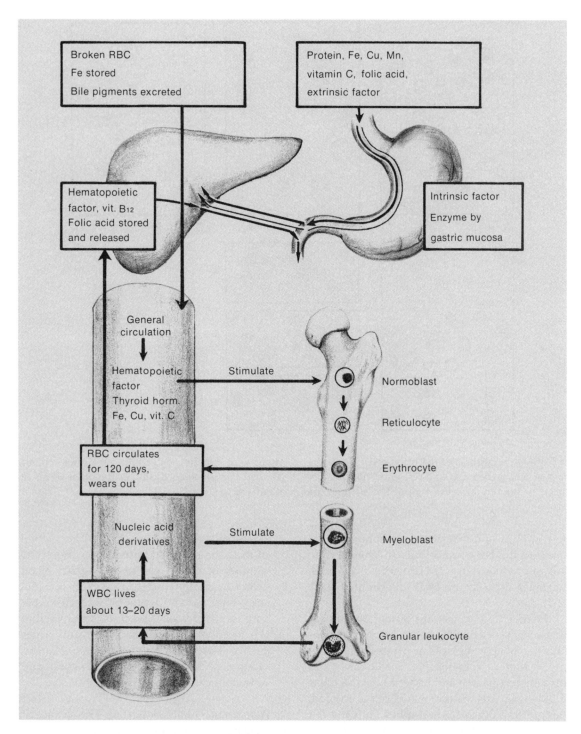

Figure 12-4 Sequence of events in, and elements essential for, normal blood formation. RBC, red blood cells; WBC, white blood cell; vit., vitamin; horm., hormone.

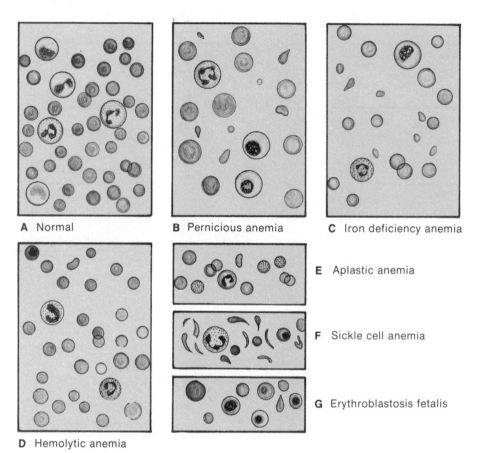

A Normal   **B** Pernicious anemia   **C** Iron deficiency anemia

E Aplastic anemia

F Sickle cell anemia

G Erythroblastosis fetalis

D Hemolytic anemia

Figure 12-5 Diseases of red blood cells. Using the normal blood smear (**A**) as a reference, note the decrease in number of cells in all of the anemia, the larger-sized cell (macrocyte) in pernicious anemia (**B**), and the odd-shaped cells in the other forms of anemia (**C** to **G**).

tion, radium, and radioactive phosphorus) tend to injure the bone marrow itself. The marrow fails to produce either red or white blood cells, causing a drop in the total cell count (Figure 12-5).

Hemolytic anemias are brought about by excessive destruction of erythrocytes. Occasionally, the phagocytic system proceeds too rapidly. More commonly, infections and infestations (malaria and bacteria) are the cause of blood cell loss. As each blood cell is entered, the infecting organism multiplies until the cell bursts (hemolysis), resulting in anemia. Hemolytic anemias are of two types:

1. *Drug-induced hemolytic anemia.* Destruction of the bone marrow causes a decrease in the number of stem cells producing red blood cells. Chloramphenicol, an antibiotic, can destroy both red and white blood cells so that corpuscular matter is markedly decreased in the peripheral blood. Radiation and drugs such as those used in the treatment of cancer often depress bone marrow function and cause reduced platelet production.

2. *Autoimmune hemolytic anemia.* The surface of human red blood cells is covered with blood group **antigens** that are capable of provok-

ing an immune reaction. Autoimmune hemolytic anemias are acquired by persons who develop specific **antibodies** (autoantibodies) to antigens on their own red blood cells. The most important groups of red cell antigens are the ABO and Rh systems. A cell membrane antigen stimulates the production of antibodies in individuals who lack that antigen's blood group. Thus, a normal immune system reacts by producing antibodies against their own antigens. However, some persons do develop autoantibodies for various unknown reasons. When this happens, red cells may become coated with antibody and complement. Mismatched blood transfusion is an example of a procedure that can result in a hemolytic reaction and massive hemolysis. The red cell count drops markedly; hemoglobin is liberated into the plasma with impairment of renal function. Less common types of anemia resulting in hemolysis include sickle cell anemia (occurring almost exclusively in the dark colored races) and a hemolysis that occurs in infants as a result of antibodies from an Rh-negative mother. This latter condition, erythroblastosis fetalis, will be discussed later.

SECONDARY ANEMIAS   Secondary anemias are due to simple loss of blood or to a defective formation of hemoglobin caused by other diseases. The average adult has about 6 liters (about 5 quarts) of blood. If 2 liters (about 2 quarts) are lost suddenly, death will result. However, a gradual loss of the same quantity over a period of weeks can be handled and withstood by the body. In chronic bleeding, it is important to locate and remedy the cause, which may be hemorrhoids, excessive menstrual flow, or ulcerations of the stomach or bowel. Malignant tumors, infections, liver cirrhosis, thyroid deficiency, and chronic kidney disease (nephritis) are other causes of secondary anemia. In one of the most common disorders, *macrocytic hyperchromic anemia*, the erythrocyte is larger than normal or **macrocytic,** and abnor-

mally high amounts of hemoglobin are present. In microcytic hypochromic anemia the reverse is true; smaller than normal or **microcytic** red cells contain decreased amounts of hemoglobin. Both of these are iron-deficiency anemias and often occur in women of childbearing age, during chronic hemorrhage, and in infants. In microcytic anemia, prolonged red blood cell production, due to loss of red blood cells in the circulation, imposes a continuous demand on the iron stores of the body, leading to iron deficiency. Consequently, the hemoglobin content of newly formed red cells is lower than normal (hypochromic), and the cells are smaller than normal. Macrocytic anemia, on the other hand, is characterized by abnormally large red cells owing to the absence of vitamin B_{12}, the intrinsic stomach factor, or both. These cells contain a greater than normal amount of hemoglobin (hyperchromic).

A person with a secondary anemia needs frequent blood checks (hemoglobin and hematocrit) and continued advice and treatment, not only to treat the cause of the disease, but also to restore the composition of the blood to normal.

Polycythemia

The condition known as **polycythemia** is characterized by the production of a greater than normal number of red cells. The erythrocyte-forming areas become overactive. Erythropoiesis is normally subject to a feedback control. The process is inhibited by a rise in the circulating red blood cell level to super-normal values and is stimulated by anemia. Erythropoiesis is also stimulated by reduced oxygen (hypoxia) situations, such as residence at high altitudes. It is thought that the low atmospheric pressure existing at high altitudes decreases the ability of hemoglobin to combine with oxygen and that this reduction tends to stimulate the formation of new red cells. As a result, viscosity of the blood is increased and blood pressure rises. The heart may become overloaded, and the danger of clotting within the circulation is increased.

Reactions of Hemoglobin

On exposure to atmospheric conditions, hemoglobin forms a covalent bond with oxygen to produce oxyhemoglobin (HbO_2), oxygen attaching to the iron (Fe^{++}) in the heme. The details of the oxidation and reduction of hemoglobin is described in Chapter 16, "Respiration."

Previously it was mentioned that the normal average hemoglobin content of blood is 15 g/100 mL, all of it in red blood cells. When old red blood cells are destroyed in the liver and spleen, the globin portion of the hemoglobin molecule is split off, and the heme is converted to **bilirubin,** which is excreted by the liver in the form of bile (see Figure 12-4). The iron is reused for hemoglobin synthesis, and if blood is lost from the body and the iron deficiency is not corrected by dietary intake, iron deficiency anemia results.

When excessive amounts of hemoglobin are broken down, or when biliary excretion of bilirubin is decreased by liver disease or bile duct obstruction, the plasma bilirubin level rises, imparting a yellow tinge to the skin (jaundice). A jaundiced condition usually accompanies an infected liver (infectious hepatitis).

LEUKOCYTES

The white blood cells of the blood, the leukocytes, are concerned primarily with the body's defense against infection (Table 12-1). Leukocytes vary in size and shape (see Figure 12-3). They are less numerous than erythrocytes, numbering 5000 to 9000/mm³. They lack hemoglobin, are colorless, contain a nucleus, and are larger than the red cells. White blood cells are extremely active, and move under their own power by ameboid motion, often against the flow of the bloodstream. Unlike erythrocytes, white blood cells pass out of the bloodstream and into the intracellular spaces to phagocytize foreign materials (such as bacteria) found between the cells (Figure 12-6).

Leukocytes fall into two distinct groups.

The first group contains granules in the cytoplasm and possesses lobed nuclei. These leukocytes, called **granulocytes,** can be divided into three types: neutrophils, eosinophils, and basophils. The second group are agranular and are referred to as **agranulocytes.** The two types of agranulocytes, lymphocytes and monocytes, develop from lymph nodes (**lymphatic** tissue) and red bone marrow (myeloid tissue) (Figure 12-3). No cytoplasmic granules can be seen, and their nuclei are usually spherical.

Granulocytes

The sequence of development from stem (embryonic) cell to mature segmented leukocytes is shown in Figure 12-3. Beginning at the myelocyte stage, three distinct cell lines are distinguishable by the appearance of specific granulation (using Wright's stain), hence the names *neutrophilic* (neutral), *eosinophilic* (eosin acid), and *basophilic* (basic).

The neutrophilic myelocyte (15 μm in diameter) shows an eccentric, condensed, round (or ovoid) nucleus with a light pink-tan cytoplasm that has numerous fine granules. As the cell matures, overall size decreases, and the nucleus changes shape from indented (metamyelocyte) to a thin band and finally becomes segmented into several lobes (two to five) connected by a fine chromatin strand. A sex chromatin body ("drumstick") may be found occasionally in neutrophils of normal females and appears as a small ovoid mass of chromatin attached to a nuclear lobe. Neutrophils are the most numerous of all the white blood cells, constituting from 60 to 70 percent of the total numbers. Neutrophils serve to phagocytize cells that are involved in inflammation. When there is a great need for them, immature forms (bands) may appear in the blood.

The eosinophil has abundant, coarse granules (up to 1.5 μm), which are blue at first but become deep orange in color. The mature eosinophil has a two-lobed nucleus, and although their function is unclear, they may be weakly

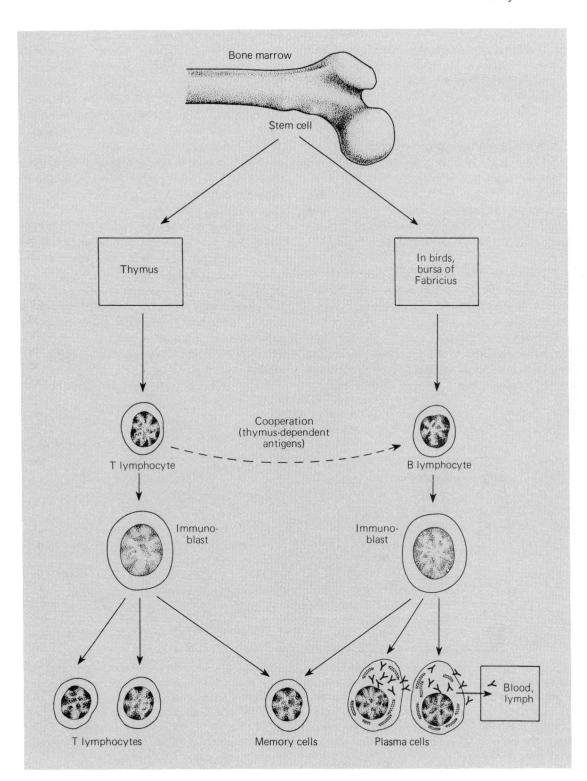

cent antibody technique. Characterization of the abnormal lymphocytes by detection of IgM has shown that certain diseases are expressions of one class of B lymphocytes.

T Lymphocytes

These constitute the greater part (65 to 80 percent) of the circulating pool of lymphocytes. They are from primitive stem cells that have circulated through the thymus gland and are capable of destroying bacteria, viruses, and fungi and of providing immunological surveillance that can eliminate cancer cells (Figure 12-11).

T lymphocytes divide about three times daily. Once the T cell has entered the bloodstream, it slips between the epithelial cells of the vascular bed or enters special regions of the lymph nodes and spleen. The T cell identifies the intruding antigen and processes its identity in a fashion to allow the B cell to form antibodies. In addition, the T cell can destroy the of-

fending agent outright. The T cell can circulate through the bloodstream via the lymphatic-general circulatory system portal in the thoracic duct and search for foreign substances or malignant cells.

Humoral immunodeficiency can be caused by a lack of either B or T cells and is especially prominent in certain malignant states such as multiple myeloma and chronic lymphocytic leukemia. Infectious problems may develop in persons deficient in these lymphocytes and their immune reactions.

Acquired immune deficiency syndrome, or AIDS, has received much public attention recently. Identified just a few years ago, this disease is fatal, and cause and cure are unknown. In AIDS, the immune system is destroyed, leaving victims with no means to fight off infectious agents. Individuals with AIDS suffer recurring infections and often develop Karposi's sarcoma, a rare cancer, and *Pneumocystitis carinii* pneumonia.

The immune system's T lymphocytes ap-

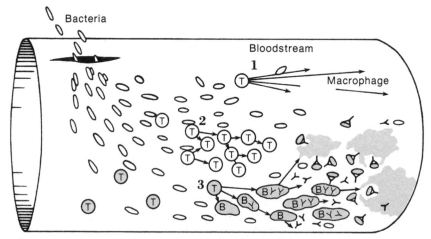

Figure 12-11 Bacteria invading bloodstream. Bacteria invade, perhaps through a tiny scratch, and begin to multiply. Some become stuck to monocytes, activating them into more aggressive macrophages. Circulating randomly, T cells adhere to macrophages and to the bacteria. This linkage stimulates T cells to send three messages: (1) Activate and summon more macrophages. (2) Produce more T cells. (3) Signal B cells to proliferate and become plasma cells that produce and release antibodies (Y-shaped molecules). Antibodies latch onto intruder molecules (or antigens) on bacteria and link the yoke of the Y to them. The shaft of the Y attaches to macrophages. They in turn envelop the bacteria, which are consumed by enzymes and become harmless debris.

pear to be affected by the disease. AIDS victims have a decreased number of these white blood cells, and those that they do have are abnormally shaped and malfunctioning.

Researchers believe that AIDS is caused by a viral agent. The disease is transmitted through blood products, such as blood transfusions and male sexual contact. AIDS has been diagnosed in Haitian immigrants, homosexuals, intravenous drug users, and persons who have received blood transfusions. The symptoms of the disease—fatigue, recurring viral infections, fever, swollen lymph glands, and night sweats—may not appear until months after exposure to the disease.

Antibodies

Antibodies are gamma globulin molecules that circulate in the plasma and are produced by the lymphoid tissues and lymphocytes. Each antigen has a specific antibody that alone can react with it. The particular configuration of the combining sites is seemingly complementary to the determining portions of the antigens that induce their formation—an arrangement comparable to a lock and key. The antigen molecule illustrated has many active chemical sites or determining groups protruding from its surface as specific configurations. These determine antigenic specificity and fit into the pocketlike-combining sites of the antibody. The number of antibody molecules produced is usually many times greater than the number of antigen molecules present.

A discussion of all immunoglobulins is beyond the scope of this book, but a few are listed below.

- IgG (75%) In maternal/fetal passage allows fetus to fight infection.

- IgM (10%) Responds to antigenic stimulation.

- IgA (12%) Neutralizes viruses and prevents attachment of bacteria to epithelial cells.

- IgO (about 2%) Minimal function found in B lymphocytes in lymphatic leukemia.

- IgE (about 1%) In allergic reactions helps to suppress foreign allergen.

Two broad groups of antibodies are recognized: *natural antibodies*, which are formed without apparent antigenic stimulation, and *acquired antibodies*, which appear only after exposure to a known antigen. The anti-A and anti-B blood group antibodies involved in blood reactions are examples of the natural type. Antibodies that are developed during the course of immunization (as for polio, measles, and typhus) are examples of the acquired type.

The combination or binding of antibodies to antigens is manifested in several different ways:

- *Agglutination*—when antigens such as large particles or whole cells (red blood cells or bacteria) are clumped together in a fluid

- *Precipitation*—when soluble antigen molecules become insoluble, and thus become more readily phagocytized by a white cell or macrophage

- *Lysis*—when red blood cells or bacteria rupture their cell membranes, releasing hemoglobin and cellular contents into solution

- *Opsonization*—when antibodies (opsonins) combine with bacteria or viruses and make them susceptible to phagocytosis

The thymus gland has an important role in antibody production. Located in the chest cavity near the lower end of the trachea in children, the thymus is a prominent mass of lymphoid tissue. During early childhood the gland produces lymphoid cells responsible for antibody production during most of the life of the individual. Removal of the thymus gland in newborn mice has proved fatal; the mice die from infection within a few weeks. Although the thymus is naturally atrophied and inactive in an adult, early removal of the gland in children can almost entirely prevent antibody production and therefore blocks the formation of an important line of defense against disease.

Allergy

Normally the antigen-antibody reaction takes place in the bloodstream, and the debris produced is removed by the reticuloendothelial system. If an acquired immunity is weak, a foreign protein may penetrate the interstitial tissue before it comes into contact with an antibody. When the immune reaction occurs in the tissues, the surrounding cells are irritated or destroyed by its side effects.

Allergies result from weak immunities, and the reactions in the tissues give rise to the allergic symptoms. In infancy, there may be allergic reaction to food proteins because the child has not yet developed the strong immunity to them possessed by adults. More commonly, however, allergic reactions are due to materials that are proteinlike and induce some antibody formation, but not enough to produce full immunity. These materials are known as **allergens** (for example, dust and flower pollens).

The deleterious effects of the immune reaction in the tissue spaces cause the cells to release **histamine,** a potent vasodilator that causes redness and swelling of the skin (hives, urticaria) and in the irritated membranes of the eyes and sinuses (as in hay fever). If the release of histamine is excessive, a sudden vasodilation can occur throughout the body, lowering the blood pressure and causing shock. A massive response to an allergen is known as anaphylactic shock. This can occur when a drug, vaccine, or any allergen to which a person is sensitive has entered the circulatory system. In other allergies, the immune reaction causes constriction of the smooth muscle fiber of the respiratory tract, producing *asthma*, a difficulty in getting enough air into and out of the lungs.

The symptoms of a hay fever type of allergy may be relieved by antihistamine drugs; the symptoms of asthma may be relieved by epinephrine. More positive treatment can be carried out following intensive skin testing with small amounts of various allergens. Serum injections are now commonly used over an extended period, either to alleviate the temporary discomfort or to eliminate the problem entirely.

BLOOD TYPING AND TRANSFUSIONS

If the amount of blood in the body is severely reduced because of hemorrhage, disease, surgical procedures, or other causes, the body cells suffer from lack of oxygen and nutrients. The usual treatment in such cases is to replenish the lost blood by instilling blood from another person. This is called a *blood transfusion.*

Such procedures may be complicated, for the blood of some persons is incompatible with that of others. The plasma proteins and red cells of each individual are slightly different from those of everyone else. Therefore a transfusion of blood from one person to another can cause an immune reaction. From birth, an infant's plasma develops antibody-like proteins called **agglutinins.** These will react specifically with proteins in the red cells from other blood, and the resulting immune reaction causes clumping (agglutination) and lysis of the red cells (hemolysis). The red cell proteins (antigens) involved are called **agglutinogens.**

ABO Blood Grouping

In 1900, Karl Landsteiner, a Viennese pathologist, discovered that human blood is not the same in all individuals. He found that human blood can be classified into four well-defined groups, based on the presence or absence of two specific antigens located on the surface of the red blood cells. These agglutinogens, labeled A and B, may be present on the erythrocytes of an individual in one of three possible combinations, or they may be completely absent. When an individual has only one of the antigens, the blood group is called A or B, depending on which is present. When both antigens are present, the blood is known as AB; when both

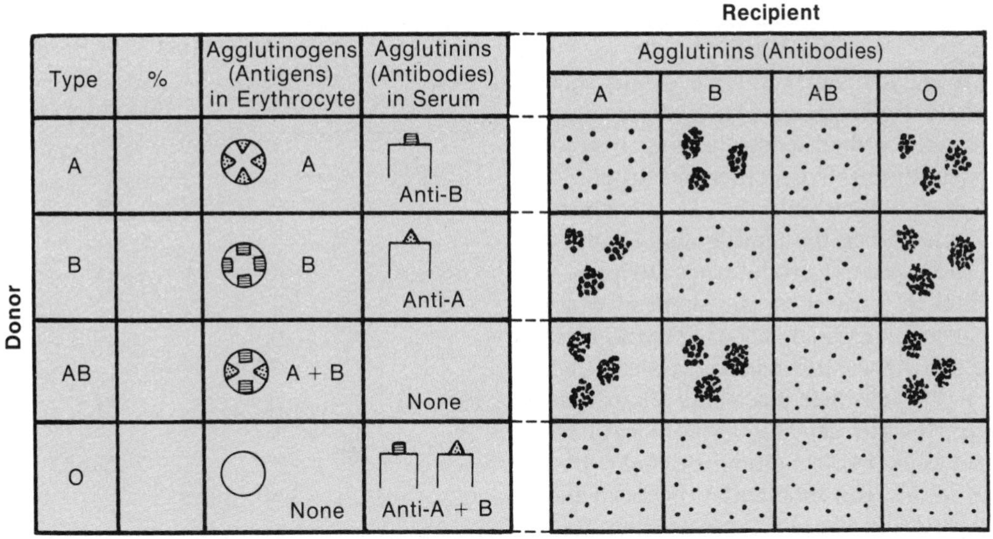

Recipient

		Agglutinogens (Antigens) in Erythrocyte	Agglutinins (Antibodies) in Serum	Agglutinins (Antibodies)			
Type	%			A	B	AB	O
A		(A) A	Anti-B				
B		(B) B	Anti-A				
AB		(A+B) A + B	None				
O		( ) None	Anti-A + B				

Figure 12-12 ABO blood group system. If a cross match of two persons is necessary, the two blood samples are mixed and observed under the microscope. If the recipient and donor are both type A, the blood cells will be evenly distributed. If, however, the recipient is type A and the donor is type B, the antibodies present in the recipient's blood will cause the donor's blood cells to clump or agglutinate, proving the two blood specimens to be incompatible.

are absent, the blood group is known as O (Figure 12-12).

Fortunately, when an agglutinogen is present in the blood, the opposite agglutinin is absent. Thus the serum of blood type A always contains the specific antibody designated as anti-B. Type B blood has an anti-A antibody; type AB blood has both antigens, and therefore has neither antibody. Type O blood has neither antigen, so it contains both anti-A and anti-B antibodies.

If an individual with type A blood should receive a transfusion of type B blood, his anti-B plasma antibodies will react with the incoming red blood cell B antigen, resulting in the clumping of the red cells of the transfused donor blood. Similarly, the A antigen of the donor's blood would react with the anti-A antibody of the patient's type B blood, causing clumping of the former's red cells. Individuals who belong to the AB group are *universal recipients*, since they do not have any antibodies in their plasma

and are able to receive any type of blood. Individuals of the type O group are called *universal donors*, since they do not have either of the antigens and therefore cannot cause agglutination in the receiver's blood. A person can usually donate blood safely to another with the same blood type. In all cases, before a transfusion is given, determination of compatibility is made (cross matching of bloods).

Rh Factor

About 85 percent of the white populace have another red blood cell protein called the Rh factor. Such individuals are said to be *Rh-positive;* the remaining 15 percent lack the protein and are *Rh-negative.* If Rh-positive blood is given by transfusion to an Rh-negative person who has previously been sensitized to the Rh factor, the blood of the recipient will produce counteracting antibodies (anti-D) that destroy the red blood cells containing the foreign Rh-positive factor.

This phenomenon can have serious consequences during pregnancy, for it can result in a blood condition fatal to the fetus or newborn child of an Rh-negative mother and Rh-positive father. If the fetus inherits the Rh-positive blood of the father, the mother may form antibodies to Rh-positive erythrocytes during or after childbirth, since there is a slight mixing of the fetal and maternal bloods (Figure 12-13). During subsequent pregnancies, these antibodies may pass from her blood into the blood of the fetus, thereby causing destruction of fetal red blood cells. This condition is called *erythroblastosis fetalis*. The infant may be born dead (stillborn); if the infant lives, replacement transfusions must be begun immediately (Table 12-2). Rh-negative blood seems to give the best results.

Erythroblastosis fetalis does not, however, invariably occur in these situations, for not all Rh-negative individuals develop anti-D antibodies in the presence of Rh-positive antigens. Furthermore, only a small percentage of Rh-negative women react in this way when pregnant with an Rh-positive fetus. Even if the mother's plasma does develop anti-D antibodies, in most cases the child will be normal. Trouble usually does not develop during the first pregnancy but rather during subsequent ones.

RhoGAM (immune globulin–human) is a commercially prepared chemical designed to suppress maternal anti-D antibody production as a result of exposure to Rh-positive fetal red cells. The principle is that passive antibody administration of RhoGAM in the proper dose to the Rh-negative mother at or soon after delivery (within 72 hours) prevents her from responding actively to the antigenic stimulus of incompatible fetal cells that entered her circulation at delivery. The protection given at the delivery of the first baby does not, however, prevent the mother from generating anti-D antibodies in subsequent pregnancies. Therefore, RhoGAM must be given immediately following each delivery.

Table 12-2   Selection of Blood for Exchange Transfusions*

Mother	Baby	Blood for Exchange
Group O	O	
	A	O
	B	
Group A	O	O
	B	
	A	A or O
	AB	
Group B	O	O
	A	
	B	B or O
	AB	
Group AB	A	A or O
	B	B or O
	AB	A, B, AB or O

*When the blood groups of the mother and baby differ, the baby's blood may not be compatible with the maternal serum because of the expected agglutinins anti-A or anti-B.

HL-A and Histocompatibility

HL-A stands for *human leukocyte antigen*. This system was first recognized in women who had given birth to several children in whom transfusion-type reactions (agglutination) occurred. However, there was no evidence of red blood cell destruction. Investigations of these reactions found that the women had antibodies to white blood cells in their plasma. These antibodies would cause agglutination of white blood cells and were therefore called *leukoagglutinins*. By studying these reactions, the scientists were able to demonstrate that only some plasma and leukocytes showed this interaction.

During pregnancy and delivery, small numbers of white cells can leak through the placenta into the mother's circulation. If the antigens on these cells are not identical to hers,

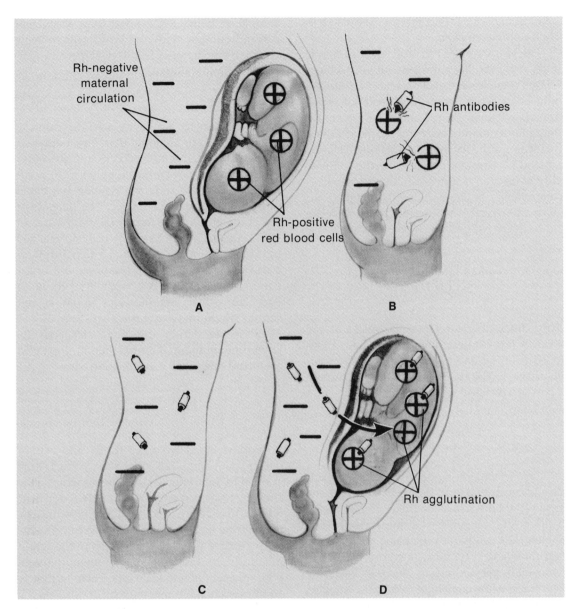

Figure 12-13 Erythroblastosis fetalis. **A,** Carriage of an Rh-positive fetus by an Rh-negative mother may not cause complications during the first pregnancy. **B,** Following delivery, however, mixing of the maternal and fetal bloods in the uterus will result in Rh antibody production (filter). **C** and **D,** In ensuing pregnancies, fetal blood cells will be destroyed.

she can produce antibodies. During each pregnancy, she would be stimulated to produce more antibodies if the same antigens were present. These antibodies are weaker than red

blood cell antibodies. Later in life, if this same woman received a transfusion containing incompatible leukocytes, her antibodies could cause leukocyte destruction and the release of chemical molecules. This reaction is called a *febrile transfusion reaction.*

The HL-A system has much greater significance, however. The antigens on white blood cells and platelets are also present on most cells in the body. When a transplant operation is performed, whether skin, kidney, or heart, the compatibility of the donor and recipient's tissue (**histocompatibility**) must be established just as in blood transfusion. Establishing the compatibility of tissues for transplantation is called *histocompatibility testing* and involves typing the donor and recipient white cells, much as donor and recipient red cells are typed.

ROUTINE BLOOD STUDIES

Many different studies may be made of the blood. A few of these have already been mentioned (clotting and bleeding time, prothrombin activity). The following are done routinely as part of a physical examination.

Hemoglobin and Hematocrit

Cellular metabolism requires an adequate supply of oxygen to assure the body sufficient energy yield and use. The hemoglobinometer (hemometer) is a device that measures the hemoglobin by comparing the blood with a standard color scale. The normal average is 15 g of hemoglobin per 100 mL of blood; as has been mentioned, a low content of iron-rich hemoglobin results in basic anemia.

The hematocrit, a measurement of the formed elements of the blood, is determined by filling a capillary tube with citrated blood (blood that has been treated with sodium citrate to prevent coagulation), packing the cells by centrifugation, and then reading the percentage of packed cells per total volume of the blood

sample. A hematocrit greater than 45 percent can indicate too many red blood cells (polycythemia); a hematocrit less than 45 percent generally indicates anemia.

Blood Cell Count

A hemocytometer is a device used for counting the blood cells in a sample. Two blood-drawing tubes are used, one for red cells and the other for white blood cells, together with an accurately ruled glass slide for viewing the blood samples under the microscope. The purpose of the gridiron rulings is to provide a fixed area in which the cells may be counted. A normal red blood cell count varies from 4.5 to 5.5 million cells per cubic millimeter; the white cell count varies from 5000 to 9000/mm³. Erythrocytosis and leukocytosis are increases in either cell count, resulting from overproduction of red blood cells and infection, respectively. Erythropenia and leukopenia reflect a decreased cell count due to disease or hemorrhage.

White Blood Cell Differential

A drop of blood is spread thinly and evenly over a glass slide. A special stain (Wright's) is applied to differentiate the colorless white cells, and the number and shape of the different white blood cells are counted and noted under the microscope. Abnormal red cells, parasitic infective stages, and abnormal shifts in the white cells indicate different types of anemia, infection, infestation, and leukemia. Such a count is an important aid in making a diagnosis.

Instruments are available (Coulter Counters, Autoanalyzers) that use a small quantity of blood and automatically perform a number of tests simultaneously, saving the technician valuable time and providing a more rapid means of obtaining the clinical results.

OUTLINE

I. Functions of blood
   A. Carries oxygen and nutrients to tissues
   B. Carries waste products away from tissues
   C. Defends against injury and disease
   D. Regulates body temperature
   E. Carries hormones to cells
   F. Maintains acid-base balance
   G. Prevents loss of blood through its clotting mechanism
   H. Maintains osmotic concentration in tissues

II. Characteristics of blood
   A. Arterial blood—scarlet red
   B. Venous blood—dark red
   C. Thicker than water (average specific gravity is 1.050)
   D. pH = 7.38
   E. Temperature = 38°C (100.4°F)
   F. Blood volume = 6 L (5 quarts) (8 percent body weight)
   G. Made up of plasma (55 percent) and formed elements (45 percent)

III. Formed elements
   A. Erythrocytes—red blood cells
      1. Biconcave, disk-shaped
      2. Produced in bone marrow
      3. Lack a nucleus
      4. Average 4.5 to 5.5 million/mm³
      5. Carry oxygen to tissues using hemoglobin
         a. Hemoglobin average is 15 g/100 mL
      6. Life span = 120 days
      7. Phagocytized by reticuloendothelial system—hemolysis
      8. Intrinsic factor, extrinsic factor (vitamin B_{12}), and iron necessary for erythropoiesis
      9. Anemias
         a. Primary—disturbance of red cell mechanism (pernicious anemia, aplastic anemia, hemolytic anemia, sickle cell anemia)
         b. Secondary—loss of blood and defective hemoglobin formation (macrocytic hyperchromic and microcytic hypochromic anemias)
      10. Polycythemia—greater than normal number of red cells
      11. Hemoglobin
         a. Carries blood gases
         b. Destruction indicated by bilirubin level
   B. Leukocytes—white blood cells
      1. Nucleated and polymorphic
      2. Granulated or agranulated
      3. Primary function—defense against infection (phagocytosis)
      4. Secondary function—antibody production
      5. Life span = about 2 weeks
      6. Formed in bone marrow and in lymphatic tissue
      7. Destroyed in the liver
      8. Inflammation
         a. Injury to tissue permits bacteria to enter the skin
         b. Bacteria release toxic substances that damage the tissue
         c. Increased blood supply to the damaged area produces reddening of the skin and increased temperature
         d. Migrating white cells ingest the bacteria and the damaged tissue cells
         e. Exuding blood plasma causes tissue edema
         f. Suppuration—formation of pus around a laceration
      9. Neoplastic blood diseases
      10. Infectious mononucleosis—possible viral etiology

11. Multiple myeloma
C. Thrombocytes—blood platelets
1. Fragments of specialized bone marrow cells
2. Formed from megakaryocytes in the bone marrow
3. Lack a nucleus
4. Active in blood-clotting mechanism
5. Thrombocytopenia—idiopathic thrombocytopenic purpura
IV. Blood plasma
A. 55 percent of total blood volume—the liquid portion approximately 92 percent water
B. Proteins
1. Serum albumin—53 percent of total plasma proteins
   a. Regulates blood volume
   b. Synthesized in liver
2. Globulin
   a. Alpha, beta, and gamma fractions
   b. Gamma fraction participates in immune reactions
3. Fibrinogen—coagulation mechanism
4. Prothrombin
   a. Synthesized with aid of vitamin K
   b. Coagulation mechanism
C. Plasma electrolytes
1. Major plasma ions—sodium, chloride, potassium, calcium, phosphate, sulfate, magnesium
2. Trace quantities—iodide, copper, iron
D. Nutrients
1. Glucose—primary carbohydrate fraction
2. Lipids (neutral fats, cholesterol, and phospholipids)—formed in liver cells
E. Waste products—lactic acid, uric acid, urea, creatine, creatinine, ammonium salts
F. Gases

1. Oxygen, carbon dioxide, nitrogen
2. Found in solution
G. Regulatory and protective proteins
1. Hormones, antibodies, enzymes, vitamins
2. Found in variable concentrations
H. Blood buffer system—prevents major shifts in acid-base balance
I. Clotting mechanism
1. Prevents fatal loss of blood
2. Clotting sequence
3. Thrombus—stationary clot
4. Embolus—clot brought by bloodstream from another site
5. Hemostasis
6. Hemorrhagic disorders
   a. Hemophilia—absence of antihemophilic factor (AHF)
   b. Disseminated intravascular clotting
   c. Purpura
   d. Excessive menstrual bleeding
7. Anticoagulants that prevent excess clotting of blood—heparin and dicumarol
J. Diagnosis of coagulation abnormalities
1. Bleeding time
2. Clot retraction
3. Prothrombin time (PT)
4. Partial prothrombin time (PTT)
5. Venous clotting time (VCT)
6. Plasma fibrinogen
V. Immunity
A. Body's defense against infection and disease
B. Antigen—foreign protein that stimulates immune reaction
C. Antibody—protein produced to react with a specific antigen
D. Antigen-antibody reaction—comparable to lock and key
E. Classes of immunity
1. B lymphocyte—humoral immunity
2. T lymphocyte—cell-mediated immunity

F. Types of antigen-antibody reactions
  1. Agglutination
  2. Precipitation
  3. Lysis
  4. Opsonization
G. Allergy—weak immunity due to proteinlike substances in air, food, and other media that fail to induce sufficient antibody formation
  1. Redness and swelling (hives)—manifestations of allergy
  2. Antihistamine drugs and epinephrine—used to relieve hay fever, asthma, and anaphylactic shock
VI. Blood typing and transfusions
  A. ABO blood grouping—based on presence or absence of two specific antigens on surface of erythrocytes
    1. Group A—only A antigen is present
    2. Group B—only B antigen is present
    3. Group AB—both antigens are present (universal recipient)
    4. Group O—both antigens are absent (universal donor)
  B. Blood type compatibility
    1. Necessary to prevent red cell hemolysis (clumping) during transfusions
    2. Cross matching
  C. Rh factor
    1. Additional red cell protein possessed by 85 percent of individuals (Rh-positive)
    2. Individuals lacking Rh factor—Rh-negative
    3. Serious complications may arise if Rh-negative mother gives birth to Rh-positive child
  D. HL-A and histocompatibility
VII. Routine blood studies
  A. Hemoglobin and hematocrit—used to detect anemia and polycythemia
  B. Blood cell count—number of red and white blood cells; important in diagnosing hemorrhage, infections, etc.
  C. White blood cell differential—shape and structure of white and red cells can indicate leukemias, infections, and infestations

STUDY QUESTIONS AND PROBLEMS

1. What are the functions of blood?
2. What are the functions of erythrocytes? Leukocytes? Thrombocytes?
3. What is the advantage of transfusing whole blood rather than simply plasma?
4. When might a physician order a transfusion of "packed cells—plasma-free" rather than of whole blood?
5. What is an "H & H"? Why would physicians order it?
6. What are the different types of leukocytes? Describe their functions and indicate which of them has a migratory capability.
7. What is the reticuloendothelial system? What is its purpose?
8. Differentiate between plasma and serum.
9. Mrs. Jones had a baby and hemorrhaged. The diagnosis was afibrinogenemia. Is this a cause for bleeding? Describe it.
10. Describe agglutination. Differentiate between coagulation and hemolysis.
11. Why don't we form clots in our blood system?
12. What is meant by the ABO blood system? List the four blood types and their frequency of occurrence.
13. Describe the "lock-and-key" theory. What does it have to do with immunity?
14. What is cross matching?
15. What is an Rh baby? Describe the probable blood type situation leading up to the birth of such a baby.

the alpha amino group of aspartic acid to glutamic acid, which results in the synthesis of oxaloacetic acid following a myocardial infarction. Levels of this enzyme begin to rise in about 8 hours, peak in 24 to 48 hours (Figure 13-11), and return to normal in 4 to 8 days. Damage to other tissues besides the heart may elevate this enzyme and cause a false diagnosis of myocardial infarction.

NERVOUS CONTROL OF THE HEART

The heart is innervated by the autonomic nervous system (Figure 13-12). The vagus nerve, also known as the tenth cranial nerve, contributes fibers to the parasympathetic portion of the autonomic system. The vagus is a mixed nerve containing both motor and sensory components. The motor parasympathetic neurons innervate the SA and AV nodes. Although the heartbeat is started by the SA node pacemaker, the frequency of the beat may vary considerably.

Stimulation of the vagal neurons results in marked slowing of the heart and decreased force of contraction. The vagus action is inhibitory, since it secretes acetylcholine at the nerve endings; this decreases the rate and rhythm at the node. It also decreases the excitability of the AV junctional fibers between the muscles of the atria and the Purkinje system, slowing transmission of impulses.

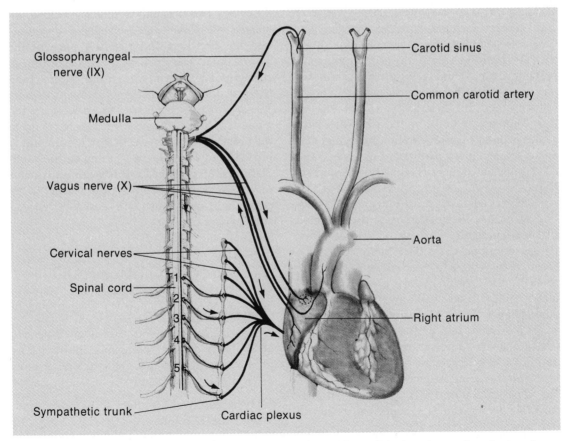

Figure 13-12 Nervous regulation of heart action. Arrows indicate direction of nerve impulse travel.

Sympathetic neurons arise from the first five thoracic segments of the spinal cord and then pass to the cervical ganglia. They innervate the SA and AV nodes and also the myocardium of the atria and ventricles.

Stimulation of the sympathetic (accelerator) nerves speeds the heart and increases the force of contraction. In general, stimulation of the sympathetic nerves increases the activity of the heart through secretion of epinephrine, thereby increasing the effectiveness of the heart as a pump.

Cardiac Reflexes

The dual control of the heart just described is coordinated by centers located in the medulla oblongata. The cardioinhibitor center gives rise to the parasympathetic neurons that travel along the vagus nerve to the heart. Activation of this center decreases cardiac action. The cardioaccelerator center sends neurons down the spinal cord to activate the accelerator (sympathetic) neurons, producing an increase in cardiac activity. The two centers also send neurons to each other, and these have an inhibitory effect. Therefore activity of the cardioinhibitor center depresses the accelerator center, and vice versa. These centers, then, coordinate vagal and accelerator function, thus regulating cardiac activity.

Reflexes adjusting heart rate may be classified according to whether they are initiated by pressoreceptors or by chemoreceptors. **Pressoreceptors,** also called *baroreceptors,* are sensitive to changes in blood pressure. They are found at the division of each common carotid artery, at the carotid sinus, and at the arch of the aorta (Figure 13-12). The carotid and aortic sinuses are innervated by the glossopharyngeal (IX) and vagus (X) nerves, respectively. An elevation in blood pressure causes the pressoreceptors to fire more frequently, which activates the cardioinhibitor center. As a result, the cardioaccelerator center is depressed, resulting in a slowing of the heart and a weakening of the force of contraction. Cardiac output falls and therefore so does the arterial blood pressure.

Chemoreceptors, which were discussed in Chapter 11, are sensitive to oxygen and carbon dioxide tension of the blood. These receptors are found in the carotid and aortic bodies and are sensitive to lack of oxygen (hypoxemia). Impulses from these receptors are conveyed to the cardiac center, and the heart rate is accelerated, increasing the cardiac output, so that more blood goes to the tissue cells. Chemoreceptors are also stimulated by an increase in carbon dioxide (hypercapnia).

The stroke volume is the difference between the volume of blood in the ventricle before systole (the end-diastolic volume) and the volume of blood in the ventricle after systole (the end-systolic volume). Any factor that increases the amount of blood in the ventricle before systole generally causes an increase in the stroke volume. Experiments with animals in which the end-diastolic volume is caused to increase have shown that the walls of the ventricles are stretched and that the force of the contractions also increases. The increase in strength of contractions as a result of stretching the walls of the ventricles is known as *Frank-Starling's law of the heart.* An increase in the amount of blood being returned by the veins to the heart will cause an increase in the amount of blood in the ventricles. According to the Frank-Starling law of the heart, this factor contributes to increasing stroke volume. Conversely, a decrease in venous return will cause a decrease in stroke volume.

CONGENITAL HEART DISEASE

Congenital lesions account for about 2 percent of all heart disease. Congenital heart disease produces symptoms and signs by one or more of the following mechanisms:

- Hypertrophy of the affected chamber, resulting in heart failure

- Diverting blood from the left atrium or ventricle to the right atrium or ventricle, increasing the work of the right ventricle and the amount of blood flow to the lungs

- Shunting of venous blood from the right atrium or ventricle into the aorta, left atrium, or left ventricle bypassing the pulmonary circulation, causing arterial unsaturation (cyanosis)

Septal Defects

Most congenital heart abnormalities are in the septum, which separates the right and left sides of the heart. The most important feature is the intermixing of the blood from the systemic and pulmonary circulations as the result of a shunt.

Atrial septal defects may be serious or trivial. An example of a trivial septal defect is the *patent (open) foramen ovale*. The opening of the septum between the two atria, which is normally present at birth, fails to close and allows blood from the right side of the heart to enter the left side directly (Figure 13-13). Mixing of venous and arterial blood produces a decreased oxygen content in the circulation resulting in a blue-gray pallor to the skin (a blue baby).

A septal defect due to a developmental irregularity is another situation. In this instance, the blood passes readily from the left to right side of the heart where the pressure is lower, so that both the right atrium and right ventricle become dilated or hypertrophied (oversized). The main concern here is not the oxygen content of the blood but the increased workload to the right ventricle. It could result in heart failure if prolonged.

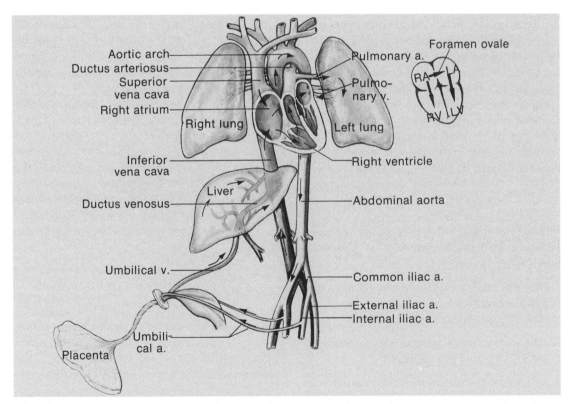

Figure 13-13 Plan of circulation in a mature fetus. Arrows indicate the direction of blood flow. The inset is a diagram of the blood flow in the fetal heart. Note the blood flow from the right atrium into both the right ventricle and left atrium through the foramen ovale. RA, right atrium; LA, left atrium; RV, right ventricle; LV, left ventricle.

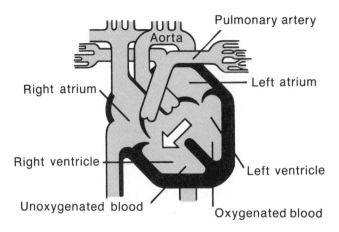

Figure 13-14 Septal defects. A large septal defect allows oxygenated blood to flow into the right ventricle, increasing the pressure on the right side of the heart while decreasing the blood pressure on the left side of the heart. Eventual reverse flow from the right to the left ventricle results in unoxygenated blood entering the aorta, and cyanosis develops.

Ventricular septal defects are persistent openings in the upper interventricular septum that allow blood to pass from the high-pressure left ventricle into the low-pressure right ventricle (Figure 13-14). The opening is usually small, cyanosis does not generally occur, and unless bacterial infection (endocarditis) occurs, the defect can be surgically repaired. With large shunts, both the right and left ventricle may become strained and fail.

Tetrology of Fallot

Tetrology of Fallot is the most serious congenital lesion of the heart (Figure 13-15). The condition is fourfold (tetra):

1. The pulmonary valve is markedly narrowed (pulmonary stenosis) so that the blood is unable to pass adequately to the lungs.

2. The aorta is out of position (**dextroposition**), overriding the right ventricle.

3. Closure of the interventricular septum is incomplete (septal defect).

4. As a result of pulmonary stenosis, the right ventricle is enlarged.

In addition to the cyanosis, red blood cell production is increased due to the inadequate oxygen supply. The fingers may become clubbed, and the fingernails are curled reportedly due to a poor oxygen supply.

Surgery can be successfully performed. An **extracorporeal** (outside the body) circulation is created; the left subclavian artery is joined to the pulmonary artery to bypass the obstructed opening of the pulmonary artery. Now the blood reaches the lungs from the left ventricle instead of the right ventricle. Repair of the ventricular septal defect will return blood to its normal flow from the right ventricle to the lungs.

Patent Ductus Arteriosus

In patent ductus arteriosus the ductus arteriosus fails to close and persists as a shunt connecting the left pulmonary artery and aorta (Figure 13-16). In this condition, the blood flows from the aorta into the pulmonary artery where the pressure is lower. The condition is compatible with life, but danger of bacterial (streptococcal) infection is high. The open ductus can be easily corrected surgically by dividing the connection between the pulmonary artery and the aorta.

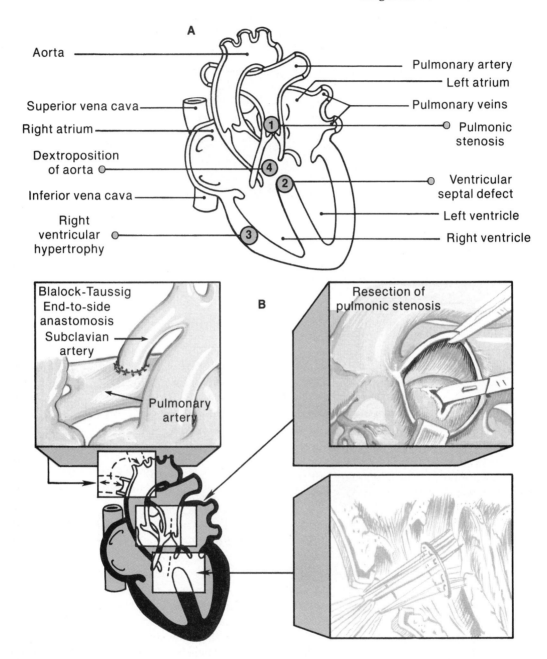

Figure 13-15 Tetralogy of Fallot. **A,** A pulmonary stenosis (1) and large ventricular septal defect (2) result in a hypertrophied right ventricle (3) and displaced (dextroposition) aorta (4). **B,** Closure of the ventricular septal defect, resection of the pulmonary stenosis, and end-to-end anastomosis of the right subclavian artery to the right pulmonary artery have proven to be an effective treatment.

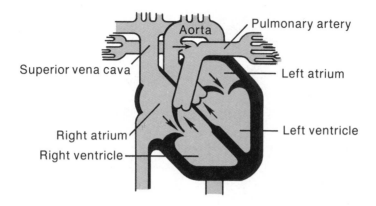

Figure 13-16 Patent ductus arteriosus. Failure of the ductus arteriosus to close results in an increased aortic blood pressure and a decreased pulmonary ar-tery blood pressure, causing the blood to flow back-ward into the pulmonary artery.

VALVULAR DISEASES

The two sets of valves responsible for maintaining correct blood flow within and out of the heart are the AV and the semilunar valves. For purposes of convenience, the two valves on the left side of the heart will be discussed in this section; they are the left bicuspid AV, or mitral, valve and the aortic semilunary valve.

Mitral Stenosis

Most cases of mitral stenosis are caused by rheumatic heart disease. Three out of four patients under 45 years of age with mitral stenosis are women. When the valve narrows to a fraction of its normal opening, the left atrial pressure must increase to maintain normal flow across the valve, resulting in a significant pressure difference between the left atrium and left ventricle during diastole. This increased atrial pressure also results in increased pulmonary venous capillary pressure, causing the patient to experience dyspnea and fatigue.

One of the more severe side effects is the possible formation of thrombi at the valve.

They in turn may travel through the vessels as emboli. The general effects of mitral stenosis are cyanosis and venous congestion. Mitral stenosis can now be treated surgically (valvotomy) or by replacement with an artificial valve (Figure 13-17).

Mitral Insufficiency

Mitral insufficiency or incompetence is the least well defined of the major valvular lesions. It is caused by sclerosis of the cusps, resulting in an inability to close completely.

During ventricular systole, the double cusps do not close normally and blood is forced back into the atrium. The clinical effect is an overworked and hypertrophied ventricle. Mitral insufficiency may lead into atrial fibrillation and eventual ventricular failure. Surgical removal and replacement with a prosthetic valve are indicated in most cases.

Aortic Stenosis

Aortic stenosis usually occurs in men over 50 years of age. In this situation, the semilunar cusps adhere together, forming a hard calcified

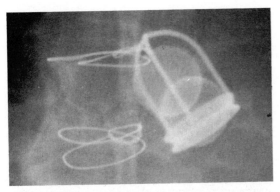

A

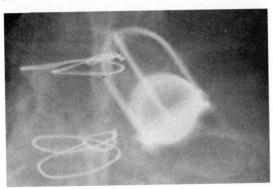

B

Figure 13-17 Starr-Edwards ball prosthesis replacement of the mitral valve. **A,** Open. **B,** Closed. (Courtesy of Salem Hospital, Salem, Massachusetts.)

rigid mass. The three most common causes are rheumatic heart disease, congenital disorder, or atherosclerosis. Because the left ventricle must pump through this valve into the aorta, the left ventricle greatly enlarges due to the increased pressure and blood volume. The most characteristic physical sign is a rough, rasping systolic murmur at the aortic area. As in the mitral valve, surgical repair or replacement is indicated.

Aortic Insufficiency

Aortic insufficiency or incompetence may be due to endocarditis (caused by rheumatic or bacterial conditions) of the valve cusps or to dilation (syphilitic) of the aorta and aortic ring. In this condition the aortic valve does not close properly, and during diastole, blood flows back into the left ventricle from the aorta. The left ventricle hypertrophies, and the regurgitant flow of blood from the aorta into the ventricle through the incompetent valve during diastole (ventricular) causes a loud diastolic murmur, which is a definite diagnostic sign. Prosthetic replacement is the indicated treatment for mitral insufficiency.

OTHER MAJOR HEART DISEASES

Congestive Heart Failure

Congestive heart failure is a common signal to myocardial infarction and occurs in 50 to 60 percent of all patients with organic cardiovascular disease. The loss of the left ventricular contraction diminishes the ability of the left ventricle to empty during systole. The right or left ventricle alone may fail initially, but combined failure is the rule in most cases. The pressure elevation is reflected backward into the left atrium and pulmonary veins and may cause pulmonary congestion and eventual pulmonary edema.

The most common causes of cardiac insufficiency are hypertension, coronary atherosclerosis, rheumatic heart disease, and bacterial endocarditis (affecting the valves). A person may experience shortness of breath, dyspnea, and cyanosis on physical exertion. The typical appearance of a patient with congestive heart failure includes **edema** (swelling caused by excessive fluid) of the extremities. Treatment involves the rapid decrease of blood volume to ease the strain on the heart and decrease the edema in the lungs and legs.

Rheumatic Heart Disease

Acute rheumatic heart disease develops from a throat or ear infection caused by group A beta-hemolytic streptococcus bacteria. The child will first show signs of a sore throat (pharyngitis), then painful joints, fever, and a rash. A few weeks later the infection develops into rheumatic fever. Rheumatic fever is an autoimmune disease resulting from a reaction between the child's own antibodies against streptococcal antigens and similar cardiac valve antigens.

The lesion of rheumatic fever is *carditis*. All three layers of the heart—endocardium, myocardium, and epicardium—can be inflamed (pancarditis) resulting in inflamed mitral valves. Blood platelets and fibrin from the blood form nodules called *vegetations*, which form a row along the margin of the cusps. The inflamed cusps become thickened and tend to adhere to one another. As the inflammation lessens, fibrous tissue replaces the inflamed tissue and tends to contract, altering the valve opening.

The effect of these events is a stenotic or incompetent valve. Serious implications of mitral stenosis and incompetency have already been discussed. Recurrence of rheumatic heart disease, with further damage to the heart, can develop years later if the individual is reinfected with streptococci. This is referred to as *chronic rheumatic heart disease*.

Bacterial Endocarditis

Bacterial endocarditis is a bacterial infection of the endocardium usually following or accompanying rheumatic heart disease, calcific valvular or congenital heart disease. The infective agents are nonhemolytic streptococci, especially *Streptococcus viridans* and *S. fecalis*, staphylococcus, and *Hemophilus influenzae*.

Bacteria lodge on the endocardium of valves (aortic and mitral) and multiply. Fibrin and platelet thrombi form vegetations similar to those found in rheumatic heart disease but much larger in size. These nodules become friable (crumble easily) and break loose, carrying bacteria throughout the body as infective emboli producing mycotic aneurysms and causing other infections (such as embolic nephritis).

A diagnosis of endocarditis is made by detecting fever, heart murmur, and bacterial growth in blood cultures. If properly treated, prognosis for this infection is good. However, acute infection produces severe destruction of

the heart valves, and heart failure is not uncommon. Treatment involves an extended use of antibiotics with possible surgical repair or valvular replacement.

CLINICAL IMPLICATIONS

Anticoagulant Therapy

Anticoagulant drugs are commonly used in the treatment of coronary thrombosis with myocardial infarction. In all cases, the use of anticoagulant drugs must be controlled by correctly chosen and carefully performed laboratory tests.

Two types of anticoagulant agents are in general use: heparin and sodium warfarin. Heparin prevents extension of existing blood clots and formation of new ones. Sodium warfarin (Coumadin) depresses the production of prothrombin and certain other blood factors.

Chemotherapy

One of the oldest chemotherapeutic agents, and still a valuable drug for many heart patients, is digitalis, which acts as a regulator in certain types of heart disease. It stimulates the myocardium directly, causing an increase in contractility and excitability. Other members of the digitalis family, digoxin and digitoxin, are also heart stimulants. They provide action of short and prolonged duration, respectively.

Nitroglycerin is used to dilate the coronary arteries, thus reducing blood pressure and cardiac work.

Heart Catheterization

Straightforward radiography can reveal the existence of a cardiac defect, but it can only hint at its severity. To answer questions concerning abnormal heart action, a measuring instrument, a catheter, is fed into the working heart itself. It is inserted through a vein in the arm or neck, fed (steered) down the vein, and then, with the aid

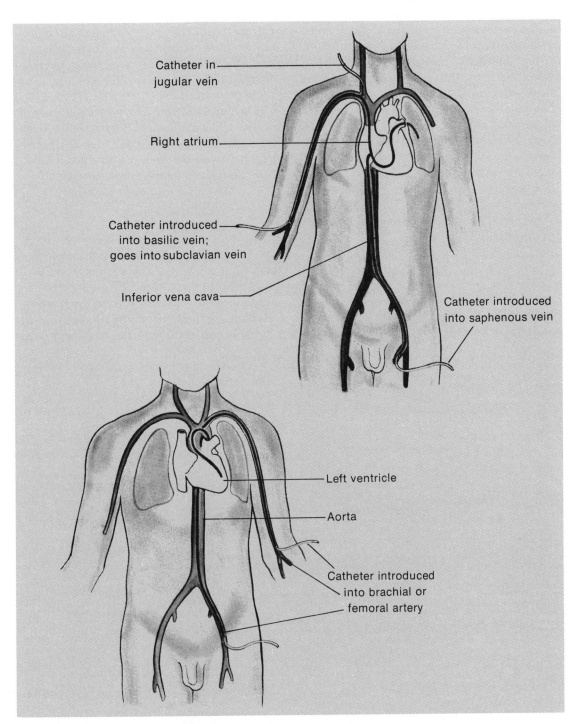

Figure 13-18 Heart catheterization. Insertion of a catheter into the right heart via the subclavian or saphenous veins or into the left heart via the brachial or femoral arteries to detect structural and other defects.

of a fluoroscopic monitor, rotated until it enters the heart. Its bent tip will lead it in different forward directions so that the desired route may be selected without the catheter being pushed into other veins. The instrument can be fed into the right atrium, down through the tricuspid valve, and into the ventricle below, where blood samples can be withdrawn for analysis. Manometers can then be used to record instantaneous pressure changes. Harmless dyes that are opaque to x-rays provide a record of the catheter's route, revealing septal defects, faulty valves, and abnormalities in the great veins and arteries (Figure 13-18).

Pacemakers

Electric battery-operated pacemakers that supply impulses to regulate the heartbeat have been implanted under the skin in the right up-

per chest area. Electrode catheters attached to the artificial pacemakers are then passed beneath the skin to the myocardium, where they are returned to the surface through a previously made incision in the chest wall (Figure 13-19). The batteries last from 3 to 5 years and are easily replaced and maintained. Many people whose hearts cannot beat effectively because their biological pacemaker does not function properly have been saved by this simple device.

Artificial Heart

The heart–lung machine has made it possible to perform many surgical operations on the heart that previously could not be undertaken. The machine siphons off blood from the large vessels entering the heart on the right side so that no blood passes through the heart and lungs.

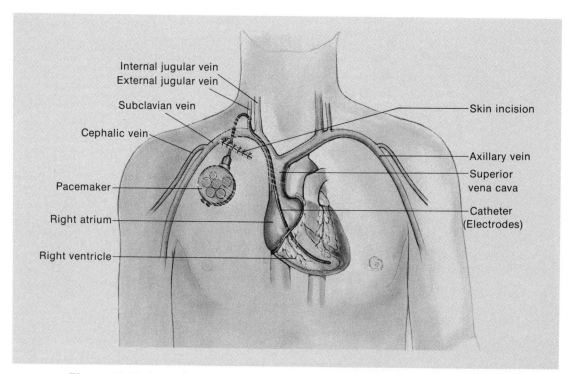

Figure 13-19 Pacemaker implantation, a current treatment for complete heart block. The electronic pacemaker replaces a nonfunctioning SA node.

While passing through the machine, the blood is oxygenated and carbon dioxide is removed. It is then returned to the general circulation through one of the large arteries. The blood is also "defoamed" to ensure that all air bubbles are removed, since this air could obstruct the blood vessels and be fatal to the patient.

A diseased valve may be repaired or replaced by an artificial valve. More spectacular has been the transplantation of a heart from the body of a person who has recently died to replace the diseased heart of a cardiac patient. Artificial hearts or parts of hearts designed to assist the ventricles in their pumping function have not proved as successful as hoped; however, research continues, and the outlook is optimistic.

OUTLINE

I. The heart
  A. Hollow muscular organ
  B. Double pump
II. Structure of the heart
  A. Muscle and membranes
    1. Myocardium
      a. Heart muscle
      b. Fibers joined by side branches
    2. Pericardium
      a. Two-layered membrane enclosing heart
      b. Outer layer—fibrous tissue
      c. Inner layer—serous membrane (epicardium)
    3. Endocardium
      a. Inner lining of heart
      b. Serous membrane
      c. Forms valves of heart
  B. Cavities
    1. Two upper atria and two lower ventricles
      a. More muscular on left side owing to greater workload
  C. Valves
    1. Right three-cusped AV valve and a left two-cusped AV valve
    2. Two semilunar valves at the origins of the aortic and pulmonary arteries
III. Cardiac cycle
  A. Action of heart
    1. Contraction of myocardium—systole

    2. Relaxation—diastole
    3. Sequence of events
      a. Blood from venae cavae and pulmonary veins enters both atria simultaneously
      b. Increased atrial pressure forces AV valves open, and blood enters ventricles by gravity flow
      c. Ventricular contraction forces blood upward, closing AV valves, opening semilunar valves, and discharging blood to arteries
      d. Relaxation of ventricles creates a backflow from arteries, closing the semilunar valves
      e. AV valves reopen, and next cycle begins
  B. Heart sounds
    1. Stethoscope
      a. Ventricular systole—closing of AV valves ("lubb")
      b. Ventricular diastole—closing of semilunar valves ("dupp")
      c. Murmurs—leakage through valves
      d. Mitral stenosis—congestion
  C. Heartbeat
    1. Depolarization spreads from specialized neuromuscular tissue
      a. Pacemaker in right atrium—SA node

b. Impulses carried by atrial muscle to AV node at ventricular junction

c. Bundle of His conducts impulses down ventricular septum to apex and up into muscle

d. Purkinje fibers attach to heart muscles directly and innervate them

2. Cardiac rhythms

a. SA node—70 to 74 beats/min

b. AV node—45 to 65 beats/min

3. Abnormal conduction

a. Heart block

b. Tachycardia—up to 300 beats/min

c. Bradycardia—lower than 70 beats/min

d. Fibrillation—complete breakdown of rhythm and impulse propagation

4. Electrocardiogram (ECG)

a. Records electrical events of heart

b. Various leads from different positions on chest

c. Normal ECG

IV. Coronary circulation

A. Arteries—right and left coronary arteries supply myocardium with blood

B. Veins—coronary sinus collects blood received from cardiac veins and returns it to right atrium

V. Nervous control of the heart

A. Heartbeat starts itself (pacemaker) but is further controlled by autonomic nervous system

1. Vagus nerve—inhibits

2. Sympathetic nerves—accelerate

B. Cardiac reflexes

1. Medulla oblongata

a. Cardioinhibitor center—parasympathetic (inhibits)

b. Cardioaccelerator center—sympathetic (stimulates)

2. Localized centers

a. Pressoreceptors—sensitive to changes in blood pressure

b. Chemoreceptors—sensitive to oxygen and carbon dioxide tension of blood

c. Counteractivity acts to regulate heart activity

d. Frank-Starling's law

VI. Congenital heart disease

A. Septal defects

1. Atrial

2. Ventricular

B. Tetrology of Fallot

C. Patent ductus arteriosus

VII. Valvular diseases

A. Mitral stenosis

B. Mitral insufficiency

C. Aortic stenosis

D. Aortic insufficiency

VIII. Other major heart diseases

A. Congestive heart failure

B. Rheumatic heart disease

C. Bacterial endocarditis

IX. Clinical implications

A. Anticoagulant therapy

1. Heparin—antithrombin

2. Coumadin—antithrombin

B. Chemotherapy

1. Digitalis family—stimulants

2. Nitroglycerin—dilator

C. Heart catheterization

1. Hollow, narrow catheter

2. Accompanying fluoroscopy

3. Venous entrance to heart

4. Blood samples analyzed

5. Pressure gradients measured

6. Radiographs detect abnormalities

D. Pacemakers

1. Electronic devices

2. Battery inserted near hip

E. Artificial heart

1. Heart–lung machine

2. Exchange of gases and removal of air bubbles

3. Effective in heart transplants and other forms of cardiac surgery

STUDY QUESTIONS AND PROBLEMS

1. Name and describe the four chambers of the heart, and discuss their roles in the cardiac cycle.
2. Describe the two sets of valves of the heart, and justify their structure and position.
3. Describe the valves of the heart, and explain their relation to the heart sounds.
4. Explain systole and diastole, and relate them to the cardiac cycle.
5. What does the pacemaker have to do with the excitation-conduction pathway?
6. Discuss the principle of the ECG and its use in diagnosis.
7. Why are so many leads used in taking an ECG?
8. What does a thrombus have to do with coronary occlusion?
9. How is blood pressure modified?
10. Where are the pressoreceptors found?
11. What are chemoreceptors and to what are they sensitive?
12. Describe the autonomic nervous system's antagonistic control of the heart muscle.
13. Why does edema occur when an individual has a congestive heart attack?
14. What part does infection play in heart disease? Give examples.
15. Of what value is heart catheterization, and how is this procedure carried out?
16. Of what value to heart patients are anticoagulants and antibiotics?

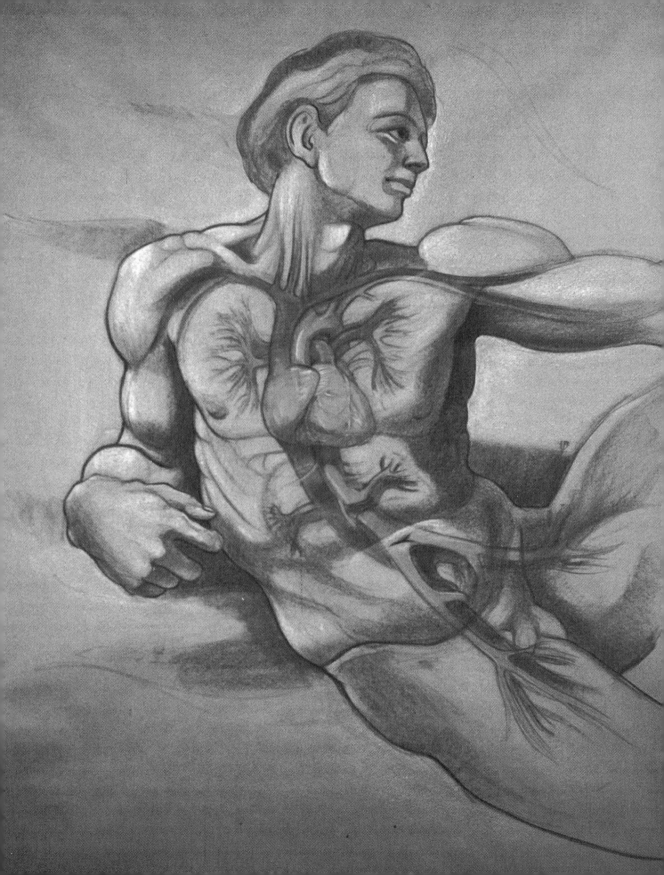

14

Circulation

LEARNING OBJECTIVES

- Identify the structural differences between an artery, vein, and capillary

- Identify the major arteries and veins of the systemic circulation

- Identify the major pressure and pulse points of the body

- Detail the formation of an arterial antheroma

- List the theories for etiology of arteriosclerosis

- Explain the significance of the HDL/LDL level in the blood

- Identify the general symptoms and causes of hypertension and arteriosclerosis

- Distinguish between primary and secondary hypertension

- Describe a varicose vein and list the common areas involved

- Differentiate between pulmonary and systemic circulation

- Describe the anatomical differences between the hepatic portal and fetal circulatory systems

- Explain the effects of vessel length, vessel diameter, and blood viscosity on peripheral resistance

- Explain the principle of a sphygmomanometer

- Describe the dynamics of capillary exchange

- Explain how exercise helps maintain a stable circulation

- Explain shock

IMPORTANT TERMS

aneurysm	artery	hemodynamics	thrombophlebitis
antheroma	atherosclerosis	hemorrhage	ultrafiltration
aorta	capillary	hypertension	varicose veins
apoplexy	cerebrovascular	oncotic pressure	vascular
arteriography	accident	pulse	vein
arteriole	coarctation	shock	venule
arteriosclerosis	cyanosis	sphygmomanometer	

Stores of oxygen and nutrients in the internal environment would be rapidly used up if they were not replenished by the turnover of the interstitial fluid, accomplished by the circulation of blood. This circulation permits the exchange of fluid and dissolved materials in the capillaries. The blood pressure provides the driving force for the **ultrafiltration** (filtration under force through minute pores) of fluid into the interstitial spaces, and the pressure of the plasma proteins causes a similar return of fluid at the distal ends of the capillaries. For this exchange to maintain the composition of the internal environment effectively, not only must the plasma composition be well regulated by the internal organs, but also the plasma must be delivered to the tissues in sufficient volume to meet their oxygen and nutrient requirements—and at sufficient pressure to ensure the all-important ultrafiltration.

GENERAL CIRCULATION

The general circulation is shown diagrammatically in Figure 14-1. As described in the preceding chapter, the heart is a muscular organ that can be thought of as two separate parallel pumps. The right heart receives blood from the veins draining the tissues and pumps it through the pulmonary arteries to the lungs. Here it passes through capillaries in which it is oxygenated and its carbon dioxide removed. The pulmonary veins return the blood to the left heart, which pumps it out again through the aorta, from which arise the various systemic arteries that distribute blood to the other tissues.

Thus there are two parallel circulations: the pulmonary circulation for gas exchange in the lungs and the systemic circulation for maintaining a constant internal environment in the other tissues. Note that all blood goes through both systems.

THE BLOOD VESSELS

Arteries carry blood away from the heart to the tissues; **veins** return blood from the tissues to the heart. The large arteries nearest the heart divide into smaller vessels. These, in turn, divide into still smaller ones, and so on. The smallest arteries are called **arterioles.** These lead into vessels of narrow caliber, the **capillaries.** The capillaries branch and form a network where the exchange of materials between blood and tissue fluid takes place. After passing through the capillary network, the blood is collected into the smallest veins, the **venules.** The venules unite to form small veins; these combine to make ever larger vessels until eventually the great veins that enter the heart are reached.

Arteries

A typical artery in cross section consists of three layers (Figure 14-2):

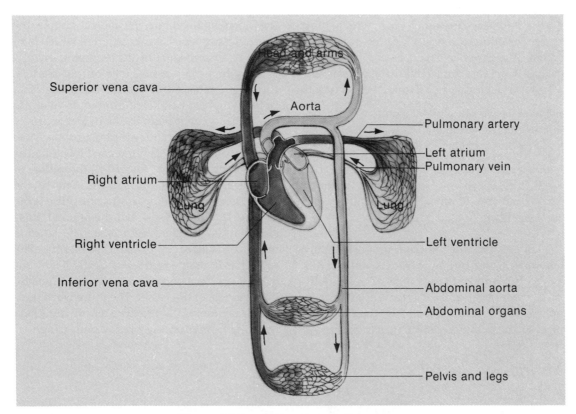

Figure 14-1 Diagram of general circulation.

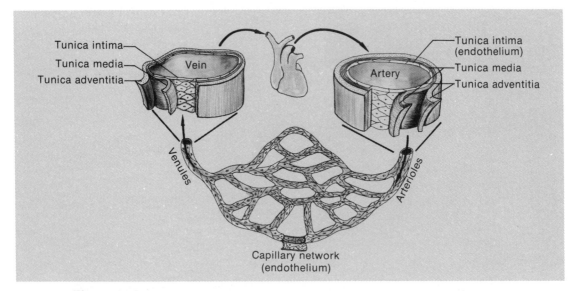

Figure 14-2 Blood vessels. Diagrammatic sketch showing the single-cell endothelium of all the vessels and the layered muscular coats of arteries and veins.

- *Tunica intima*—the inner coat, which itself consists of three layers: a layer of endothelial cells lining the arterial passage (lumen), a layer of delicate connective tissue, and an elastic layer made up of a network of elastic fibers

- *Tunica media*—the middle coat, which consists mainly of smooth muscle fibers with elastic and collagenous tissue

- *Tunica externa or adventitia*—the external coat, which is composed of loose connective tissue with bundles of smooth muscle fibers and elastic tissue

The smaller arteries have proportionately less elastic tissue and more smooth muscle; this is especially true of the arterioles. The elasticity of the large arteries is of great importance in maintaining a steadier blood flow than that initiated by the pulsating pumping of the heart. As the blood is forced into the aorta at systole, the great pressure stretches the arteries, which temporarily absorb some of the pumping energy. Then, when the pressure falls during diastole, the elastic rebound of the arterial walls compresses the blood and continues to drive it into the tissues, even though the heart is relaxed.

The distention of the arteries during systole extends to the periphery and can be detected as a throbbing sensation, known as the **pulse,** in several arteries near the surface. One of the most useful areas for measuring the pulse is the radial artery at the wrist, but the femoral pulse also may be palpated with ease. The latter is more convenient in examining infants. The pulse has the same frequency as the heart, but its propagation along an artery is determined by the structure of the vessel. The pulse wave travels in a manner similar to the wave of vibrations resulting from plucking a taut violin string.

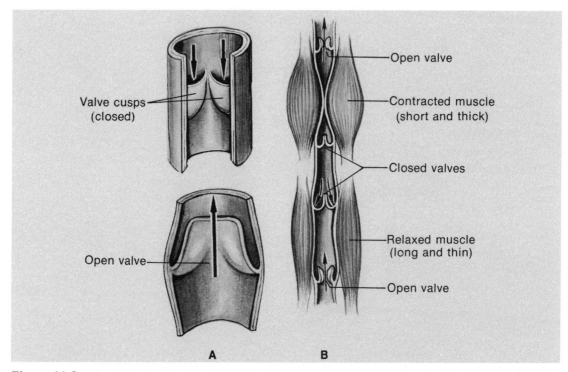

A   **B**

Figure 14-3 Valve action in the veins. **A,** Section of a vein, showing an open and a closed valve. **B,** Action of a contracted muscle forces the blood in a vein up toward the heart. This squeezing or "milking" action moves blood along the veins back toward the heart.

Veins

The veins have the same three layers as the arteries but differ in having a much thinner tunica media (see Figure 14-2). They have little elastic tissue or muscle and, therefore, when dissected, appear collapsed.

The unique feature of the veins is the presence of valves made of loose pockets of endothelium, which permit the blood to flow in one direction only—toward the heart (Figure 14-3). Veins have a much larger total diameter than arteries, and the blood moves back to the heart rather sluggishly compared with the surging force of the arterial circulation. The return of venous blood is helped by the "milking" action of muscular contractions (Figure 14-3). When a vein is compressed by movement of a limb, its valves allow the blood to be squeezed only in the proper direction.

Capillaries

The walls of the capillaries consist of one layer of endothelial cells continuous with the layer that lines the arteries, veins, and heart (see Figure 14-2). The capillaries form an interlacing network of variable size.

As noted, it is in the capillary beds that the main work of blood is done—the exchange of materials between cells and blood. A molecule diffuses through the capillary wall into the tissue fluid and from the tissue fluid into the cell, or in the reverse direction as the case may be (Figure 14-4). In addition to diffusion, some of the fluid part of the blood filtrates through the capillary walls, especially at the arterioles where the blood pressure is high, so that the blood continually contributes to the tissue fluid.

Capillaries in glands supply the substances needed for secretion; in the ductless glands they

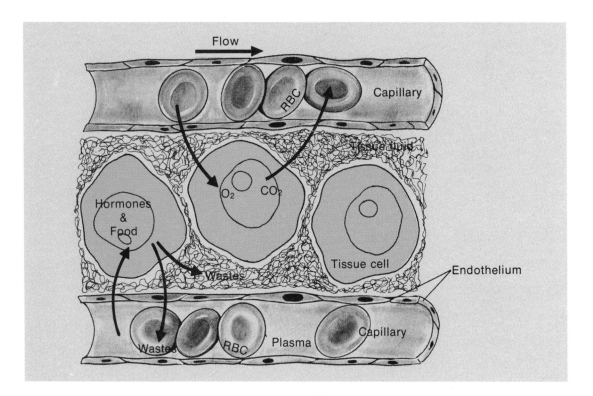

Figure 14-4 Diagram illustrating some of the physiologic exchanges between capillaries and body cells. RBC, red blood cell.

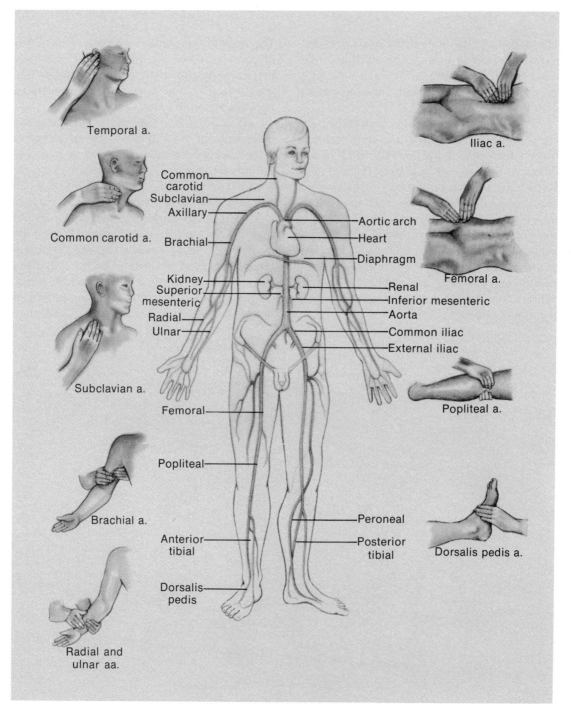

Figure 14-5 Major arteries and pressure points of the body.

(LDL), and high-density lipoproteins (HDL). Recently, most attention has been focused on the LDL (which carry most of the cholesterol found in blood) as playing a key role in both the development of atherosclerotic lesions and in the regulation of cholesterol metabolism in cells.

It is not just the total amount of cholesterol in a person's blood that determines susceptibility to coronary heart disease but also the type of cholesterol. By measuring the two major kinds of cholesterol—LDL and HDL—and calculating the ratio of one to the other, one can predict more accurately whether a person is at risk for developing coronary heart disease or having a heart attack.

LDL transport cholesterol in the bloodstream and, when levels are high, deposit it on artery walls. HDL may perform an opposite function by removing cholesterol from the artery walls and carrying it to other locations, such as the liver, to be metabolized and excreted. In blocking the deposit and reducing buildup of fats that clog blood vessels, HDL may reverse this main cause of heart attacks. The higher the content of HDL in total blood cholesterol, then, the lower the risk of coronary heart disease.

Investigations have shown that human fibroblasts and other kinds of cells contain specific receptors for LDL, and that interaction of the lipoprotein with the receptors is a necessary first step before the degradation of LDL can occur. If this event is prevented from happening, then large quantities of cholesterol-bearing LDL may accumulate in the bloodstream. This is apparently what happens to persons with the genetic disease called *hypercholesterolemia*. It is an autosomal dominant disorder in which the mutation causes a deficiency of LDL receptors; not only is the cholesterol elevated, but atherosclerosis and myocardial infarction also may occur.

Although evidence favors the hypothesis that cholesterol, especially that carried by LDL, is somehow involved in the formation of lesions in the arterial walls (atherogenesis), the question remains to be answered. Will lowering the concentration of cholesterol in the blood prevent or reverse the process of plaque formation and prevent strokes or heart attacks?

The clinical signs of arteriosclerosis (atherosclerosis) are directly related to vascular insufficiency. Ischemia (due to lack of oxygen) of vital organs such as the brain produces blackouts, loss of memory, and associated neuron damage. Poor circulation in the renal system results in a buildup of toxic nitrogenous compounds in the blood and possible toxemia. As mentioned before, coronary artery insufficiency can result in myocardial infarction and death. A debilitating disease, *gangrene*, is a serious result of arterial insufficiency to the extremities. Progressive narrowing of the arteries of the legs is detected by a cramping of muscles after standing or walking for a period of time. Pain results from the mild hypoxia and resulting ischemia of the exercised muscles.

Hypertension

Hypertension, which afflicts 24 million people in the United States, is the most common chronic disease. It is a major health problem because people with high blood pressure are more likely to have strokes, heart disease, or kidney failure than are people with lower blood pressure. The risk of kidney failure and strokes falls with appropriate therapy to lower the blood pressure. The factors that can aggravate this disease are well indicated and their avoidance should be practiced. Some common behavioral factors are excessive dietary salt intake, excessive weight, and smoking.

Primary Hypertension

Hypertension, as defined by the American Heart Association, is any blood pressure that exceeds 140/90 mm Hg. Approximately 14 percent of adults are hypertensive; 90 percent of those have no identifiable cause (primary or essential hypertension). Key factors include heredity, environment, job-related stress, and race. For instance, black people have a higher incidence of hypertension than other races. The following discussion will concern the remaining 10 percent of people who have an identifiable reason for hypertension (secondary hypertension).

Secondary Hypertension

Causes of secondary hypertension include renal disease, coarctation of the aorta, and endocrine disorders. Hypertension due to renal disorders is caused by narrowing (stenosis) of the renal arteries or damage to the tissue itself. Chronic destructive renal disease is the most common cause of secondary hypertension. In this disease the renal artery is partially occluded, causing increased secretions of renin, an enzyme that controls blood pressure. The stenosis can be surgically corrected and reversal of renal hypertension occurs. Other diseases (chronic pyelonephritis or chronic glomerulonephritis) that damage the kidney tissues with infective or immune responses are not as easily reversed. The diseased kidney may have to be removed (nephrectomy).

Coarctation of the aorta is a congenital disorder involving a constriction (stenosis) of the aorta as it leaves the heart. This structural problem causes hypertension proximal to the constriction and is generally corrected by surgery.

Endocrine (hormonal) causes of hypertension usually relate to increased secretion of adrenal hormones. The renin-angiotensin-aldosterone hormonal system regulates arterial blood pressure and sodium balance (see Figure 10-18).

Control Mechanisms

Although the cause of essential hypertension is not known, three control mechanisms—neural, endocrine, and renal-electrolyte—are probably involved.

The neural control of arterial blood involves the use of mechanoreceptors responding to pressure or volume stimuli (Figure 14-14). These receptors are found throughout the circulatory system and in skeletal muscle. A fall in circulatory pressure or increased skeletal mus-

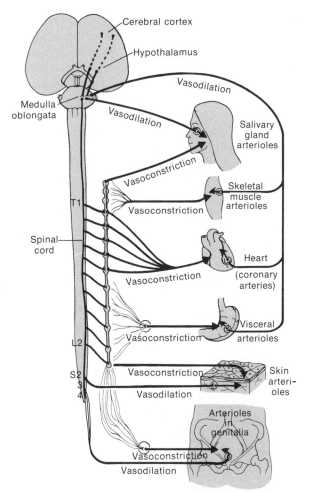

Figure 14-14 Nervous regulation of visceral and muscular arteries and arterioles. Stimulation of sympathetic nerves induces vasoconstriction of the vessels, whereas stimulation of the parasympathetic nerves produces vasodilation of the vessels.

cle tension stimulates the vasomotor center in the medulla to increase cardiac output and constrict peripheral arterioles, thereby increasing blood pressure. A rise in blood pressure or skeletal muscle relaxation diminishes vasoconstriction and reduces blood pressure. The mechanism is a reflex action.

The hormonal (endocrine) and renal-electrolyte control mechanisms have received considerably more attention. Renin's role in renovascular hypertension and primary aldosteronism is unquestioned.

Renin is a proteolytic enzyme released by the kidney. It reacts with angiotensin I in the blood in the lung to form a potent vasoconstrictor, angiotensin II, which increases the blood pressure. Angiotensin II also stimulates the adrenocortical hormonal synthesis and release of the sodium-retaining hormone, aldosterone, thereby promoting increased volume and contributing to increased pressure.

A third action of the angiotensins involves the nervous system. They stimulate the release of the neurotransmitter norepinephrine by the nerves of the sympathetic nervous system and by the adrenal medulla. Norepinephrine constricts arterioles, increasing the blood pressure.

A drop in blood pressure through the kidneys stimulates renin release, resulting in activation of the renin-angiotensin mechanism and an increased blood pressure. Conversely, high blood pressure activates a negative feedback mechanism to inhibit renin release and lower the blood pressure. When these feedback mechanisms do not work normally, chronic hypertension may result.

PRINCIPAL SYSTEMIC VEINS

In general, the total diameter of the veins returning blood from any organ is about twice that of the arteries carrying blood to the organ. Blood from the systemic veins returns to the heart through three main channels (Figure 14-15):

1. Blood from the wall of the heart returns through the coronary sinus.

2. Blood from the head, neck, upper extremity, and part of the trunk returns through the superior vena cava.

3. Blood from the remainder of the trunk and lower extremities returns through the inferior vena cava.

The coronary sinus receives most of the veins of the heart. It lies on the posterior aspect of the organ and terminates by emptying into the right atrium (see Figure 13-2).

Veins of the Neck and Head

The external jugular veins are small vessels that lie superficially in the lateral aspect of the neck, one on each side (Figures 14-15 to 14-17). They descend over the sternocleidomastoid muscle beneath the platysma and empty into the subclavian vein.

An abundant blood supply reaches the brain through the internal carotid arteries; numerous veins on the brain surface and within it are concerned with gathering this blood and returning it by way of the internal jugular veins. Deep within the substance of the brain are many veins that drain into large sinuses. These veins and sinuses have very thin walls and lack valves.

The internal jugular veins are the principal vessels draining the head and neck (Figures 14-15 to 14-17). They arise from the transverse sinus and descend laterally beside the common carotid artery. At their base they join with the subclavian veins to form the brachiocephalic (innominate) veins (Figure 14-17). These receive numerous tributaries from the face, neck, and cranial cavity. The right brachiocephalic vein is almost vertical and only about 1 inch long. The left brachiocephalic vein is almost horizontal and about 3 inches long. It passes above the heart to join the right brachiocephalic vein and then form the superior vena cava.

The superior vena cava extends downward about 3 inches to drain into the right atrium of the heart (Figures 14-15 and 14-17). The azygos vein is a long, slender, unpaired vessel that passes upward along the right posterior abdominal and thoracic walls and receives blood from the intercostal and lumbar veins. It terminates by emptying into the superior vena cava.

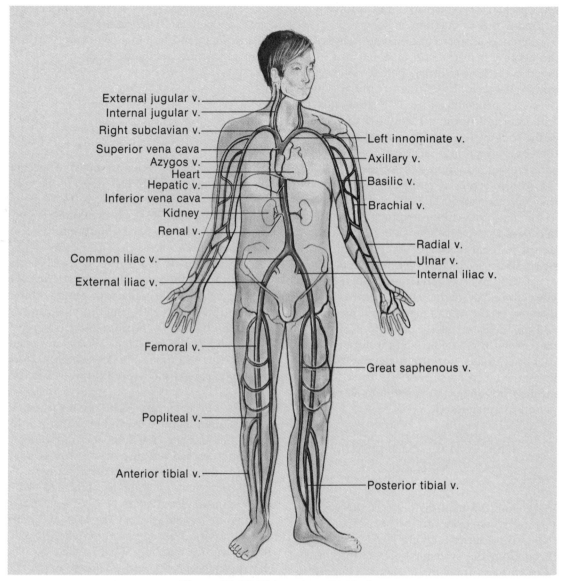

Figure 14-15 Major veins of the body.

Veins of the Upper Extremity

Blood from the upper extremity is returned by deep and superficial sets of veins. The deep veins are called by the same name as their corresponding arteries. They frequently anasto-mose with one another and the superficial veins. There are valves in both sets of veins, and both groups finally open into the axillary and subclavian veins.

The superficial veins lie just beneath the skin and are easily seen on the back of the hand and the front of the elbow (Figure 14-18). They

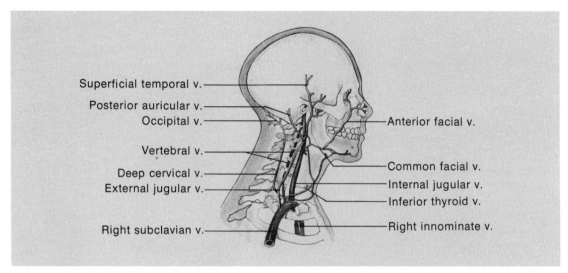

Figure 14-16 Veins of the head and neck.

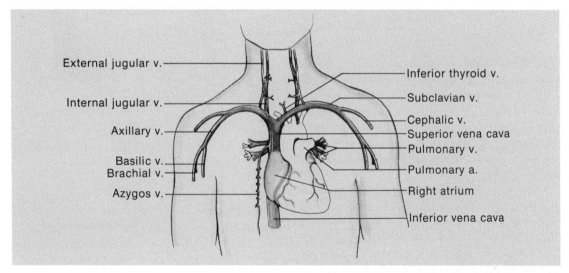

Figure 14-17 Veins of the upper extremity.

are larger than the deep veins and return more blood. They include (starting from the distal end of the limb) the cephalic vein (dorsal surface of the hand), the accessory cephalic vein (a lateral branch), the basilic vein (medial side of arm), the median cubital vein (at the bend of the elbow), the axillary vein (continuation of the basilic), and, finally, the subclavian vein that drains the axillary and cephalic veins.

The large superficial veins in front of the elbow (basilic, median cubital, and cephalic) are commonly used when drawing blood from the body or when injecting medication into the bloodstream (intravenous injection).

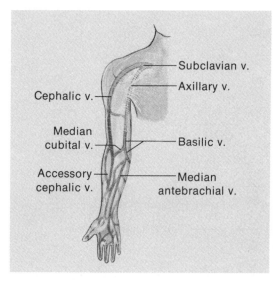

Figure 14-18 Superficial veins of the anterior surface of the right arm.

Veins of the Lower Extremity

Like those of the arm, the veins of the lower extremity are both superficial and deep. The superficial veins lie beneath the skin between the layers of fascia. The deep veins are covered by deep layers of muscle and follow the large arteries, usually enclosed in the same sheath. Both sets have valves, which are more numerous in the deep veins.

The superficial veins include the great saphenous vein, which begins in the marginal vein of the foot, extends up the medial side of the leg and thigh, and ends in the femoral vein just below the inguinal ligament. It receives blood from all aspects of the superficial layers of the leg and foot. A major tributary is the small saphenous vein, which passes up the back of the leg and frequently empties into the popliteal vein at the knee (Figure 14-19).

Veins of the Pelvis and Abdomen

The internal and external iliac veins unite to form the common iliac veins. The common iliac

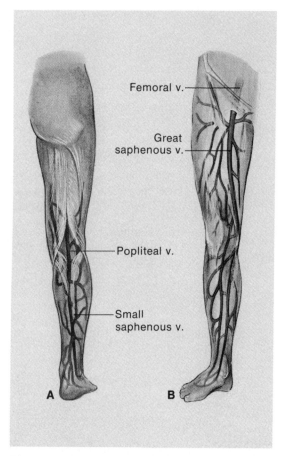

Figure 14-19 Superficial veins of the right lower extremity. **A,** Posterior aspect. **B,** Anterior aspect.

veins progress from the sacral region to form the inferior vena cava in the lumbar region.

The inferior vena cava ascends along the right side of the aorta, passes through the diaphragm, and terminates by entering the right atrium of the heart (Figure 14-15). It receives veins that have the same names as the peripheral branches of the aorta (lumbar, renal, suprarenal, inferior phrenic, superior phrenic, spermatic or ovarian).

The hepatic veins arise in the liver and flow into the inferior vena cava (Figure 14-20). These carry the return flow from the hepatic artery and the blood that comes to the liver through the portal system.

3. Filtration based on difference between opposing pressures in any given area
   a. Arteriolar feeding—greater hydrostatic pressure than oncotic pressure
   b. Venular draining—greater oncotic pressure than hydrostatic pressure
XIII. Clinical considerations
   A. Exercise
   B. Shock
   C. Aneurysm
   D. Arterial degeneration

STUDY QUESTIONS AND PROBLEMS

1. Discuss the differences between arteries, veins, and capillaries according to their structure, function, and location.
2. If a vessel were severed, how would you know whether it was a vein or an artery?
3. What is meant by the pulse? Where and how would you take a pulse?
4. Where and how would you place pressure to stop bleeding from an artery?
5. What are the symptoms of arteriosclerosis, and how are they produced?
6. Discuss the various theories proposed to explain the etiology of arteriosclerosis.
7. Define hypertension, and list possible causes.
8. Trace a drop of blood through the pulmonary circulation.
9. Indicate the gaseous content along with the structural anatomy.
10. Where is the carotid sinus, and what does it do?
11. Which veins are commonly used to obtain a sample of blood?
12. How does the portal vein differ from other veins? Why?
13. Outline the distinctions between fetal circulation and circulation after birth.
14. Explain the principle of the sphygmomanometer. Why is the stethoscope placed distal to the inflated cuff?
15. Correlate the two sounds heard through the stethoscope during the taking of the blood pressure with the cardiac cycle.
16. Define systolic and diastolic blood pressures in relation to the structural anatomy of the heart and arteries.
17. Differentiate hydrostatic pressure of tissues, blood pressure, colloidal osmotic pressure, and pulse pressure.
18. What are some factors that may cause an increase in the pulse rate?
19. Describe nervous and endocrine effects on the blood vessels and heart.
20. What is hemorrhage? If it occurred in the left region of the circle of Willis, what part of the body would be affected?
21. Define shock and indicate its seriousness.

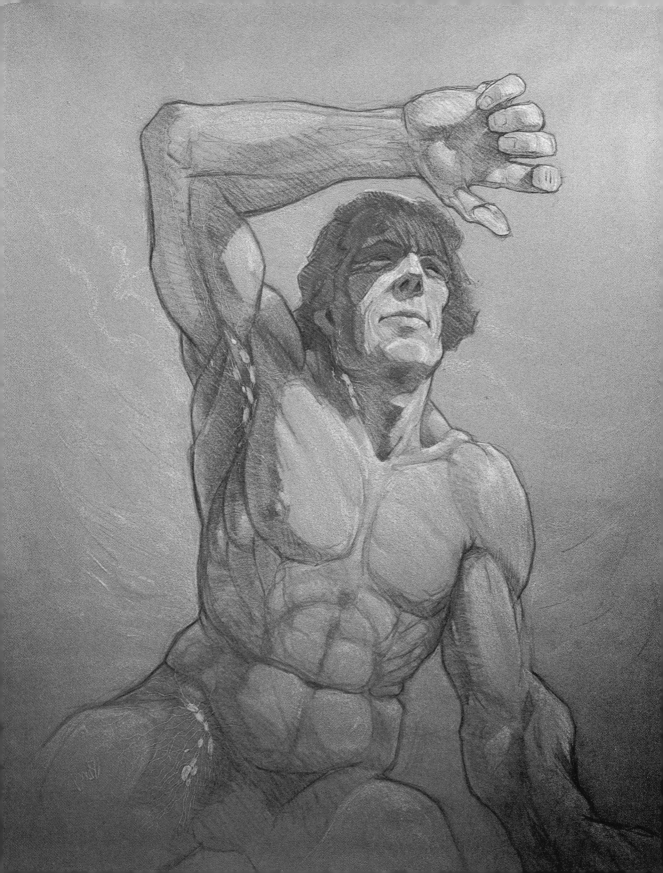

15

The Lymphatic System

Learning Objectives

- Describe and identify the anatomical arrangement of a capillary bed

- Compare the structure of a vein and lymph vessel

- Identify the major vessels of the lymphatic system

- Locate the major clusters of lymph nodes in the body

- Explain how a superficial lymph node protects the body from infection

- Explain the function of Peyer's patches in the intestinal mesentery

- Explain how lymph fluid is produced

- Identify, histologically, Hodgkin's disease

- Identify the causative agent and explain the progress of elephantiasis

- Explain tissue drainage

- Describe edema and explain a number of ways it can occur

IMPORTANT TERMS

chyle
elephantiasis
excretia
lacteal

lymph
lymphangiogram
lymphatic tissue
lymph nodes

lymphoma
mastectomy
splenomegaly

GENERAL ANATOMY

Lymphatic Vessels

The lymphatic system is a one-way collecting system that gathers and drains filtered fluid and cellular constituents that accumulate in the spaces between the cells. The larger lymphatic vessels drain into veins, which return the lymph to the blood circulation.

Lymphatic capillaries resemble blood capillaries in structure. Both consist of a single layer of endothelial tissue. The major difference is that the lymphatic capillary has a closed terminal end; lymph fluid is absorbed from the tissue spaces through the endothelial membrane (Figure 15-1). Most of the tissues of the body, with the notable exception of the central nervous system, are drained by lymphatic.

Larger lymphatic vessels drain the capillary network. The walls of these vessels resemble the walls of veins in structure. The muscle fibers in both the middle and outer layers are longitudinal and oblique, and therefore these larger vessels are contractile; lymphatic capillaries are not.

The larger lymphatics not only have the same basic three layers as veins; they also have valves on their inner surfaces to prevent backflow (Figure 15-2). The valves are bicuspid or tricuspid and occur at shorter intervals than do those of veins. Lymphatic capillaries do not contain valves.

Lymph Nodes

Lymphatic tissue filters and removes bacteria. Along the course of the lymphatic vessels are small bodies of lymphatic tissue called lymph nodes (Figure 15-3). These usually are oval and are commonly referred to as *microkidneys*.

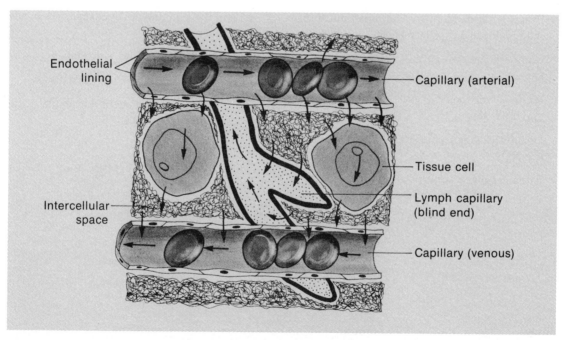

Figure 15-1 Diagram of a capillary bed, showing how materials diffuse between arterial capillaries and venous capillaries. Materials that are trapped in the intercellular tissue spaces are collected by lymphatic capillaries and returned to the blood system.

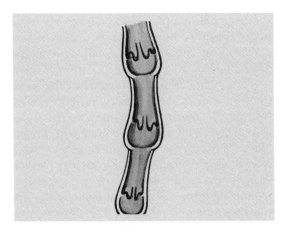

Figure 15-2 Diagram of valves in a lymph vessel.

sponge. Sinuses are spaced throughout the tissue and provide a filtering network for the lymph stream (Figure 15-3). These sinuses consist of a maze of passageways lined with phagocytic cells that engulf bacteria or other foreign products from the lymph fluid.

Experimental evidence indicates that lymph nodes may afford some protection against infection or disease. Lymphocytes (leukocytes) of the lymph nodes and spleen actively destroy such foreign substances as bacteria and antigens that are introduced into the body (Chapter 12). Lymph nodes commonly become swollen and inflamed during severe bacterial infections.

Lymphatic Tissues

The following organs containing lymphoid tissue perform somewhat different functions, particularly with respect to the substances that they filter:

Afferent lymphatic vessels enter the node at different points, but the efferent vessels leave only at a small depression, the hilus (Figure 15-3). Blood vessels also enter and leave from the hilus. Internally the lymph node resembles a

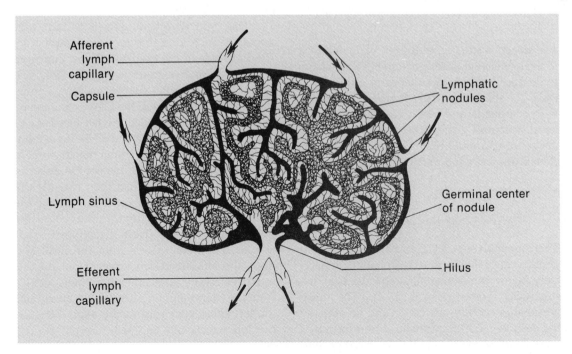

Figure 15-3 Anatomy of a lymph node. Lymphocytes develop in the germinal centers of the nodules.

Tonsils

These masses are found just beneath the epithelium of the mouth and throat and include the following: the *palatine tonsils* (at each side of the soft palate), the pharyngeal tonsils (behind the nose on the back wall of the upper pharynx, sometimes called the *adenoids*); and the *lingual tonsils* (back of the tongue). Any or all of these tonsils may become so infected with bacteria that their surgical removal may be indicated.

Thymus

The thymus gland, usually comprising two lobes, plays a key role in the formation of antibodies in the first few weeks of life and the development of immunity. It is located at the base of the neck and manufactures lymphocytes. Removal of the thymus causes a decrease in lymphocyte production throughout the body and also a decrease in size of the spleen and lymph nodes. The thymus is most active during early life; after puberty it undergoes degenerative changes and is replaced by adipose tissue.

Spleen

The spleen is situated in the left abdomen next to the stomach. It is designed to filter blood and to produce lymphocytes that carry on phagocytosis. The spleen has many functions concerned with lymphatic and blood circulation. Prominent structures inside the spleen are round masses of lymphoid tissue, and these justify its classification as a lymphatic organ.

Lymphatic Circulation

There are both superficial and deep sets of lymphatic vessels. The surface lymphatics, immediately below the skin, often continue near the superficial veins (Figure 15-4). The deep lymphatics usually are larger and accompany the deep veins. All lymphatic vessels form networks and, at certain points, carry lymph into the regional lymph nodes. Groups of large nodes are in the neck (cervical), under the arm (axillary), at the elbow (cubital), and in the groin (inguinal).

The lymphatics in the region of the small intestine, called *Peyer's patches,* have the special function of absorbing digested fat. The inner wall of the small intestine is lined with fingerlike projections, the villi. Each villus contains a central lymph capillary, or **lacteal,** which absorbs the end products of fat digestion. Lymph carrying absorbed fat has a milky appearance and is called **chyle.** The cisterna chyli is concerned with the collection of lymph from the lower limbs, intestine, and kidneys (Figure 15-5).

Lymphatic vessels are named according to their location. Those on the thumb side of the forearm are called the *radial lymphatic vessels;* those on the medial part of the forearm are the *ulnar lymphatic vessels.* Nearly all the lymph from the upper extremity and from the breast is carried into the axillary lymph nodes and eventually to one of the two large drainage ducts that empty into the bloodstream.

The right lymphatic duct is a short tube about 0.5 inch long that receives lymph from the right side of the head and thorax and from the right upper extremity (Figure 15-4*A*). It empties into the right subclavian vein. The remaining three-fourths of the body is drained by a 16-inch long tube called the *left lymphatic* or *thoracic duct* (Figures 15-4 and 15-5). It begins at the back of the abdomen below the attachment of the diaphragm, and its first section arises in the cisterna chyli. The duct then extends upward through the diaphragm and along the back wall of the thorax up into the root of the neck on the left side. Here it receives the left jugular lymphatic vessels from the head and neck, the left subclavian vessels from the left upper extremity, and other lymphatic vessels from the thorax and its parts. It then opens into the left subclavian vein in the angle between the left subclavian and left internal jugular veins.

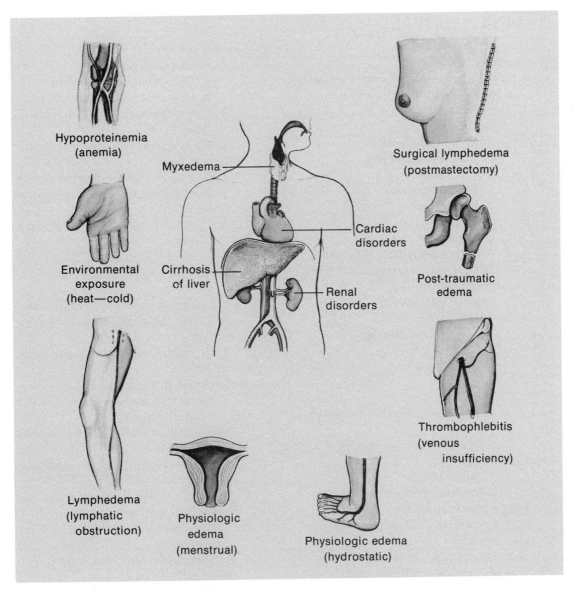

Figure 15-9 Some causes of peripheral edema.

In congestive heart failure venous pressure is elevated, and the capillary pressure is consequently elevated. In cirrhosis of the liver, oncotic pressure is low because liver synthesis of proteins is depressed; in nephrosis, oncotic pressure is low because large amounts of protein are lost in the urine (proteinuria).

Another cause of edema is inadequate lymphatic drainage. Following the removal of the axillary networks of lymph vessels in a radical mastectomy, drainage from the arms is disrupted. In elephantiasis, filarial blockage of the inguinal lymph nodes creates an accumulation of tissue fluid in the lower extremities.

OUTLINE

I. General anatomy
   A. Drainage system that returns extracellular fluid to the blood circulation
   B. Lymphatic vessels
      1. Capillaries
         a. Single endothelial layer
         b. Closed terminal end
      2. Larger vessels
         a. Resemble veins
         b. Have valves
   C. Lymph nodes
      1. Oval-shaped
      2. Afferent vessels drain into node
      3. Efferent vessels leave only at hilus
      4. Spongelike sinuses
      5. Produce lymphocytes
   D. Lymphatic tissues
      1. Tonsils—protect oral region
         a. Palatine
         b. Pharyngeal
         c. Lingual
      2. Thymus
         a. Produces antibodies before puberty
         b. Manufactures lymphocytes
      3. Spleen
         a. Filters blood
         b. Produces lymphocytes
   E. Lymphatic circulation
      1. Superficial and deep vessel networks
      2. Groups of nodes at key locations
         a. Cervical, axillary, cubital, and inguinal
         b. Intestinal nodes (Peyer's patches)—involved in digestion
      3. Right lymphatic duct—drains upper right quadrant of body
      4. Left lymphatic duct (thoracic duct)—drains remaining three-fourths of body
II. Physiology of lymphatic tissue
   A. Lymph formation
      1. Extracellular fluid accumulation

      2. Similar to blood plasma
         a. Decreased protein quantity
         b. Higher albumin ratio
      3. Low leukocyte content; no erythrocytes
      4. Many digestive enzymes
   B. Lymph flow
      1. Very slow under resting conditions
      2. Flow a result of muscular contractions
      3. Valves in large vessels prevent backflow
   C. Function of lymphatics
      1. Defense mechanism—phagocytosis
      2. Immunity—antibody production
      3. Tissue drainage—edema prevention
      4. Excretion—removal of waste products
      5. Digestion—absorption and circulation of end products of fat digestion
III. Clinical considerations
   A. Lymphangitis—inflammation of lymphatic vessels
   B. Adenitis—enlarged lymph nodes or glands
   C. Hodgkin's disease—infected and enlarged lymph nodes
   D. Lymphosarcoma—malignant lymphatic tumor
   E. Splenomegaly—enlarged spleen following infectious disease
   F. Elephantiasis
      1. Parasitic worms (filariae) grow in lymph vessels
      2. Blockage of lymph vessels causes lower extremities to swell
   G. Edema—excess accumulation of interstitial fluid
      1. Increased accumulation of metabolites

2. Gravity effect on dependent parts of body
3. Salt retention
4. Systemic failures
   a. Congestive heart failure
   b. Nephrosis—kidney
   c. Cirrhosis—liver

5. Inadequate drainage
   a. Surgical removal of diseased lymph nodes and vessels
   b. Parasitic blockage of lymph network

STUDY QUESTIONS AND PROBLEMS

1. Explain why lymph capillaries, vessels, and nodes are important in everyday living.
2. Why are tonsils and adenoids important in early childhood?
3. Why are tonsils and adenoids not removed the first time they become infected?
4. What is the function of the thymus gland? What is its longevity?
5. List the networks of nodes that are found at junctions of the appendages to the main body trunk. Why do you think they are primarily located in those regions?
6. What is lymph? How does it originate?
7. What are antibodies? Where are they produced in the lymphatic system?
8. Distinguish between Hodgkin's disease and a lymphosarcoma.
9. Can Hodgkin's disease be treated? Explain.
10. Most parasites use the general circulatory system as a means of transportation in the larval stage. What system does the filaria worm use?
11. Explain why elephantiasis is the result of a mechanical blockage of vessels and nodes.

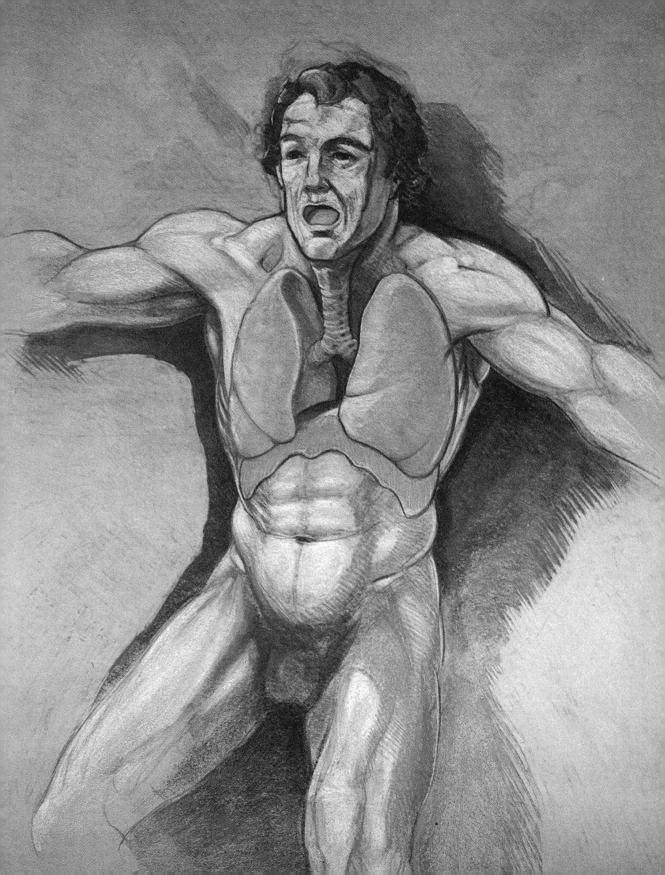

16

The Respiratory System

LEARNING OBJECTIVES

- Label the structures of the respiratory system

- Identify two functions of the cilia and mucus cells

- List the steps in inspiration and expiration

- Compare intrapulmonary and intrathoracic pressures

- Define respiration, ventilation, inspiration, and expiration

- Identify the different volumes of air used in vital lung capacity

- Describe the vagal Breuer control of breathing

- Explain the oxygen-associated chemoreceptor mechanism for the control of ventilation

- Explain the process of gaseous exchange in the alveoli and in the tissues

- Describe how oxygen and carbon dioxide are transported in the blood

- Specify the consequences of hypoventilation and hyperventilation

- Describe the cause, effects and prevention of "decompression sickness"

- Describe CPR and identify its advantages over previous methods of resuscitation

- Define respiratory acidosis and alkalosis and give examples

- Define pleura, pneumothorax, atelectosis, thoracentesis, and asphyxia

- Define COLD and explain the difference between asthma and emphysema

- List at least two pneumoconioses and postulate an EPA corrective measure

- Explain the significance of surfactant coating the alveoli

- Define hyaline membrane disease and differentiate it from sudden death syndrome

- Describe how a pulmonary embolism could form and be detected

IMPORTANT TERMS

alveolus	emphysema	hyperpnea	pleura	ventilation
apnea	epistaxis	inspiration	pleurisy	vital capacity
asphyxia	eupnea	intrapulmonary	pneumothorax	
aspiration	expiration	intrathoracic	respiration	
bronchus	hyaline	mediastinum	resuscitation	
dyspnea	hypercapnea	perfusion	stretch receptors	

Most animal cells require a continual supply of oxygen. They also need a means of getting rid of the surplus carbon dioxide that results from cellular activity. Any full consideration of respiration must include the processes involved in utilizing oxygen and producing and disposing of carbon dioxide. In the broadest sense, **respiration** refers to all the oxidative processes concerned with the release of energy from nutrients; in its narrowest sense, the term sometimes is used merely as a synonym for breathing. Oxidative processes within the cells are discussed in Chapter 17; this chapter will cover the delivery of oxygen to the cells and the removal of carbon dioxide from them.

GENERAL ANATOMY

Upper Respiratory Tract

Air drawn in through the nose passes over a complex series of surfaces formed by the nasal septum and the nasal conchae (Figure 16-1). These surfaces not only warm and moisten the air but also act to baffle its flow, causing dust particles to settle out on the mucous films that line the nasal passages. Their ciliated epithelium acts to move dust-laden mucus backward to the pharynx, where it is either swallowed or expectorated. Breathing through the mouth allows relatively cold and unfiltered air to reach the lungs.

Both the nasal and buccal cavities open into the **pharynx,** and the esophagus and larynx descend from the lower part of the pharynx. The trachea (windpipe), in turn, descends from the larynx. Mucous membranes of the superior conchae contain the endings of the olfactory nerve fibers. The nerve fibers for the muscles of the nose come from the facial nerve (cranial nerve VII), and the skin receives fibers from branches of the trigeminal nerve (cranial nerve V).

The partition separating the two nasal spaces from each other is called the nasal septum. The nasal septum is rarely situated exactly at the midline. If it is marked to one side, the anomaly is described as a *deviated septum.* Attacks of hay fever or colds, accompanied by swelling of the nasal mucosa, make breathing extremely difficult in the smaller of the two nasal cavities. Such occurrences may lead to surgical repair of the septum, a *submucous resection,* in which the cartilaginous septum is removed entirely, allowing the opposing mucous membranes to act as a septum and to be more flexible.

Nosebleeds (**epistaxis**) can result from an injury or blow to the nose. Growths and, occasionally, high blood pressure also can cause nosebleeds. Pressure applied to the nostril of the bleeding side and cold compresses over the nose usually alleviate the problem. If not, cotton packing dipped in epinephrine has proved helpful as an emergency measure.

Lower Respiratory Tract

In addition to the nose and pharynx, other parts of the respiratory tract are the larynx, trachea,

tery divides into smaller and smaller branches until it finally resolves itself in capillary networks around the bronchioles and alveoli (Figure 16-5). Pulmonary capillaries anastomose to form small venules as the blood leaves the alveoli. These then become the pulmonary veins, which carry freshly oxygenated blood back to the left atrium of the heart.

If a foreign body enters the smaller air passages of the lungs, it may obstruct the bronchi, causing a blockage of air in the portion of the lung they supply. This produces a partial collapse of the lung, causing symptoms of coughing and expectoration. In many patients it is possible to remove the foreign body following examination by bronchoscopy. With this method, a rigid, lighted tube is inserted through the mouth, pharynx, and trachea into the obstructed bronchus. After **aspiration** (suctioning) of the secretions, the foreign body may be removed with lung forceps (Figure 16-6).

Bronchoscopy is also used in the diagnosis and treatment of many intrathoracic diseases. Specimens from tumors of the air passages can be removed and examined (biopsy); secretions can be aspirated; and bacterial studies made of them. In the study of some diseases of the lungs, such as bronchiectasis, liquids are injected to outline the air passages and an x-ray film taken (bronchogram). Esophagoscopy is the viewing of the interior of the esophagus through a lighted tube. Foreign bodies may lodge in the esophagus just as they do in the bronchi; and disease processes of the esophagus are similar to those of the bronchus.

The thoracic cavity is divided from the top to bottom by a double wall, the **mediastinum.** The heart, great vessels, esophagus, and many nerves lie within this area; the lungs are in pleural cavities on each side.

The two lungs vary in shape and size. The right lung is the larger and has three lobes. It is wider than the left lung but a little shorter. The left lung is divided into two lobes and is somewhat narrower and longer than the right (see Figure 16-2).

A thin serous membrane covers each lung

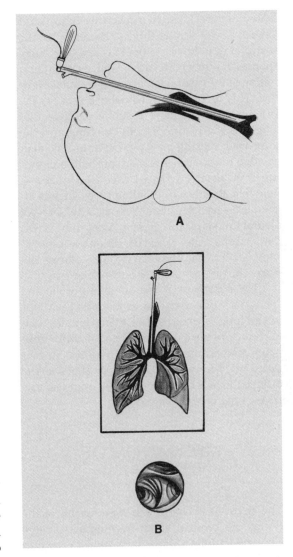

Figure 16-6 Introduction of the bronchoscope. **A,** The bronchoscope is a device used to explore the trachea and bronchi. **B,** View through the bronchoscope as it approaches the bifurcation of the trachea into left and right bronchi.

and continues over the thoracic wall, the diaphragm, and the mediastinum. This membrane is called the **pleura** and has two layers. The pulmonary (visceral) pleura covers the lung; the parietal pleura lines the thoracic cavity. The space between them is referred to as the **intra-**

thoracic space. These two layers are in constant contact, and a serous fluid is secreted by the membranes to lubricate the surfaces that move upon each other with each respiration. Inflammation of the pleural membranes is called **pleurisy.** This condition can be painful because the inflammation produces a sticky exudate that roughens the pleura of both the lung and the chest wall, causing an acute irritation. In pleurisy due to bacterial infection, pus will accumulate in the pleural cavity—a condition known as *empyema.* Removal of the fluid by syringe is often necessary. An abnormal accumulation of pleural fluid is *pleural effusion.* Aspiration of this fluid is done under closed conditions so as not to allow air into the chest cavity and collapse the lung. This process is called a *thoracocentesis.*

The substance of the lungs is porous and spongy because of the air in the sacs. Each lobe of the lung is composed of many lobules, and into each lobule a bronchiole enters and terminates in an atrium. Inflammation of the lungs is called **pneumonia.** Surgical removal of a single diseased lobe is called a *lobectomy;* removal of a whole lung is a *pneumonectomy.*

MECHANISM OF BREATHING

The rhythmical changes in the capacity of the thorax are brought about by muscular action. Changes in lung volume with intake or expulsion of air follow passively.

Lung tissue is elastic but is unable to expand or contract of itself. The mechanism by which the lungs are filled is best illustrated by a model like that shown in Figure 16-7A. Balloons are connected to the outside of a bell jar by a Y-shaped tube that passes through the stopper of the jar. The space between the outside of the balloons and the inside of the jar contains air but is closed from the outside. The floor of the jar is made of rubber, and when it is pulled down, the same amount of air in the jar fills the greater space, thereby creating a negative pressure on the outside of the balloons. The inside of

each balloon is still subject to the external atmospheric pressure. Since the external pressure is now higher than before, air enters the balloons and inflates them until the pressure in the jar and the expanded elastic tension of the balloons are equal to the atmospheric pressure. When the rubber floor is allowed to return to its original position, the increased pressure in the jar squeezes air out of the balloons.

The filling and emptying of the lungs are similar to but more complex than the model just described (Figure 16-7B). Two pressures are involved here. The intrapleural space is a potential space between the lungs and the chest wall. The pressure exerted there normally is less than atmospheric pressure to permit full expansion of the lungs. The intrapulmonary (intra-alveolar) pressure is that pressure in the alveolar sacs that rises and falls above or below atmospheric pressure according to the muscular action of breathing.

The rib cage, or thorax, is airtight and analogous to the jar. The diaphragm closes off the floor of the thorax as does the rubber in the model, and the bronchi and trachea connect the lungs with the outside atmosphere. At rest, the diaphragm is curved and bowed into the chest cavity. When it is stimulated to contract, the muscle tightens, forming a relatively straight line—thus increasing the air space in the thorax and decreasing the intrathoracic pressure (Figure 16-7B). The intercostal muscles are located between the ribs. Contrary to popular belief, the ribs are not rigidly fixed to the spinal column. When the intercostal muscles are stimulated, the ribs pull together collectively, working in conjunction with the diaphragm to increase the thoracic volume.

Recognizing that the combined efforts of the diaphragm and intercostal muscles result in an increased thoracic volume and decreased intrathoracic pressure and considering that the thorax is a closed chamber, the only way that the increased chest space can be filled is to draw air in from the outside, filling the lungs. This is called **inspiration.** Relaxation of the diaphragm and intercostal muscles returns them to their normal dome-shaped and extended positions,

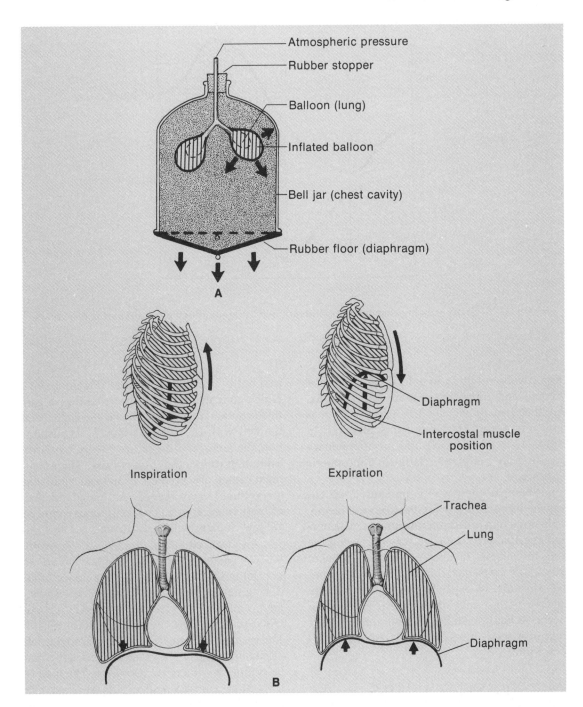

Figure 16-7 Mechanism of breathing. **A,** Bell jar apparatus used to simulate the action of the diaphragm. As the rubber floor (diaphragm) is pulled down, the balloons (lungs) are forced to inflate owing to the closed air cavity in the jar. **B,** Chest and diaphragm positions front and side views, in breathing. *Left,* the end of a normal inspiration, with chest expanded and diaphragm down; *right,* the end of a normal expiration, with chest cavity contracted and diaphragm relaxed.

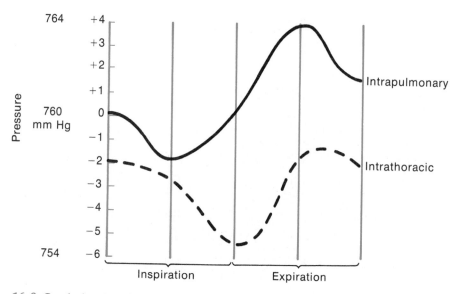

Figure 16-8 Graph showing changes in intrathoracic and intrapulmonary pressures during respiration.

respectively, resulting in a decrease in thoracic volume and increased intrathoracic pressure. Air that had previously been "sucked" in is now forced out. This is called **expiration.** In Figure 16-8, intrathoracic and intrapulmonary pressures are compared relative to atmospheric pressure.

When the chest wall is punctured, air enters the intrathoracic space between the pleura, a phenomenon called **pneumothorax.** The normal negative pressure is decreased, causing the lung to collapse. In cases of multiple gunshot wounds of the thorax, the patient can die of anoxia (insufficient oxygen) unless the holes are covered with a tight dressing. Artificial pneumothorax has proved beneficial in treating such lung diseases as tuberculosis by resting the infected lung.

LUNG VOLUMES

Much can be learned about lung function by measuring the volume of air taken in and exhaled at each breath. This is conveniently car- ried out by using a spirometer (Figure 16-9A), which consists of a drum inverted over a gas-filled container. A tube connects the airspace in the bell with the subject's mouth. At inspiration, the removal of air causes the drum to sink; expiration causes the drum to rise. The movement can be observed on a scale and calibrated for volume of air moved. An added rotating drum (kymograph) gives a continuous record of volume changes within the drum.

An idealized record is shown in Figure 16-9B. The small oscillations at the left are due to normal quiet breathing, and the amount of air during normal breathing is called the *tidal volume.* A maximal inspiration followed by a maximal expiration gives the next part of the record. The greatest volume that can be inspired is the *inspiratory capacity;* the greatest volume that can be expired is the *expiratory reserve.* The total of inspiratory capacity and expiratory reserve is the **vital capacity.**

On the whole, athletes tend to have larger than normal vital capacities. Small vital capacities are indicative of pulmonary disease. Residual volume is that air still contained in the lungs

DISORDERS OF THE RESPIRATORY SYSTEM

Atelectasis

Atelectasis is the collapse of an expanded lung. Congenital (primary) atelectasis is the only form of collapse that deserves to be called atelectasis, for the alveoli fail to expand (Figure 16-14). It is seen in the stillborn child, who has never breathed, and in children who live only a few days and never breathe well.

Secondary atelectasis occurs in adults and is a result of an inadequate expansion of the lungs. Alveolar collapse occurs in obstructive disorders (tumors, fluid) where the flow of inspired air is partially or completely obstructed with the subsequent reabsorption of alveolar air. This atelectasis may be caused by pleural effusion, empyema (pus in the pleural cavity), or the presence of air under pressure in the pleural cavity (pneumothorax). Obstruction of the bronchus may also be caused by foreign bodies such as food, coins, or a tooth that becomes lodged in a bronchus. No air can enter the lung and the air already there is absorbed into the blood, and the lung collapses.

Pneumothorax

Pneumothorax is air in the pleural cavity, the region between the parietal pleurae, which lines the chest cavity, and the visceral pleura, which covers the lungs. Air pressure in the cavity less than that within the lungs keeps the lungs expanded. A wound, spontaneous rupture (tension pneumothorax), or rupture of an emphysematous lung can cause air to enter the pleural cavity. Air enters the pleural cavity upon inspiration but cannot escape, and the lung gradually collapses. The patient experiences sudden, severe chest pain, breathing be-

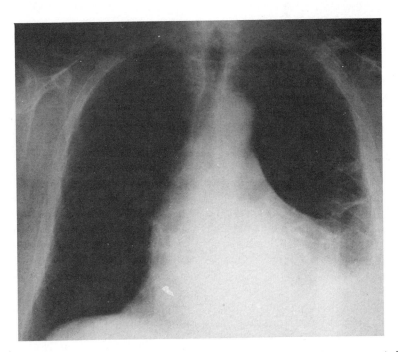

Figure 16-14 X ray of a patient with a left lower-lobe atelectasis accompanied by pleural effusion. (Courtesy of Salem Hospital, Salem, Massachusetts.)

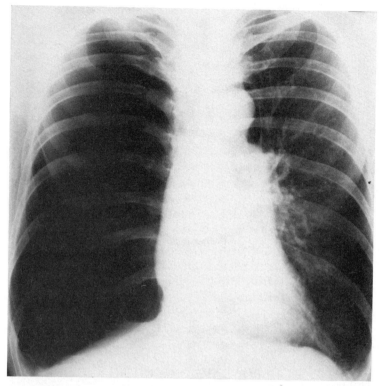

Figure 16-15 Right spontaneous pneumothorax. Lobes of the right lung have collapsed due to entrance of air into the pleural cavity. Contrast right side of chest (dark) with normal left side (gray), in which lung fills the cavity. (Courtesy Salem Hospital, Salem, Massachusetts.)

comes difficult (dyspnea), and the pulse becomes weak and rapid. Limitation of motion and hyperinflation on the affected side are revealed by physical examination. Unilateral pneumothorax can be tolerated fairly well, but bilateral pneumothorax may be fatal. The presence of air can easily be detected by the physical signs and x-ray examination (Figure 16-15).

Bronchiectasis

Bronchiectasis is a dilation of the bronchi, either local (saccular) or generalized (cylindrical). The smaller and peripheral bronchi and bronchioles may become chronically dilated (Figure 16-16). This dilation can result from two main factors—infection or obstruction.

Children who have had infections such as influenza or pneumonia may develop chronic inflammation of the bronchial walls, with destruction of the musculoelastic tissue resulting in dilation of the part of the bronchial tree. The resultant dilation favors future accumulation of secretion, with added infection and injury to the bronchial wall. An abscess may result. With the destruction of the smooth muscle in the walls, the patient is unable to cough up the purulent mucus. This infection can spread to the lung (pneumonia), to the pleural membrane on the lung (pleurisy), or invade the pleural cavity (empyema).

One major sign of bronchiectasis is the coughing up of foul-smelling, pus-containing material. If detected in the early stages, bronchi-

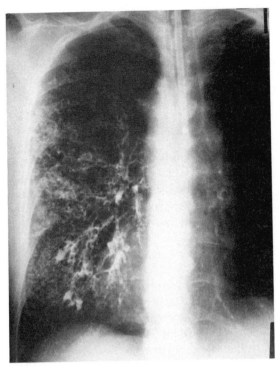

Figure 16-16 Bronchogram demonstrating bronchiectasis in the left lower lobe. Note the expanded white sacs that appear to be larger than the remaining lung tissue. (Courtesy Salem Hospital, Salem, Massachusetts.)

ectasis is treatable and reversible; however, chronic bronchiectasis develops into irreversible widening (either cylindrical or saccular) of the bronchi.

Pneumonia

Pneumonia (pneumonitis) is an inflammatory condition of the lung that may be caused by pathogenic bacteria, viral agents, or other agents. The lung is the one organ in the body that is in direct contact with the outside air and is, consequently, a prime target for airborne organisms. The onset in bacterial pneumonias is abrupt and is characterized by severe chills, fever, dyspnea, and sharp chest pain. In contrast, viral pneumonias are gradual, and the chief symptom is a hacking cough.

Approximately 90 percent of cases of lobar pneumonia are caused by the pneumococcus *Streptococcus pneumoniae*. Lobar pneumonia involves one lobe of the lung and is generally caused by organisms found in the flora of the upper respiratory tract. They are nonpathogenic.

Chronic Obstructive Lung Disease (COLD)

Chronic obstructive lung disease (COLD), also known as *chronic obstructive pulmonary disease (COPD)* refers to a group of lung disorders in which resistance to airflow is increased, especially on expiration. COLD is comprised of three distinct diseases: bronchitis, asthma, and emphysema. Characteristic manifestations are dyspnea and decreased exercise tolerance.

Cigarette smoking is the most common cause of emphysema and bronchitis. Cigarette smoke inhibits the action of cilia that line the respiratory tract and causes inflammation of the respiratory mucosa and hypertrophy and excessive secretion of the mucous glands. Exposure to industrial pollutants is an additional but less significant factor.

In obstructive lung disease the time required for forced vital capacity is increased, and the forced expiratory volume is decreased.

Bronchitis

Bronchitis is inflammation of the bronchus. The condition may be either acute or chronic. The acute form may also involve the trachea and is called *tracheobronchitis*. In this instance, the responsible irritant may be bacterial (pyogenic), mechanical (airborne dusts), or environmental (toxic chemicals or gases). In any case, the mucous membrane becomes red and swollen, and the lumen is filled with pus. Acute bronchitis is most serious in small children and those with chronic illness. The tiny bronchioles of small children may become obstructed while individuals with chronic illness are susceptible to secondary infections.

Chronic bronchitis is a prolonged and often slowly progressive condition of bronchial inflammation and hypersecretion of mucus. The mucous glands enlarge, and the mucosa becomes thickened and inflamed. Interference in the air passages due to mucosal swelling and excess mucus reduces the oxygen intake (hypoxia) of the patient and results in dyspnea and wheezing. When the obstruction becomes widespread, carbon dioxide is retained, resulting in cyanosis, and circulatory resistance is increased. Demands on the heart may be increased, leading to hypertrophy and possibly failure.

In a majority of cases, chronic bronchitis can be attributed to cigarette smoking and atmospheric pollution (sulfur dioxide). There is no known cure for chronic bronchitis. It is treated with antibiotics and moist vapors to relieve the symptoms. Preventive therapy is indicated for smokers and people who work in areas where the quality of air is poor.

Asthma

Bronchial asthma is a generalized spasm of the bronchioles resulting in marked wheezing and difficulty with expiration. The patient with asthma is subject to acute attacks of extreme dyspnea. Bronchial asthma is not inherited, but the associated hypersensitivities to various allergens is genetic. Common allergens are house dust, pollen, animal dander, food, molds, perfumes, cosmetics, solvents, and aerosol cleaners. Inhalation of an allergen by an asthmatic person results in constriction of the bronchioles due to a reaction of the antigen with immunoglobulin (IgE) antibodies coating mast cells in the lung. The mast cells and basophils release histamine, which constricts the smooth muscles of the bronchial tubes. Also, due to narrowing of the bronchial tubes, mucous plugs form and obstruct the airways. Expiration of air is reduced if not totally obstructed, and the patient wheezes or pants for air.

Immediate relief occurs with administration of epinephrine (a vasodilator). At present,

the maintenance of adequate blood levels of theophylline is the prescribed long-term treatment. Theophylline indirectly supports the increase of cyclic AMP in the tissues, which results in the relaxation of bronchial smooth muscles and inhibition of histamine release from the mast cells and basophils. Adrenal corticoid steroids have proved to be somewhat effective antiasthmatic drugs. However, their potential side effects restrict their use.

The cause of asthma is still a question. A family history of allergy and an individual history of hypersensitivity are prominent indications. Psychological factors also play a major role in some cases. The emotional factor has been known to initiate an asthmatic attack.

There is no cure for asthma. It is important to identify the precipitating allergen and avoid its contact. Skin tests (scratch tests) can isolate and identify the allergen or allergens. Allergy shots, with resultant desensitization, can reduce the incidence of severity of the attacks.

Emphysema

Emphysema is a pathological enlargement of the alveoli and air passages due to the rupture and tearing of the elastic alveolar walls. This results in a lung filled with pockets that cannot collapse. The disease may be a consequence of bronchitis, pulmonary tuberculosis, whooping cough, or bronchial asthma.

As the disease progresses, shortness of breath increases, and the thorax becomes enlarged and barrel-shaped as the patient attempts to inhale. The injury may be limited to a primary lobule distal to the respiratory bronchiole (panlobular), it may be limited to the center of a secondary lobule (centrilobular), or very large balloons in the lung called *bullae* may occur. The exchange of gases in these alveoli becomes markedly impaired. This inability to exchange lung gases leads to dyspnea, wheezing, and exhaustion. Resultant acidosis, cyanosis, and failure of the right side of the heart (cor pulmonale) can be fatal.

To properly understand the entire process

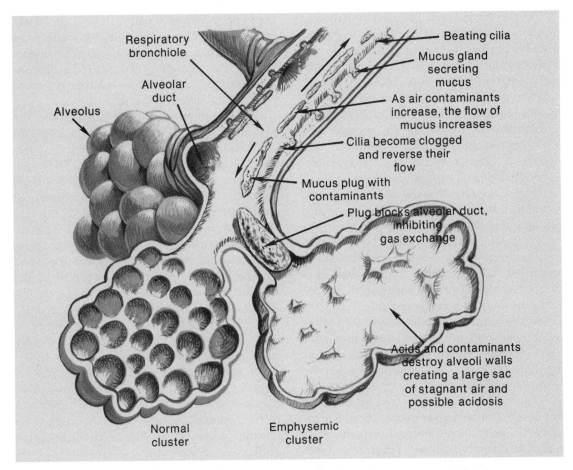

Figure 16-17 Diagrammatic representation of the development of emphysema.

of lung obstruction and destruction in emphysema, it is best to describe the process anatomically. Deeply imbedded in the wall are mucous glands. The glands produce a large amount of mucus. The mucus is transported up the inside surface of the airway where tiny hairs called cilia propel it upward toward the throat where it may be expelled or swallowed. At the surface, goblet cells also produce a small amount of mucus (Figure 16-17). The mucus is continually propelled upward against gravity by the swimming motion of the cilia. This is one of the lung's important defenses. When smoke or dust particles enter airways, the defenses are ready to repel them. The falling smoke and dust particles are trapped in the mucus layer. If the contamination is not too severe, cilia and mucus can keep the airways clean. Heavy contamination over the years imposes a severe strain on the lung's defenses. The most common lung contaminant is tobacco smoke. Under severe conditions the cleaning processes begin to break down. The dust and smoke eventually overload the cilia, and they can no longer remove the dirt-filled mucus. The body must now rely on coughing to clear the airways, hence the smoker's cough. The uncleared mucus sinks more deeply into the lower airways, however, and as the ciliated cells are shed, they are replaced by skinlike cells.

Deep in the airways, the constant irritation causes the mucous glands to increase in size and produce even larger amounts of mucus. This excessive production of mucus results in coughing, a symptom of bronchitis. Mucus that is not coughed up may clog the smaller airways. When a mucus plug cuts off the supply of fresh air to an alveolar cluster, the air exchange process is stopped, and inflammation and infection may result. Earlier when air was available, blood cells released carbon dioxide and picked up oxygen. But now that the air supply to this part of the lung is cut off, the air exchange process fails, and the blood cannot obtain oxygen for the body's tissues. Alveoli break down, forming large useless sacs (bullae). This is the pattern of fully developed emphysema. In a lung where this disease is widespread, the patient's breathing is severely limited.

As the lungs become less efficient in moving air in and out, the heart tries to pump more blood to compensate for the inadequate oxygen supply to the body. This overwork causes the heart to hypertrophy.

Diagnosis of emphysema can be achieved by physical tests and x-ray examination. The total vital capacity should normally be exhaled in 3 seconds. If emptying time is longer than 5 seconds, significant bronchial obstruction is present. The match test is helpful. If the patient is unable to blow out a match held 6 inches from the face, ventilatory obstruction is evident. X-ray examination reveals increased volume of the lungs. The anteroposterior chest diameter is increased. Bullae may appear, and tissue destruction and fibrosis can be seen. Scintillation scanning and pulmonary angiography are also helpful in evaluation.

Tuberculosis

Tuberculosis is a communicable disease caused by *mycobacterium tuberculosis.* Although many strains of the tubercle bacillus (TB) exist, it is believed that only the bovine (ox or cow) and the human varieties are pathogenic for humans. Tuberculosis is spread by coughing and sneez-ing. The causative organism has a protective coat (capsule) that enables it to live outside the body for prolonged periods of time. This disease used to be a major cause of death; today, it ranks twentieth. Inborn resistance is important; some races are more susceptible than others.

There are two forms of tuberculosis. The first infection with tubercle bacilli occurs in children following initial contact with the bacillus. This is called *primary tuberculosis.* The lesion is a small caseous spot or focus called a *Ghon lesion.* Initially neutrophils surround the bacilli in the alveoli and the bronchioles. The bacilli then drain into lymph nodes where T lymphocytes react with the bacilli (cell-mediated immunity) and activate neighboring macrophages. The bacilli are engulfed and destroyed by the macrophages, calcium salts and fibers encompass the bacilli, and the infection is arrested until such time as the person's immunity is decreased. With the production of antibodies by the tubercle bacillus and its toxin, the cells become hypersensitive to the bacillus, and reinfection must proceed along other lines. This hypersensitivity is the basis for the tuberculin test. Injection of the tubercle protein triggers an immune response (within 48 to 72 hours) if antibodies are present.

Secondary tuberculosis also relies on the lymphatic system. Reinfection normally occurs in adults and is basically a reactivation of the bacilli that were arrested during the primary infection. Infection spreads to the regional lymph nodes, which become enlarged and caseous. If the infection is not checked and it invades the bloodstream, other organs can become involved, resulting in fatal miliary tuberculosis or invasion of a bronchus with rapidly fatal bronchopneumonia.

Diagnosis of tuberculosis is made by microscopic detection of bacilli in purulent sputum. Lesions can be detected through x-ray examination. A person's immune response should be tested using the tuberculin (Tine) test. Positive swelling and redness at the site of the injection after 48 hours indicates an exposure to the bacillus and antibodies have been produced.

Pneumoconioses

Pneumoconioses are a group of diseases caused by longstanding lung irritation due to inhalation of certain toxic dust or metallic materials. The particles are inhaled into the lungs where they are engulfed by phagocytes and deposited in lung and lymphoid tissue. This may cause fibrosis and destruction of lung tissue.

Silicosis is the most widespread and best known pneumoconioses. Silica (silicon dioxide) is a fine dust particle contained within many of the hard surface materials that are mined or shaped today. Coal, gold, and tin miners and granite, stone, and quartz cutters are constantly inhaling silica. These fine particles are inhaled into the lungs and produce inflammation. Fibrosis occurs and, in most cases, develops into emphysema. Silicosis is diagnosed by clinical history and x-ray examination.

Anthracosis ("black lung") is a coal miner's condition. After many years of exposure to carbon particles, the coal miner retains coal dust containing silica. This results in shortness of breath and cough due to the progressive fibrosis. All city dwellers have a minor degree of anthracosis from the soot and polluted air they breathe.

Asbestosis is caused by the inhalation of asbestos or magnesium silicate. Recently asbestos was declared a carcinogenic agent. Individuals who inhale asbestos and smoke cigarettes are at a greater risk of developing cancer than if exposed to one or the other. Asbestos inhalation causes diffuse fibrosis of the lungs and impaired respiratory function. Asbestosis may be acquired in the process of the manufacture of asbestos or from handling the insulation product. A simple act of cutting the board insulation creates a fine dust that can be inhaled and cause lung injury.

Lung Neoplasms

Lung neoplasms may be primary, developing from the lung or mediastinum, or secondary, as a result of a primary tumor elsewhere. Primary tumors may be benign or malignant (bronchogenic carcinoma). The tumor may result in obstructive problems (emphysema or atelectasis) and, with continued growth, may require surgical removal to remedy the situation.

Lung cancer has accelerated to the point where it is the number one cause of death from cancer in the United States, especially in men 50 to 60 years of age. The incidence of malignant lung tumors has increased markedly over the past 25 years. The rate is considerably higher in urban and industrial areas than in rural districts.

Bronchogenic carcinoma, a common form of lung cancer, originates in a bronchus. The tumor may grow until a major bronchus is blocked, cutting off the air supply to that lung. The lung then collapses, and secondary infection results in pneumonia or the formation of a lung abscess. Symptoms of lung cancer invariably begin with a cough, pain, and shortness of breath due to a decrease in pulmonary function. About half of the patients have bloody sputum (hemoptysis), which results from erosion of blood vessels by the growing malignancy. If a bronchus becomes occluded, there may be diminished expiratory function and dyspnea heard as wheezing. In some cases loss of appetite, loss of weight, and general body weakness accompany the growth of the tumor. Metastasis to the brain, liver, and bone may occur in advanced stages.

The lung is a common site of metastatic cancer spread by the bloodstream. Tumor cells from other parts of the body are carried to all areas of the lung by vascular channels. Some grow to produce tumor nodules (sarcomas). Many advanced carcinomas also spread to the lung.

Diagnosis of lung carcinoma is made by x-ray, cytological, and bronchoscopic examination. Early x-ray diagnosis offers the greatest possibility of detection and cure. Unfortunately, only lung cancers that are of sufficient size and maturation can be seen. Cytological examination of sputum can reveal bronchogenic carcinoma cells (Figure 16-18), since the tumor

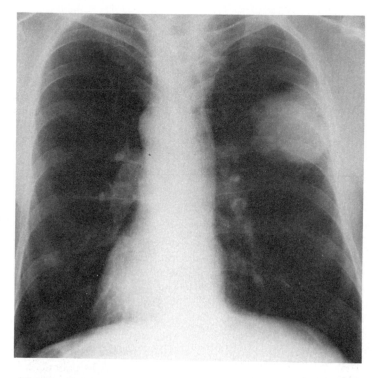

Figure 16-18 X ray of a right upper lobe bronchiogenic carcinoma. (Courtesy of Salem Hospital, Salem, Massachusetts.)

commonly exfoliates into sputum. Bronchoscopic examination with biopsy of the tumor is the most successful technique in the diagnosis of carcinoma. Examination of the pleural fluid (thoracentesis) is also used in diagnosis.

Respiratory Distress Syndrome

Respiratory distress syndrome (RDS), or *hyaline membrane disease,* is an abnormality of the lungs in newborn infants. If an infant is born prematurely before the lung tissue is sufficiently developed, a membrane of protein (**hyaline**) may form on the lining of the alveoli. The thick hyaline membrane coats the alveoli, blocking the passage of gases through the alveolar capillaries.

It takes a few hours after birth for the membranes to form and dyspnea occurs within 36 to 48 hours. Production of the membrane is related to a lack or insufficient quantity of surfactant fluid containing lecithin within the alveoli. This fluid is produced by the lungs and acts to bathe the alveoli, thereby preventing their collapse. It is now possible to withdraw amnionic fluid (by amniocentesis), to chemically determine whether enough surfactant is being produced to prevent this disease.

Hyaline membrane disease is a major cause of death in newborns. Death may be prevented by connecting the infant to a mechanical respirator for a period long enough for the lungs to have time to mature and produce adequate quantities of surfactant. The same hyaline membrane disease may also be caused by prolonged ventilation using respirators.

Sudden Infant Death Syndrome

Sudden infant death syndrome (SIDS) is defined clinically as the sudden, unexpected death of an apparently healthy infant for whom an autopsy fails to identify the cause of death. It is often called *crib death*, since it usually happens while the baby is sleeping. SIDS occurs in over 7000 infants every year in the United States, making it the most frequent cause of death between the ages of birth and 1 year.

The initial hypothesis for this problem was underventilation. Over half of the victims showed signs of increased muscle structure, or hyperplasia, in the pulmonary arteries. Underventilation reduces the level of oxygen in the alveoli, causing nearby arteries to constrict. Hyperplasia increases the resistance of the vascular system to blood flow to the lungs, which in turn produces hypertrophy of the right side of the heart.

A second physical sign to support the underventilation or hypoxia theory was the presence of abnormal quantities of brown fat found in more than half of the victims. Normally, brown fat, which is rich in mitochondria, is lost following the first year of life, but will persist if the infant is in a state of hypoxemia.

Another supportive finding was the production of red blood cells by the liver. Normally, the liver does not make red blood cells after the first week of life. This condition, however, is caused by hypoxemia, which stimulates the kidney to produce and release erythropoietin, which stimulates the manufacture of red blood cells.

Recently evidence has been presented to relate the hypoxemia to abnormal respiratory control centers in the brain. The carotid body, a small sensory organ in the neck, has an important role in detecting oxygen content of the blood and controlling the reflex activity of respiration. More than half of the victims of SIDS had an underdeveloped carotid body. If this organ is not operating correctly, the infant may not be able to restart its breathing during a prolonged period of apnea.

The long-term prospect for preventing SIDS is promising. Extensive investigations are now in progress to determine an exact cause. Several drugs that stimulate the respiratory control centers might serve to prevent future episodes if such a situation is indicated.

Pulmonary Embolism

Pulmonary embolism causes over 200,000 deaths per year. It is second only to coronary artery disease in cardiovascular mortality. The majority is not detected except by autopsy, thus recognition and diagnosis are difficult.

A clot which forms abnormally in a blood vessel is called a thrombus. Slow stagnant venous blood flow, abnormalities in the endothelial lining of blood vessels (atherosclerosis, infections, trauma), and alterations in the various components of the blood-clotting mechanism are predisposing factors to blood clot formation. If the clot is broken off from its point of origin and floats through the bloodstream, it is called an embolus. Depending on the number and size of the clots, one or more branches of the pulmonary artery may become occluded. This is called a pulmonary embolism (Figure 16-19). The immediate result is an obstruction of the blood flow to that portion of the lung which it supplies. The consequences may be respiratory and circulatory; either or both are fatal.

Pulmonary embolism may be present as a sudden onset of dyspnea and anxiety, with or without substernal pain, followed by acute right heart failure and circulatory collapse. Blood studies may be normal or slightly increased (sedimentation rate and leukocytosis). Blood gases are decreased. ECG is often normal except for an elevated heart rate, and chest x-ray examination may be normal.

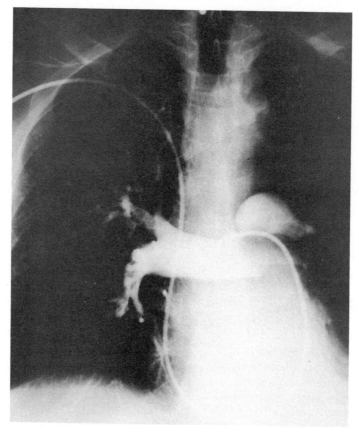

Figure 16-19 Pulmonary embolism. Note the expanded spread of dye in the upper and middle right lobes of the lung. A blockage (embolism) has prevented the dye from spreading into the right lower lobe. (Courtesy of Salem Hospital, Salem, Massachusetts.)

OUTLINE

I. Introduction
  A. Purposes of respiration
    1. Delivers oxygen to the tissues
    2. Removes carbon dioxide from the tissues
II. General anatomy
  A. Upper respiratory tract
    1. Nose
      a. Ciliated epithelium
      b. Nasal septum
      c. Nasal conchae
      d. Mucous membranes containing olfactory nerve fibers
    2. Pharynx—membranous sac between the nasal and oral apertures and the esophagus
    3. Abnormalities
      a. Deviated septum
      b. Nosebleeds
  B. Lower respiratory tract
    1. Larynx—voice box
      a. Guards opening of trachea

b. Contains vocal folds—speech

c. Laryngoscopy

d. Laryngospasm—croup

2. Trachea

  a. C-shaped hyaline cartilage rings

  b. Connects larynx with bronchi

  c. Lined with ciliated mucous epithelium

  d. Tracheotomy–tracheostomy

3. Bronchi

  a. Supported by overlapping cartilaginous rings

  b. Ciliated mucous epithelium

  c. Bronchoscopy

4. Bronchioles

  a. Small bronchi

  b. No supporting structure

5. Atria

  a. Elongated sacs at ends of bronchioles

  b. Lined with alveoli

    (1) Thin-walled air cells

    (2) Highly vascular

    (3) Gas exchange occurs here

6. Mediastinum

  a. Divides thoracic cavity into two parts

  b. Right lung—larger, three lobes

  c. Left lung—smaller, two lobes

  d. Pleura—serous membrane

    (1) Parietal pleura—lines thoracic cavity

    (2) Pulmonary pleura—covers lungs

    (3) Membranes secrete mucus to avoid friction between pleura

    (4) Pleurisy—inflammation of pleural membranes

III. Mechanism of breathing

  A. Anatomy

    1. Respiratory muscles

      a. Diaphragm—dome-shaped at rest

      b. Intercostal—between the ribs

    2. Nerve supply

      a. Phrenic—controls diaphragm

      b. Intercostal—controls intercostal muscles

    3. Thoracic pressures

      a. Intrathoracic pressure—in thoracic cavity

      b. Intrapulmonary pressure—in lungs

  B. Physiology

    1. Stimulation of respiratory muscles

      a. Contraction results in increased thoracic volume and decreased intrathoracic pressure

      b. Decreased intrathoracic pressure causes air to be sucked in—inspiration

    2. Relaxation of respiratory muscles

      a. Thoracic cavity returns to normal shape and volume

      b. Increased intrathoracic pressure forces air out—expiration

    3. Pneumothorax—air in thoracic cavity

IV. Lung volumes

  A. Spirometer and kymograph

    1. Tidal volume

    2. Vital capacity

    3. Inspiratory capacity

    4. Expiratory reserve

    5. Residual air

    6. Total lung capacity

  B. Ventilation

    1. Eupnea—normal

    2. Hyperpnea—increased

    3. Apnea—decreased or discontinued

    4. Dyspnea—difficult

    5. Cheyne-Stokes—hyperpnea-apnea

V. Physiology of respiration

  A. Neural control

    1. Brain—medulla

      a. Inspiratory center—stimulates

      b. Expiratory center—inhibits

      c. Pneumotaxic center—in pons

        (1) Safety mechanism

    2. Sensory receptors

      a. Stretch receptors—in lungs

b. Chemoreceptors—in alveoli and aortic and carotid bodies

B. Chemical regulation
1. Increased CO_2 in bloodstream stimulates inspiratory center to promote increased inspiration
2. Increased ventilation permits blood to lose CO_2 more readily

VI. Respiratory function of the blood
A. External respiration—exchange of O_2 and CO_2 between air and blood
B. Gaseous exchange
1. Exchange regulated by diffusion pressure gradients
2. In lungs
a. O_2 content higher in alveoli than in capillaries; therefore O_2 diffuses into capillaries
b. CO_2 content higher in capillaries; therefore, CO_2 diffuses into alveoli
3. In tissues—reverse of lungs
a. CO_2 diffuses into capillaries
b. O_2 diffuses into tissue cells
4. Small quantities of nitrogen gas remain in inert form
C. Gaseous transport
1. Hemoglobin in erythrocytes acts as carrier
2. Oxygen transport
a. $O_2 + Hb \rightleftharpoons HbO_2$ (oxyhemoglobin)
3. Carbon dioxide transport
a. $CO_2 + H_2O \rightleftharpoons H_2CO_3$ (dissolved CO_2)
b. $Hb + CO_2 \rightleftharpoons HbCO_2$ (carboxyhemoglobin)
c. $HbCO_2 + NH_2 \rightarrow HbNH_2COOH$ (carbamino Hb)
d. $CO_2 - H_2O \rightleftharpoons H_2CO_3 \rightleftharpoons H^+ + HCO_3^-$ (bicarbonate ion)
4. Chloride shift
a. Carbonic anhydrase aids bicarbonate ion formation in red blood cells but not in plasma

b. Bicarbonate ions diffuse from red blood cells into plasma
c. Chloride ions diffuse from plasma into red blood cells in compensation
5. Cyanosis—lack of oxygen and increased carbon dioxide

VII. Clinical considerations
A. Oxygen therapy
1. Hypoxia—deficiency of oxygen
2. Anoxia—absence of oxygen
3. Utilized in lung diseases, carbon monoxide poisoning, etc.
4. Restores oxygen tension of blood plasma
B. The bends
1. Nitrogen gas is found in solution in blood under normal atmospheric conditions
2. Increased external pressures produce nitrogen gas bubbles
3. Normal reduction of increased external pressure returns N_2 bubbles to solution
4. Too rapid a decompression traps N_2 bubbles, and they circulate in the body, causing pain and ultimate death
C. Artificial respiration
1. Support of normal breathing
2. Schafer method—prone pressure
3. Holger-Neilsen method—back pressure–arm lift
4. Mouth-to-mouth method—cardiopulmonary resuscitation (CPR)

VIII. Disorders of the respiratory system
A. Atelectasis
1. Primary—congenital
2. Secondary—alveolar collapse
B. Pneumothorax
C. Bronchiectasis
D. Pneumonia
E. Chronic Obstructive Lung Disease (COLD)
1. Bronchitis

2. Asthma
3. Emphysema
F. Tuberculosis
1. Primary—Ghon lesion in children
2. Secondary—infected lymph nodes
G. Pneumoconioses
1. Silicosis—silica
2. Anthrocosis—coal dust
3. Asbestosis—asbestos (insulation fibers)

H. Lung cancer
1. Primary—develop from lung
2. Secondary—metastasis from another organ
I. Respiratory distress syndrome (RDS)
1. Hyaline membrane disease
J. Sudden infant death syndrome (SIDS)
1. Hypoventilation
2. Abnormal medullary control
K. Pulmonary embolism

STUDY QUESTIONS AND PROBLEMS

1. What is the purpose of respiration?
2. Differentiate structurally between the trachea and the bronchi.
3. How do the vocal folds produce sound?
4. Trace the passage of air from the nose to the alveoli.
5. If a tracheotomy had to be done as an emergency measure, how would you do it?
6. What part do the diaphragm and intercostal muscles play in the mechanism of breathing?
7. Describe the mechanism of breathing. What pressures are involved?
8. What is meant by inspiration and expiration?
9. Mike had chest surgery and was sent back to his room with an underwater seal suction coming from his chest. Why?
10. Discuss the Hering-Breuer reflex.
11. Why do we need a pneumotaxic center?

12. Define chemoreceptor and pressoreceptor and describe their roles in respiration.
13. Mr. Jones is stabbed in the chest by a neighbor. What will happen to his respiratory mechanism?
14. What is the function of the pleura? What is pleurisy? Pneumothorax?
15. Discuss the reflex control of breathing. Can Mary sit in the corner and hold her breath until she dies? Why?
16. What is meant by a diffusion pressure gradient?
17. What body organ is most sensitive to lack of oxygen?
18. If you saw a man choking on something, what would you do?
19. Differentiate among asphyxia, emphysema, atelectasis, and bronchiectasis.
20. What are the effects of rapid decompression? Of hyperventilation?

17

The Alimentary Tract: Metabolism

LEARNING OBJECTIVES

- Define metabolism and state where and how it occurs

- Identify the basic structure of the three foodstuffs

- Identify the common carbohydrates, fats, and proteins in the diet

- Differentiate between a fat-soluble and water-soluble vitamin and give examples

- Describe the effect of minerals on the body

- Identify the anatomy of the alimentary tract

- Visualize the parts of a typical tooth

- Explain how saliva is produced and its effect on carbohydrates

- Describe peristalsis and its control mechanism

- Define varicose veins and give examples of occurrence along the alimentary tract

- Describe a hiatus hernia

- Differentiate between the cardiac and pyloric stomachs in reference to position and function

- Define peptic ulcer and where it is found

- Identify the sites and formation of bile

- Describe the role of bile in digestion

- Identify the function of hormones in digestion and give examples

- Identify the enzymes and steps involved in the digestion of the foodstuffs

- Explain how the end products of digestion are absorbed by the villi

- Describe the mixing and propulsive movements of the large intestine and the act of defecation

- Describe jaundice and explain how it originates

- Detail the steps in aerobic metabolism in the cell

- Explain the role of ATP in bodily functions

- Define calorie and relate its content to the three basic foodstuffs

- Relate the importance of a substrates enzyme,

pH, and proper temperature to provide a complete reaction

- Explain how each of the three foodstuffs is metabolized for energy, synthesis, and storage

- List different techniques used to diagnose an alimentary tract disorder

IMPORTANT TERMS

bile	coenzyme	emulsification	mastication	transamination
bolus	constipation	endoscope	metabolism	ulcer
calorie	deamination	endothermic	nutrition	varices
carbohydrate	defecation	enzyme	pepsin	vitamin
caries	deglutition	exothermic	peritoneum	
cholecyst	diarrhea	fat	protein	
chyme	diverticulum	intussusception	sterol	
cirrhosis	dysphagia	jaundice	substrate	

The composition of an organism's internal environment depends ultimately on what is eaten. **Nutrition** is the utilization of food for growth. Nutritional substances required by the cells must be provided for them dissolved in interstitial fluid. A state of adequate nutrition exists when the internal environment contains all the materials the cells require in the correct proportions and concentrations.

CELLULAR REQUIREMENTS

The absolute necessity of water is self-evident. It is consumed as such or in beverages; it is also present in most foods.

In a resting state, the very existence of a cell as an island of highly organized material requires energy. Any form of work, such as muscular contraction, adds to this energy requirement. The most important source of energy for the cells is oxidized glucose (simple

sugar); therefore the diet must contain materials that will yield glucose. These materials are the carbohydrates.

Much of the cell's structure is protein. Because protein molecules are extremely large, they cannot readily diffuse out of the blood capillaries or across cell membranes. The internal environment therefore must contain units from which proteins can be constructed by the cell. These building blocks, the amino acids, are obtained by the digestion of dietary protein in the intestine, where they are readily absorbed and carried to the cells.

Biochemical catalysts, the **enzymes**, facilitate the many chemical reactions that proceed in living cells. Enzymes themselves are protein in nature, but many of them require the presence of other organic compounds, **coenzymes**, before they can perform their tasks. The essential dietary constituents from which many of the coenzymes are replenished are the vitamins. They are required only in minute quantities, unlike the proteins and carbohydrates.

Vitamins

Vitamins, once known as "accessory food factors," are chemical substances found in fresh foods. They are of great importance for the maintenance of health but are required by the body only in minute quantities. Vitamins act as organic catalysts in a manner somewhat similar to that of the digestive enzymes: they assist in chemical reactions in the body without being destroyed in the process. The vitamins constitute a *nutritional factor*, which has been defined as a single substance performing a vital function in nutrition. In humans, absence or insufficiency of certain vitamins in the diet results in various deficiency diseases (Table 17-4).

Vitamins may be classified into two large groups, those soluble in fat and those soluble in water.

Fat-Soluble Vitamins

Fat-soluble vitamins require the emulsifying action of bile in the intestinal tract for their absorption.

Vitamin A and its precursor substance, carotene, provide a nutritional factor essential to growth and to the maintenance of epithelial tissues in particular. Vitamin A is found in animal fats, milk products, and the oils of fish livers (for example, cod liver oil). Carotene is found in green vegetables, carrots, and some fruits; it is converted into vitamin A in the body.

Two forms of vitamin A are detectable—an acid and an aldehyde. Vitamin A acid is responsible for the maintenance of the epithelium; vitamin A aldehyde makes up part of the molecule of rhodopsin, the retinal pigment essential for night vision (see Chapter 11). Vitamin A is not destroyed by ordinary methods of cooking and is therefore stable. Lack of the vitamin in the diet leads to growth failure, night blindness, a thickening and roughening of the skin, and, in severe cases, destruction of the cornea and blindness.

Vitamin D is the *antirachitic* (rickets preventing) *factor*. A lack of it causes rickets, a disease characterized by softening and deformation of bone. In adults, vitamin D deficiency leads to a form of osteomalacia in which the bones become soft and fragile. This disease is due to a deficiency of calcium and phosphorus in the bones; vitamin D assists in the absorption of these salts into the body from the contents of the small intestine.

Vitamin D is found in fish liver oil, milk products, and eggs. It is not destroyed by heat or by the canning process.

Vitamin E prevents sterility in experimental animals but not necessarily in humans. This vitamin is found principally in vegetable oils obtained from the germ of wheat and rice.

Vitamin K influences the clotting of blood. It promotes the synthesis of prothrombin by the liver. Its absence or deficiency in the diet leads to a delayed clotting time and hemorrhage (see Chapter 12). Vitamin K is found in a variety of foods, particularly such vegetables as cabbage, spinach, and tomatoes.

Diseases caused by deficiencies of each vitamin are listed in Table 17-4. It is important to remember, however, that very large doses of vitamins A, D, and K are toxic. Hypervitaminosis A is characterized by gastrointestinal disturbances, loss of hair, patchy skin, and bone pain. Hypervitaminosis K also results in gastrointestinal disorders and can cause anemia.

Water-Soluble Vitamins

The most significant problem associated with water-soluble vitamins is that they may be lost during the preparation of common foods. The water used to boil potatoes or to cook vegetables commonly is thrown out after the vegetables have been strained and served. Thus a significant quantity of the water-soluble vitamins is poured down the drain.

Vitamin B is not a single vitamin but a group of vitamins. Therefore the term *B complex* is used. The group includes vitamins B_1 and B_2, nicotinic acid (niacin), folic acid, vitamin B_{12}, and pyridoxine (vitamin B_6).

Table 17-4   Principal Vitamins and Their Functions

Name	Source	Function	Effects of Deficiency
Fat-Soluble Vitamins			
Vitamin A (precursor: carotene)	Animal fat, milk, fishliver oils	Vision (rhodopsin)	Skin disorders, night blindness, bone and nerve disorders
Vitamin D	Fish liver oils, milk, eggs	Ca and P regulation	Rickets, osteomalacia
Vitamin E	Rice oils, wheat germ	Reproduction (in experimental animals)	Sterility
Vitamin K	Green vegetables	Blood clotting	Hemophilia
Water-Soluble Vitamins			
Thiamine (B_1)	Peas, yeast, beans, nuts	Metabolism of energy-rich carbohydrate molecules	Beriberi, neuritis
Riboflavin (B_2)	Liver, milk, meat, eggs	Oxidation	Skin lesions
Nicotinic acid (niacin)	Meat, fish	Cellular oxidation	Pellagra
Folic acid	Liver, cereals, green vegetables	Red cell maturation	Anemia
Pyridoxine (B_6)			
Vitamin B_{12}			
Vitamin C (ascorbic acid)	Citrus fruits, tomatoes	Capillary function	Scurvy

In its active form, vitamin B_1 (thiamine) is concerned with the metabolism of energy-rich carbohydrate molecules. Effects of vitamin B_1 deficiency are far-reaching because it causes a peripheral neuritis leading to paralysis and loss of appetite—a disease called *beriberi*. This disease is especially common in the Orient, where polished (white) rice is used extensively as the principal food. Beriberi can be prevented by using unpolished rice instead of polished rice or by supplementing the diet with such other foods as peas, yeast, beans, ham, and nuts.

Vitamin B_2 (riboflavin) is present in eggs, liver, milk, and meat. It is a constituent of a series of enzymes, the flavo proteins, which serve vital roles in the oxidation of foods. A deficiency of riboflavin leads to lesions around the mouth, eyes, and nose.

Nicotinic acid (niacin) is a B vitamin widely distributed in several foodstuffs, including meat, fish, liver, yeast, and whole-meal flour. The vitamin is part of the structure of the coenzymes important in many different enzymic processes. Niacin deficiency leads to pellagra, a condition characterized by rough skin and lesions of the mouth and tongue, followed by a degeneration of the myelinated nerve fibers in the body.

Folic acid, vitamin B_{12}, and pyridoxine (vitamin B_6) collectively provide the *antianemic factor*. Folic acid is found in liver and leafy vegetables. Vitamins B_6 and B_{12} are found in liver, cereals, and green vegetables. This group is essential because it provides a metal, cobalt, necessary for the maturation of red blood cells. A deficiency causes pernicious anemia. Liver is a potent source of vitamin B_{12}, and patients of an earlier day were treated with crude liver extracts or "tonics." Today vitamin B_{12} may be administered directly by injection.

Vitamin C (ascorbic acid) is found in fresh fruits and vegetables. Prolonged cooking and overheating destroys the vitamin, and deficiency causes bleeding from the subcutaneous tissues, especially in the gums and other mucous membranes—a condition called *scurvy*. In children, deficiency of ascorbic acid causes the teeth and bones to develop in an abnormal fashion. The chief lesion in scurvy is a reduced tissue strength, leading to capillary fragility and hemorrhage.

Although one vitamin may have an effect on the action of another, one vitamin cannot replace another. Each has its own function, and absence of any one leads to a sequence of signs and symptoms. A balanced diet allows the body to select what it needs at any given moment.

Minerals

The principal mineral salts are the chlorides, phosphates, and carbonates of sodium, potassium, calcium, and magnesium. Their widespread occurrence in all foods means that dietary deficiencies are noted only under bizarre conditions. It is estimated that about 5 percent of the body weight is made up of mineral salts, which are found mainly in the bones and teeth as salts of calcium and phosphorus.

Calcium is the most abundant mineral in the body. Salts of calcium make up the hard parts of the skeleton and teeth; plasma calcium is important in the blood clotting mechanism; and myofibril calcium plays a major role in the contraction of skeletal muscle.

A fetus requires a large quantity of calcium for developing bones, and the young child requires it for growth. Dietary deficiencies of calcium lead to decalcification of bone (rickets in youth, osteoporosis in the adult) and, in women, to possible dental conditions in the year or two following childbirth.

The best sources of calcium are milk, beans, wheat, and other cereals. Calcium absorption by the intestine, however, depends on the availability of vitamin D. Also, any defect in fat digestion causes calcium to be lost with the undigested fat in the feces.

Phosphorus is the body's second most abundant mineral. It is widely distributed in the soft tissue cells and occurs in relatively low concentrations in the extracellular fluid. Nucleic acids in the genetic materials of the cells contain phosphorus, as does the most important energy carrier, ATP.

Iron is present in red blood cells as hemoglobin and is also a constituent of a whole series of respiratory enzymes responsible for controlled oxidation in the cells. The simplest symptom of anemia is a shortage of iron demonstrated by a low hemoglobin count. Eggs, meat, liver, and fruits are good sources of iron.

Some mineral elements, such as copper, manganese, and cobalt, are known as *trace elements* because they are required in such minute amounts. They are, however, essential to health. Iodine is the major component of the thyroxine produced by the thyroid gland. Copper, cobalt, and manganese act as catalysts in the chemical activity of the blood. The gaseous element fluorine plays an important role in preventing tooth decay.

DIGESTIVE ORGANS AND PROCESSES

Before food can be absorbed into the body, it must undergo a number of changes. Some of these are purely mechanical; others are of a highly complex chemical nature. Most food-stuffs are constructed of molecules too large to pass through the walls of the intestine, and the purpose of digestion is to convert these into smaller molecules.

During digestion, the chemical process of hydrolysis occurs. Complex molecules are split into simple molecules by the action of digestive enzymes, known collectively as the *hydrolases*. These enzymes are present in the secretions of the several digestive glands of the alimentary tract: the salivary glands, the gastric glands, the pancreas, and the intestinal glands.

The alimentary tract or canal extends from the mouth to the anus, and different names have been given to its various parts (Figure 17-1). The greater part of the tract has to do with the actual digestion of food, but it is also concerned with absorption of the nutrient part of food and with the excretion of waste products.

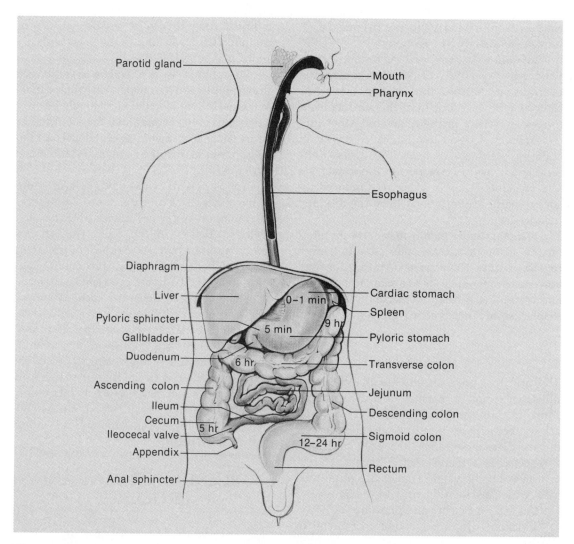

Figure 17-1 Human digestive tract. Times indicated along the tract represent how long it takes food to pass through each area during the process of digestion.

ric sphincter. The middle or main part of the stomach is the body. The left border of the stomach forms a large curve called the *greater curvature*. The right border, which is much shorter, is called the *lesser curvature.*

The wall of the stomach is composed of three layers of smooth muscle lined by mucous membrane and covered by peritoneum (Figure 17-6). The fibers of the outermost layer of muscle run longitudinally and are continuous with the esophagus. This layer is very thin at both extremities of the stomach but relatively thick along the curvatures. The middle layer is composed of circular fibers that encircle the stomach; the innermost layer of muscle fibers runs obliquely and does not form a complete coat, being confined largely to the fundus. The mucous membrane of the stomach is covered with nonciliated columnar epithelium, many cells of which secrete mucus for protection. At the lower end of the stomach the fibers of the circular muscle layer form the strong ring of the pyloric sphincter; at the upper end these fibers form the cardiac sphincter.

Three kinds of cells are in the mucosa of the stomach:

1. *Mucous cells,* located in the neck region of the stomach, secrete mucin, which protects the mucosa (inner lining) from the other highly acid gastric juices.

2. *Chief cells* (or *peptic cells*) are located along the tubule. These cells form glands for the secretion of pepsinogen.

3. *Parietal cells* (or *oxyntic cells*) located alongside the chief cells secrete hydrochloric acid (2 percent). These cells do not actually store and secrete hydrochloric acid; rather they extract the component chemicals, H^+ and Cl^-, from the parietal cells and blood cells, respectively. These then combine to form hydrochloric acid.

Hydrochloric acid secreted by the parietal cells combines with pepsinogen from the chief cells to form the enzyme **pepsin.** Pepsin is re-sponsible for the chemical breakdown of proteins to proteoses and peptones (peptides).

Rennin is an enzyme that acts upon milk proteins and is thought to be produced by the chief cells in small children. It is not found in the adult. Rennin coagulates milk, producing a white mass called *curd* and a fluid material called *whey.* The action of rennin is to convert milk protein, casein, into a form suitable for digestion (Table 17-5).

Peristaltic contractions initiated in the esophagus are continued through the stomach. Within minutes after the bolus has entered the stomach, it begins to soften, mixed with the acidic gastric juices, and reduced to a thin liquid mass called **chyme.** As the chyme is gradually propelled into the duodenum of the small intestine, the pressure of the fundus forces the food into the pyloric end.

The secretion of gastric juice is constant. It is a thin, colorless fluid with an acid pH ranging from 1.0 to 4.0. The quantity of acid secreted depends on the amount and type of food digested.

There is a correlation between enzyme activity and the relative acidity of the digestive juices (see Table 17-1). Salivary amylase, secreted in the mouth by the salivary glands, is basically neutral in reaction. The enzymes pepsin and rennin, secreted by the stomach's glandular cells, are effective in an acid medium. Understanding enzymatic activity is important because it is fundamental to the digestive process. Enzymes are specific in that they are effective only in their own pH range. Each enzyme works best in catalyzing a single chemical reaction.

Enzymes are frequently proteins. They act upon a substance called a **substrate.** Each enzyme catalyzes a chemical reaction most rapidly with a particular substrate but also catalyzes reactions with other substances of the same general type. A means of identifying the enzyme is the suffix *-ase* at the end of the substrate's name; for example, the enzyme maltase breaks down maltose into simple sugars.

Enzymes are affected by temperature as

Table 17-5   Digestive Enzymes

Location of Digestive Activity	Digestive Secretion	Source	Enzyme	pH	Enzyme Action
Mouth	Saliva	Salivary glands	Amylase (ptyalin)	6.8–7.0	Starch → maltose
Stomach	Gastric juice	Stomach lining	Pepsin	1–4	Proteins → proteoses and peptones
					Nucleoproteins → peptones and nucleic acids
			Rennin	1–4	Casein → calcium paracaseinate
Small intestine	Bile	Liver	—	8–10	Emulsifies fats; no enzymatic activity
	Pancreatic juice	Pancreas	Trypsin	8–10	Peptones → polypeptides
			Pancreatic amylase (amylopsia)	8–10	Starch and glycogen → maltose
			Pancreatic lipase	8–10	Fats → fatty acids and glycerol
	Intestinal juice	Small intestine (crypts of Lieberkühn)	Enterokinase	8–10	Polypeptides → amino acids
			Sucrase	8–10	Sucrose → glucose and fructose
			Maltase	8–10	Maltose → glucose
			Lactase	8–10	Lactose → glucose and galactose
			Lipase	8–10	Fats → fatty acids and glycerol
			Nuclease	8–10	Nucleic acids → nucleotides

well as by pH. Their activity increases twofold with every 10°C rise in temperature. This rate of increase in activity continues until temperatures of 40 to 45°C are reached; then the peptide bonds rupture and activity ceases. It is therefore necessary to maintain the normal body temperature (37°C or 98.6°F) to ensure an efficient digestive process. Lowering the body temperature decreases the rate of enzyme activity and energy production. Increasing the body temperature above 40°C (above 106°F) may be fatal if the condition is allowed to persist.

Hyperacidity of the stomach is usually a temporary condition resulting from overeating. Strong acids convert tissue protein to acid proteinase, which dissolves in concentrated acid. A common example of necrosis (tissue death) from acid is a peptic ulcer. Most peptic ulcers are located in the funnel-shaped pyloric antrum. Ordinarily, peptic ulcers result from oversecretion of hydrochloric acid, but the condition may occur if insufficient protection is provided by the mucus secreted by the stomach cells for just that purpose. If the necrosis is su-

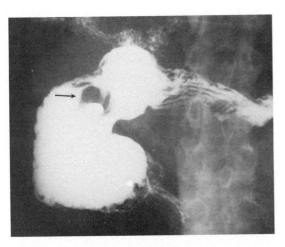

Figure 17-7 X ray of a sharply defined punched out peptic ulcer in the stomach lining (arrow). (Courtesy of Salem Hospital, Salem, Massachusetts.)

perficial, failing to extend completely through the mucosa, the lesion is termed an *erosion*. If the defect extends through and beyond the mucosa, it is an **ulcer.** Bleeding ulcers indicate that a vessel has ruptured in the defect, and blood is evident in the vomitus or feces. In perforated ulcer (see Figures 17-5 and 17-7), all the layers of tissue from the inside to the outside of the stomach wall are penetrated, causing extreme abdominal cramps and eventual collapse.

Surgical removal (gastrectomy) is indicated in perforated ulcer, whereas medicine and controlled diet may eventually cure general peptic ulcer. Regenerative activity is always present and may at any time lead to healing of the ulcer.

A common congenital anomaly of the stomach is obstructive narrowing of the pyloric valve, causing eventual closure of the opening or *pyloric stenosis* and preventing further passage of food (see Figure 17-5). Infants who repeatedly vomit their feedings may have this defect. Prolonged stenosis can result in constipation and rapid loss of weight. Dilation of the valve sphincter is indicated to relieve this problem.

Acute inflammations (gastritis) may be due to exposure to alcohol, caffeine, aspirin, bacte-

rial toxins, viral infections, and various allergens. The inflammation is generally short-lived as the mucosal lining cells of the stomach are normally replaced every 36 hours or so, and this rapid turnover allows any superficial damage to be repaired when the causative agent is removed.

Chronic gastritis, on the other hand, involves cellular changes that may be seen in biopsy specimens obtained through gastroscopy. The changes include accumulations of inflammatory cells or atrophy of glandular tissue. Pernicious anemia is an example of a result of chronic gastritis, caused by inefficient production of the necessary quantities of the intrinsic factor for red blood cell production.

Small Intestine

The food leaves the stomach through the pyloric sphincter and enters the small intestine, where most digestion and absorption take place. The first section of the small intestine is the *duodenum* (Figures 17-6 and 17-8). The duodenum leads into the middle portion, the *jejunum*, and the small intestine finally terminates in the *ileum* (Figure 17-8). The entire small intestine of an adult male is about 7.5 meters (23 feet) long and 2.5 cm (1 inch) in diameter.

Partially digested food from the stomach passes into the duodenum, where the acidity of the food stimulates the intestine to secrete intestinal hormones. These in turn stimulate the pancreas to secrete pancreatic juices and the liver to secrete bile.

Bile salts are produced by the liver and stored in concentrated form in the gallbladder during periods when the intestine is devoid of food. When the chyme reaches the intestine from the stomach, the major function of bile salts is the digestion and absorption of fats. Bile salts emulsify the fats into minute fat particles, thus providing a larger surface area for the action of pancreatic lipase. Bile (pH 7.0 to 8.0) also aids in the conversion of the acid chyme to an alkaline medium (pH 8.0 to 10.0), which is the necessary pH range for intestinal enzyme activ-

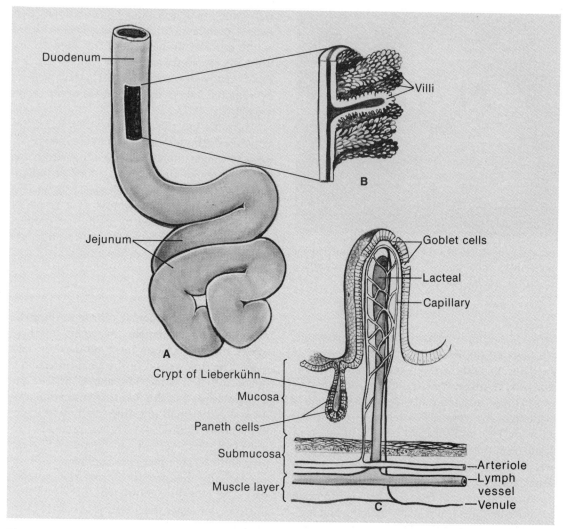

Figure 17-8 Small intestine. **A,** The long muscular duodenum and jejunum, two of the three parts of the small intestine. **B,** A section cut out of the wall of the duodenum, showing the many villi. **C,** Midsagittal section of a villus, showing the arrangement of its vascular elements and digestive cells.

ity. The intestinal juices secreted by the crypts of Lieberkuhn and the succus entericus glands are always alkaline (pH 7.0 to 10.0) and assist the bile salts in maintaining a proper pH for intestinal digestion.

Since the small intestine is the place where absorption of the end products of digestion occurs, special structural adaptations increase the absorptive surface area (Figure 17-8). The surface of the inner mucous membrane is made up of tall columnar epithelium and contains an abundance of goblet cells. The middle layer of the intestine, the *submucosa*, consists of loose connective tissue lying beneath the mucous membrane and uniting it to the outer muscular layer. This third layer contains blood vessels, lymphatics, and nerve networks (Auerbach's plexus) dispersed among layers of circular and longitudinal muscles. A fine serous coat (*serosa*) covers the entire intestine on the outside. This layer is made up of fibrous tissue and is formed by a visceral layer of peritoneum.

The innermost lining of the small intestine is not smooth, but has numerous fingerlike projections, the *villi*, which greatly increase the absorptive surface (see Figure 17-8). In addition, electron microscopy has shown that this surface is further increased by numerous microvilli to such an extent that the total surface area of an intestinal lining is estimated to be five times the surface area of the exterior of the body.

The muscularis mucosa continually contracts and relaxes rhythmically (peristalsis) during digestion. This causes corresponding shortening and lengthening movements of the villi that act to stir the intestinal fluids, thereby facilitating absorption of the digestive products.

Within each villus is a network of capillaries and a central lymph vessel or lacteal. Simple sugars, amino acids, vitamins, minerals, and water are absorbed by the capillaries. Fatty acids and glycerol are absorbed by the lacteal, and after absorption they again combine to form the triglycerides found in the bloodstream.

The intestinal glands are Paneth cells in the crypts of Lieberkuhn (see Figure 17-8). These secrete the following enzymes, which assist the pancreatic enzymes to a marked extent:

1. *Enterokinase,* a protein-splitting enzyme, activates trypsin and completes the breakdown of peptides to amino acids.

2. *Maltase,* a carbohydrate-splitting enzyme, hydrolyzes maltose into glucose.

3. *Sucrase,* a carbohydrate-splitting enzyme, hydrolyzes sucrose into glucose and fructose.

4. *Lactase,* a carbohydrate-splitting enzyme, hydrolyzes lactose into glucose and galactose.

5. *Lipase,* a fat-splitting enzyme, splits emulsified fats into fatty acids and glycerol.

6. *Nuclease* acts on the nucleic acid component of available nucleoproteins.

Patches of lymphoid tissue, known as Peyer's patches or lymphatic nodules, are found in the mucosa of the small intestine.

The jejunum succeeds the duodenum. It is about 2.5 meters (8 feet) long and constitutes two-fifths of the coiled part of the intestine. The ileum, the remainder of the coiled part, joins the large intestine at the ileocolic valve (Figure 17-9). Although the duodenum is the major portion of the small intestine concerned with digestion, the entire small intestine is lined with villi and acts to absorb end products of digestion.

The first part of the duodenum is the commonest site for peptic ulcers, which form 2 inches from the pyloric sphincter.

The patient with duodenal ulcers secretes an abnormally high amount of hydrochloric acid. This acid secretion is carried into the duodenum where ulceration occurs. The major complications are bleeding, perforation with possible peritonitis, and obstruction. The patient with a duodenal ulcer experiences sudden excruciating pain as a result of either perforation or obstruction. Bleeding from the ulcer is common and may appear as hematemesis or melena.

Diagnosis and treatment include use of barium enema and x-radiation. Medication, change of diet, and alkalis (antacids) have proved beneficial to mild cases. If the ulcer does perforate, surgical resection is generally indicated.

A **diverticulum** of the intestine is an outpouching of the wall to form a sac whose mucosal lining is continuous with that of the intestine. The usual size of a diverticulum is that of a large pea. Its contents are fecal and are subject to bacterial infections. If multiple diverticula are present, the condition is called *diverticulosis.* If one or more diverticula become inflamed, the process is called *diverticulitis.* Congenital diverticula occur chiefly in the small intestine in the mesenteric border of the duodenum or jejunum. Diverticula may occur in any part of the intestine, but the most common area is the sigmoid colon of the large intestine between the descending colon and the rectum.

Acute diverticulitis is similar in symptoms to appendicitis, except that it occurs on the left side of the abdomen. Diagnosis with barium and x-ray examination demonstrate the classic picture of small white balls lining the outside of the intestine due to the collection and latent emptying of each pouch as the barium is excreted.

Most patients with diverticular disease can be managed conservatively and lead a relatively normal life. Acute diverticulitis, on the other hand, may require surgery to prevent or alleviate complications of hemorrhage and obstruction. Meckel's diverticulum is an outgrowth or outpouching of the ileum and has been referred to as a secondary appendix.

Regional enteritis (Crohn's disease) is a chronic inflammatory disease of the small intestine causing fever, weight loss, and disturbed bowel function. It generally occurs in young adults and runs an intermittent course with mild to severe complications. The changes seen are those of chronic inflammation, with the formation of numerous granulomas and fibrous tissue in the wall of the small intestine. A unique feature of the disease is that some areas of the small intestine become affected while others do not. Each of the affected patches becomes thick walled and rigid, with intervening areas of normal, delicate, and flexible intestinal wall.

The cause of regional enteritis is unknown, but there does appear to be a psychogenic element involved. Patients with the disease have had frequent emotional upsets. There is a direct relationship to onset or relapse of the disease to stressful situations.

An **intussusception** is a prolapse of the intestine or other related organs (esophagus or stomach) in which the lumen folds over or under the adjoining portion, creating an overlapping of the organ and thus an obstruction. It occurs mostly in children and generally affects the ileum just proximal to the ileocecal junction.

Intestinal obstruction prevents the onward movement of the intestinal contents. The pathological effects are fluid and electrolyte loss, strangulation (Figure 17-5), vomiting, and probably acidosis. Malignant tumors of the small in-

testine (adenocarcinoma) can cause obstruction and jaundice. Removing the involved segment of the intestine (resection) may be curative.

Gallbladder

The gallbladder, or cholecyst, is a pear-shaped sac lying on the underside of the liver.

Bile passes from the liver where it is produced by way of the hepatic duct into the common bile duct and the gallbladder (Figure 17-9). The gallbladder stores the bile in the form of bile salts until needed for digestion, at which time they are released into the small intestine. The color of bile depends on the type and concentration of pigment that it contains. If the pigment is bilirubin, the bile is yellow; if the pigment is biliverdin, the color varies from green to black.

An important constituent of bile is cholesterol, which is also the chief constituent of gallstones (Figure 17-5). Many people who show a tendency to form gallstones are fed bile salts. These help to maintain cholesterol in solution and to stimulate liver bile flow, thereby reducing the possibility that the cholesterol concentration will rise to a level sufficient to cause gallstone formation.

Acute *cholecystitis* is inflammation of the gallbladder that may be caused by bacterial infection or by an increased retention of bile salts. The infection starts in the mucosa and extends into the wall of the gallbladder. Cholecystitis is generally responsible for the formation of gallstones in the gallbladder *(cholelithiasis)*. Pure cholesterol gallstones, described above, develop because of metabolic anomaly. Bilirubin gallstones develop as a result of hemolytic jaundice. The most common complications of cholelithiasis include obstruction, perforation of the ileum or colon, and formation of a fistula (ulceration).

A cholelithotomy is an operative procedure whereby the cystic duct or common bile duct is opened to remove gallstones; the ducts may be dilated to allow passage of the stones into the duodenum for eventual expulsion in the feces. If the cystic duct is damaged or if the stones

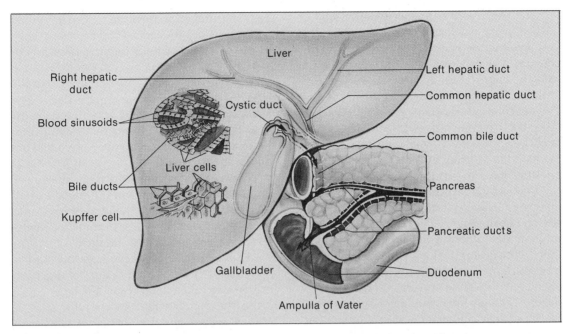

Figure 17-9 Liver and its interrelationship with the gallbladder, pancreas, and duodenum. A section has been removed from the liver and the area enlarged to show the arrangement of liver cells, bile ducts, Kupffer cells, and blood sinusoids to one another. Arrows indicate the direction of flow of bile from the gallbladder and liver and of digestive juices from the pancreas into the duodenum.

present are of a sufficient size and number to indicate future problems, removal of the gallbladder (cholecystectomy) may be indicated.

Removal of the gallbladder diminishes markedly the quantity of bile that will enter the duodenum; therefore, a low fat or nonfat diet is indicated to prevent digestive discomfort.

Pancreas

The *pancreas* is the most active and versatile of the digestive organs (Figure 17-9). In the absence of other digestive secretions, its enzymes alone are capable of almost completing the digestion of all foodstuffs. The pancreatic secretion into the duodenum is a complete digestive juice, since it contains all the enzymes needed to break down foodstuffs:

1. *Carbohydrate enzyme*, or *pancreatic amylase* (amylopsin), acts upon starch and glycogen, producing maltose.

2. *Fat enzyme*, or *pancreatic lipase* (steapsin), is the main fat digestive enzyme in humans. Each fat molecule is broken down into a molecule of glycerol and three molecules of fatty acid.

3. *Protein enzyme*, or *trypsin*, is secreted as an inactive precursor, trypsinogen, which is converted to active trypsin by enterokinase, an enzyme found in the intestinal juices. Trypsin digests proteins and peptides partially digested in the stomach. The end products of trypsin digestion are polypeptides and amino acids. A second proteolytic enzyme, chymotrypsin, is secreted by the pancreas as inactive chymotrypsinogen, which is converted to *chymotrypsin* by trypsin. Chymotrypsin acts in a manner similar to that of trypsin.

4. *Nuclease* is a nucleic acid–splitting enzyme that results in the production of mononucleotides.

The pancreas lies on the posterior wall of the abdomen; its expanded portion, or head, is positioned in the concavity formed by the duodenum. The pancreas is sometimes called the *salivary gland of the abdomen,* because microscopically it closely resembles the salivary gland. The secretions of the gland exit through a long tube, the pancreatic duct (Figure 17-6). This duct opens into the duodenal mucosa, often at the ampulla of Vater, and its contents join the intestinal juices to further the digestive process.

Inflammation of the pancreas *(pancreatitis)* involves a lack of coordination between the secretion of pancreatic juice and its ability to flow through the pancreatic duct into the duodenum. Obstruction of flow, whatever its cause, results in a rupture of the cells within the gland owing to the increased pressure of the enzymes (Figure 17-5). The enzymes amylase and lipase are measurable, and determination of their levels is clinically useful.

Interference with the secretion of the hormone insulin from the islets of Langerhans in the pancreas will result in diabetes mellitus, which is discussed in Chapter 10.

Large Intestine

In humans the junction between the small intestine and the large intestine (colon or large bowel) is usually in the lower right portion of the abdominal cavity. A blind sac, the *cecum,* projects from the large intestine near the point of puncture, the *ileocecal (ileocolic) valve* (Figure 17-10). Here a small, fingerlike process, the *appendix,* projects from the cecum. The appendix frequently becomes infected and sometimes must be removed surgically (appendectomy).

The cecum and appendix serve no purpose except as storage centers for microorganisms that aid digestion. Humans cannot digest the complex cellulose found in plant cells without assistance; the stored microbes digest this cellulose for them but most of the nutritional value of the material is lost. The duodenum, not the cecum, is the principal site of digestion.

The large intestine is about 2 meters (5 to 6 feet) long, and its sections are named according to the directions they travel and the shapes they assume. The cecum leads into the *ascending colon,* which is followed by the *transverse colon, descending colon, sigmoid colon* (shaped like the letter "S"), *rectum,* and *anus* (Figure 17-10). The muscular coat of the large intestine contains three flat longitudinal bands that extend from the vermiform appendix to the rectum and are known as the *taeniae coli.*

The major function of the large intestine is the reabsorption of water into the body. If all the water in which enzymes are secreted into the digestive tract were lost in the waste material (feces), dehydration would occur.

A second function of the large intestine is the removal of the waste products of digestion by a process called **defecation.** The large bowel is also the region where excess salts, such as those of calcium and iron, can be excreted from the body when their concentration in the blood is higher than normal. The last portion of the large intestine, the rectum, stores the waste materials until defecation occurs. The feces are expelled through the anus.

Occasionally, the colon becomes irritated, and peristalsis moves the material through it faster than normal. As a result, not enough water is reabsorbed, and a watery waste material, **diarrhea,** persists until the cause of the irritation is remedied. Conversely, if material moves too slowly, too much water is reabsorbed, and constipation results. Diarrhea may be the result of a bacillary dysentery or of many other causes. **Constipation,** on the other hand, results from an overstimulation of the intestinal muscle, narrowing the canal and preventing passage of fecal material. Nervous tension, excessive amounts of bulky foods in the diet, and use of laxatives increase the muscle tone of the intestines.

The wall of the large intestine closely resembles that of the small intestine in structure, except in its mucosal layer. This is covered with columnar epithelium, but there are no villi; the glands are longer, contain larger numbers of goblet cells, but do not secrete digestive en-

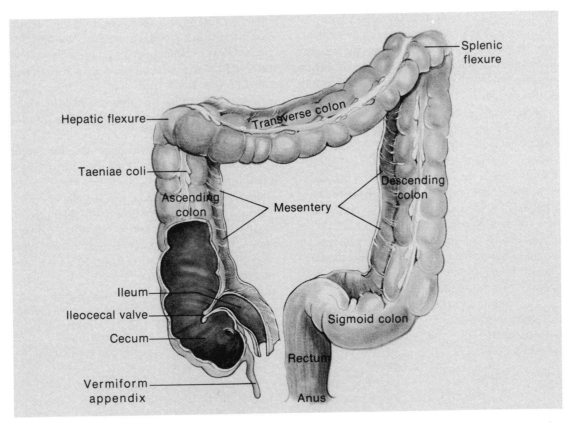

Splenic
flexure

Hepatic flexure

Transverse colon

Taeniae coli

Descending
colon

Ascending
colon

Mesentery

Ileum

Ileocecal valve

Cecum

Sigmoid colon

Vermiform
appendix

Rectum

Anus

Figure 17-10 Large intestine. Each segment is named according to the direction it travels or according to its shape. Note the junction of the small intestine (ileum) with the cecum of the large intestine (bowel). It is here that the appendix is found.

zymes. Veins in the mucosa of the anal canal may become dilated (varicose), forming small swellings when a strain is exerted on the anal region. These swellings are called *internal hemorrhoids* or *piles*. Similar varicose swellings in the skin-lined lower part of the anal canal are called *external hemorrhoids* (Figure 17-5). Medication, heat treatments, or surgery can be used to relieve the discomfort caused by this condition.

Inflammatory diseases of the large bowel (*colitis*) may be caused by tuberculosis, typhoid fever, or bacillary and amebic dysentery. Ulcerative colitis is a specific inflammatory disease of the colon that occurs most commonly in middle-aged people (see Figure 17-5). It may be caused by allergy, infection, or cellular destruction (necrosis) of the colon. Because circulating antibodies to the patient's own colonic cells have been demonstrated, an autoimmune etiology has been suggested. Hypersensitivity to milk, foods, or eggs has been implicated also as a pathogenic factor.

Characterized by severe bloody diarrhea, abdominal pain, weight loss, and fever, the disease may have many remissions over many years. The colon of a chronic ulcerative patient has a characteristic appearance on x-ray examination. The normal white puffy areas ("cloud" haustra) are not evident, being replaced by a straight and rigid, granular color.

Malignant tumors (adenocarcinoma) of the

large bowel and rectum grow around the lumen, decreasing its size and causing an obstruction (Figure 17-5). The colon above the obstruction becomes dilated, and a resection eventually is indicated. Massive tumor growth in the rectum and anal canal may necessitate surgical removal of the rectum, closure of the anus, and protrusion of the colon through the abdominal wall (colostomy) (Figure 17-5). Defecation is then performed through the colostomy opening into a sterile bag strapped to the abdomen. Metastasis (spread of cancer cells to other organs) from adenocarcinoma of the large intestine commonly occurs in the regional lymph nodes and in the liver. Malignant growths of the abdomen spread via the circulation, resulting in liver involvement. The disease may spread as far as the lung and the brain.

Gastritis and enteritis are inflammations of the stomach and intestinal linings. Commonly both areas are involved, and the condition is called *gastroenteritis.*

Liver

The liver, the largest of the glandular organs of the body, is situated in the upper right quadrant of the abdominal cavity. It lies to a large extent under the shelter of the lower ribs, its upper surface conforming to the lower surface of the diaphragm. The liver is divided into a large *right lobe* and a smaller *left lobe* by a fold of peritoneum called the *falciform ligament* (Figure 17-11). Two smaller lobes, the *quadrate* and *caudate lobes,* are marked off on the posterior and inferior surfaces of the right lobe. The quadrate lobe acts to cushion and enfold the gallbladder.

Microscopically the liver is composed of rows or cords of cubical cells that radiate from a central vein. On one side of each cord lies a blood vessel, considerably wider than a capillary, called a *sinusoid.* Blood from both the portal vein and hepatic artery is brought into the sinusoid; it in turn empties into the central vein. On the other side of the single row of cells is a bile capillary. Bile pigment and other materials derived from phagocytosis are removed from the

blood by the hepatic (Kupffer) cells and deposited into the bile capillaries as bile. (Bile and bile pigment were described on p. 448.) The blood received by the central veins of a number of neighboring liver units empties into larger veins, which unite to form the hepatic veins that in turn drain the blood into the inferior vena cava (Figure 17-11).

Located in the sinusoids are the larger phagocytic *Kupffer cells.* These are macrophages capable of ingesting bacteria and other foreign materials from the blood. These cells are commonly tested with a dye, bromsulphalein (BSP), to determine the extent of possible damage caused by liver disease. If the cells are damaged, they tend to absorb and retain larger than normal quantities of the dye. Therefore high retention of BSP can indicate liver cell damage.

The liver has numerous functions including the following:

1. Production
   a. Blood plasma proteins
   b. Antibodies to fight off diseases
   c. Heparin, a substance that prevents blood clotting
   d. Bile pigments from red blood cells in the form of bilirubin and biliverdin

2. Storage
   a. Vitamins and minerals
   b. Glucose in the form of glycogen

3. Conversion and utilization
   a. Fats (desaturation)
   b. Carbohydrates (glycogenesis)
   c. Proteins (deamination)

4. Removal
   a. Old blood cells and toxins by phagocytosis (Kupffer cells)
   b. Waste products (urea and ammonia) from amino acids

Blood from the digestive tract has a special pathway for return to the heart. Blood from the gastric and mesenteric arteries enters the capillaries of those organs, and digested food materials are absorbed into it. These capillaries

calories (cal) by 180, the heat per gram equals approximately 4 cal.

Heat/g = 680 cal/180 = approximately 4 cal/g

Fats have a heat per gram value of 9 cal. Proteins have a heat per gram value of 4 cal.

If the heat production of an individual is known from either direct or indirect measurement, the overall rate of chemical action, the person's basal metabolic rate (BMR), may be compared directly with the rate of another person with the same body dimensions. A set of values, or standards, can be derived and comparisons made to determine a person's ability to utilize the energy in the food consumed. In persons who are gaining weight, oxidizable material is stored, not used, and the energy in food eaten exceeds the metabolic rate. Conversely, in persons losing weight, only part of the heat produced is obtained from the food eaten, and there is a net decrease in stored oxidizable material.

As mentioned in Chapter 10, thyroxine is the main oxidative agent that regulates metabolic rate. Clinical tests utilizing isotopes (T_3 and T_4) and protein-bound iodine (PBI) studies have proved beneficial in diagnosing potential metabolic problems.

The term metabolism encompasses all changes that occur in the body involving foodstuffs. Thus metabolism includes not only the breakdown of food (catabolism) but also its synthesis (anabolism). Energy derived from a catabolic reaction is **exothermic,** or released from the reaction. An anabolic reaction in which energy is incorporated or stored is **endothermic.**

Carbohydrate Metabolism

The pancreas plays an important role in the regulation of carbohydrates. The endosecretion of insulin and glucagon from the islets of Langerhans provides a definite cellular control.

Following their absorption from the intestine, nutrient sugars are transported to the liver by the hepatic portal system. The liver cells can either convert most of the sugars into glycogen (glycogenesis) for storage, or if levels of available glucose in the blood are lower than normal, the liver can break down the available glycogen back into glucose (glycogenolysis) to maintain a normal blood sugar level.

The greatest and most important body use of glucose is in oxidation. When glucose is oxidized by tissue cells, carbon dioxide and water are formed and energy is released. Most tissues, apart from the liver, utilize glucose for their supply of energy. As soon as the glucose permeates the cell membrane, it is trapped in the cytoplasm. Phosphate bonds are attached to the glucose, forming a phosphate ester that cannot diffuse through the cell membrane.

Insulin promotes the oxidation of glucose by the tissue cells. Most combustion and extraction of energy occur in the mitochondria of the cell, and it is here that the released energy is stored as ATP (Figure 17-14). The mitochondrial membrane is permeable only to small molecules, however. Thus a problem exists for nutrients larger in chemical structure than a three-carbon molecule. To facilitate entry of the larger-sized glucose (six-carbon molecules), lysosomes in the cytoplasm release phosphorylating enzymes that reduce the size of the six-carbon glucose molecule to two three-carbon molecules, with a net gain of two ATP molecules or 20,000 Kcal of energy. The products of this anaerobic process of glycolysis (sugar breakdown) are two molecules of a three-carbon pyruvic acid that, after an intermediate conversion to acetic acid, may enter the mitochondria, where each combines with a four-carbon acid to form a six-carbon citric acid.

The Krebs citric acid cycle is a series of reactions involving the breakdown of six-carbon citric acid to a four-carbon acid, with the release of carbon dioxide and energy in the form of hydrogen molecules (Figure 17-14). This is the principal energy cycle in the body. The quantity of energy released by the cycle is enough to operate almost all the other body cycles.

The next sequence of reactions, called the electron transfer chain, involves the transfer of electrons from the hydrogen bonds through a

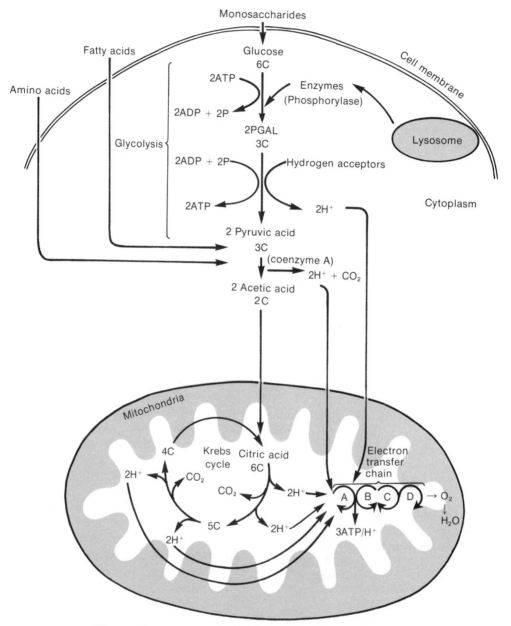

Figure 17-14 Metabolism of carbohydrates, proteins, and fats.

series of electron acceptors, each acceptor being at a progressively lower energy level. The enzymes are called *cytochrome oxidases*, or *respirases*, and effect the conversion of twelve hydrogen bonds into thirty-six molecules of ATP (Figure 17-13). The ATP molecules from the Krebs cycle added to the two ATP molecules from glycolysis produce a total of thirty-eight ATP molecules, formed by the complete oxidation of one molecule of glucose. This amounts to a considerable quantity of stored energy, because one molecule of ATP stores about 20,000 calories of free energy.

Glycogen is also formed and stored in the muscles. Such glycogen is used in muscles and not ordinarily returned to the blood.

Fat Metabolism

As previously discussed, fat enters the system in the form of fatty acids, glycerol, phospholipids, and cholesterol. A small amount of fat is synthesized in the liver, but most is synthesized in adipose tissue. Because 1 g of fat yields about 9 cal/g as compared with 4 cal/g of both protein and carbohydrate, fat deposits in the body provide a concentrated source of energy. These fat concentrations furnish about 40 percent of the energy used by a normal individual. The absorbed fats are acted upon by special cells in the liver (lipolysis), and the resultant products are channeled into the Krebs citric acid cycle (Figure 17-14). Synthesis of fats (lipogenesis) results from an excess of fatty acids and glycerol, which combine to form triglyceride. In addition to serving as a source of energy, stored fats act as cushions for some internal organs. Phospholipids are used in the formation of plasma membranes.

Fats are completely oxidized to carbon dioxide and water with a release of energy. This process occurs mainly in the liver by a mechanism that involves the breakdown of long-chain fatty acids. The end products of fatty acid oxidation are called *ketone* or *acetone bodies.* Under normal conditions, the quantity of ketones produced is excreted in the urine. When an excessive amount of fat is catabolized, ketone bodies are formed in massive quantities, and a condition known as *ketosis* results. This can occur in a person undergoing a complete fast, where there are insufficient carbohydrates and excess fats in the diet, and in severe diabetes. An excess accumulation of ketone bodies in the system requires neutralization with strong alkalis to preserve the pH of the blood and tissue fluids. An increased production of ketone bodies, as in

diabetes, will result in depletion of the alkali reserve, producing acidosis.

Cholesterol, as noted earlier, is a sterol found in fats derived from such animal foods as egg yolk and meats. Increased cholesterol levels in the blood (hypercholesterolemia) produce degenerative changes in the arterial walls (atherosclerosis).

Protein Metabolism

Amino acids absorbed from the intestinal tract are used by the metabolic mill for the synthesis of proteins, converted to fatty acids and glycogen, or oxidized as an energy source. All or a portion of the amino acid may be converted to glycogen or ketone bodies. Those amino acids that yield pyruvic acid are channeled into the Krebs cycle as an energy source (Figure 17-14).

An important role played by amino acids is the donation of their amine group (NH_2), a process called **deamination,** and transfer of this group to other substances, a process called **transamination.** Deamination and transamination are important because they are the first stop in the metabolic breakdown of these compounds and in the synthesis of amino acids from other compounds.

The primary use of amino acids is in the formation and repair of cells. During growth and pregnancy, additional proteins are formed from the amino acids taken from the blood.

The major end product of protein metabolism in humans is urea. Other waste materials that are closely correlated are ammonia, creatinine, and uric acid.

Basal Metabolic Rate

The BMR is the amount of energy expended by the body to maintain its vital functions when it is at rest. Although the needs of each individual vary, the average person requires about 3000 calories daily in the diet. Of this total, 1500 to 1800 calories are needed to maintain vital functions. The average person obtains about two-thirds of the energy needed from the oxidation of carbohydrates. From 45 to 50 percent of the total calories of the diet should be furnished by carbohydrates, 35 to 40 percent by fat, and 12 to 15 percent by protein.

A calorie diet breakdown for a young man of average size who expends about 3000 calories daily could be as follows:

- Basal metabolism (70 cal/hour for
  16 waking hours   1120

- Metabolism while asleep (64 cal/hour
  for 8 hours)   512

- Allowance for week   1500

                                         3132

DIAGNOSTIC PROCEDURES

Liver Function Tests

A serum screening analysis includes the enzymes alkaline phosphatase, lactic dehydrogenase (LDH), serum glutamic pyruvic transaminase (SGPT) and serum glutamic oxaloacetic transaminase (SGOT). These are naturally occurring compounds that catalyze biochemical reactions of the body. Alkaline phosphatase is found mainly in bone and liver; SGPT is found mainly in the liver; SGOT is found mainly in the heart, skeletal muscle, and liver; LDH is present in all metabolizing cells, especially in the liver. When an elevated alkaline phosphatase is associated with liver disease, it indicates an obstructive process (biliary tract) due to hepatitis, cirrhosis, or a space-occupying lesion of the liver that may be metastatic. The increased enzyme is presumably coming from the damaged cells lining the bile ducts. An increased alkaline phosphatase in the absence of any indicated liver disease is suggestive of bone pathology.

Bilirubin is formed from the hemoglobin of destroyed erythrocytes by the liver. It is a waste product that must be excreted by the biliary system. A markedly elevated total bilirubin level along with a drop in hemoglobin level and reticulocytosis is indicative of gross hemolysis.

Elevation of the direct bilirubin level indicates obstructive jaundice (hepatitis, cholangiolitis, and lower biliary obstruction by calculus or tumor).

Liver Scanning

Radioactive agents that emit low-energy gamma rays may be used in the diagnosis of many pathological conditions. Use of radioactive material that distributes itself in normal tissue may be used to detect abnormalities or pathological conditions. These localized agents can be detected by scanning devices, and the distribution of substance in various organs can then be determined.

The liver scan is used to detect structural damage and to evaluate the size, shape, and position of the liver and spleen. The radioactive substance commonly used is a colloidal sulfur compound tagged with radioactive ^{99m}Tc, which is picked up by the reticuloendothelial cells of the liver and spleen. Liver scans can demonstrate the location of space-occupying lesions (Figure 17-15) and can be of value in the differential diagnosis of jaundice.

Ultrasound

Ultrasound (echogram) of the liver can differentiate pleural effusion, ascites, cysts, abscesses, and tumors. Liver ultrasonograms are extremely useful in conjunction with hepatic radioisotope studies. A cirrhotic liver can be detected because it contains more echoes (reflects sound waves) than a normal liver. Serial scans can be used to determine the volume of the liver and may detect an adenocarcinoma, which will have a dense central echo pattern surrounded by a less echo-producing halo. The image pattern is called a *bulls eye*.

Laboratory Tests for Biliary Tract Disease

The most common gallbladder disorders to be tested are gallstones and cholecystitis. Radio-

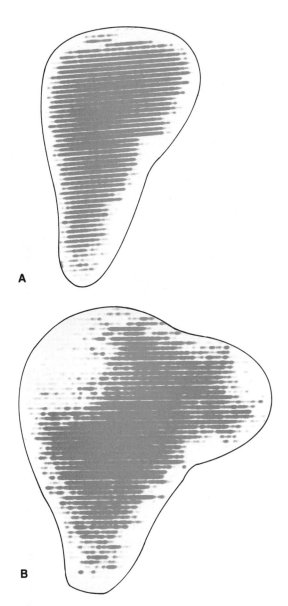

A

B

Figure 17-15 Liver scans. **A,** Normal scan using 99mTc, anterior aspect. The distribution of radioactivity indicates normal structures. **B,** Abnormal scan, anterior aspect, indicating an enlarged liver with a lesion in the right upper lobe.

graphs in combination with radiopaque dyes are used to demonstrate the presence of gallstones, tumors, or other obstructions. These dyes may be administered orally or by injection.

Oral Cholangiography (cholecystography)

The night before the test is to be performed, an iodine dye (Telepaque) is given orally to allow absorption and concentration of the dye in the gallbladder. Since gallstones are not usually radiopaque, it is necessary to fill the gallbladder with the dye, which causes stones to show up on x-ray film as shadows. This test is effective only if the liver cells are functioning normally and are capable of excreting the dye into the intestine.

Intravenous Cholangiography (IVC)

When an iodine contrast medium is given intravenously, both the biliary tree and the gallbladder can be visualized (Figure 17-16). This test is usually done after nonvisualization of the gallbladder following an oral cholecystography. A contrast dye is injected intravenously and followed by radiographic and tomographic evaluation.

Diagnostic Procedures For the Alimentary Tract

Radiographic Procedures

Gastric radiography and fluoroscopic examination are done to visualize the form and position, mucosal folds, peristaltic activity, and motility of the stomach. Preliminary film without contrast media is useful in detecting perforations, presence of metallic foreign substances, thickening of the gastric wall, and displacement of the gastric air bubble, indicating a mass outside the stomach wall.

The use of oral contrast substances can demonstrate a hiatal hernia, pyloric stenosis, gastric diverticulitis, gastritis, congenital anomalies, and diseases of the stomach such as ulcer, cancer, and polyps. If the examination includes the esophagus and upper part of the jejunum, it is called an *upper GI (gastrointestinal) series.*

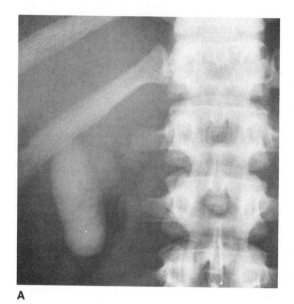

A

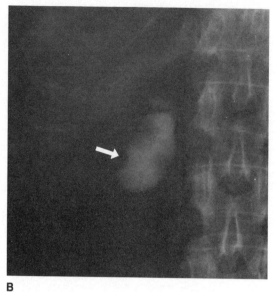

B

Figure 17-16 X ray of an intravenous cholangiogram. **A,** Contrast medium demonstrates a normal gallbladder. **B,** A gallbladder with gallstones (arrow). (Courtesy of Salem Hospital, Salem, Massachusetts.)

Small intestine radiography is done to diagnose diseases of the small bowel such as ulcerative colitis, tumors, active bleeding, or obstruction. A contrast medium is used to aid in the diagnosis of Meckel's diverticulum, congenital atresia, obstruction, regional enteritis, intussusception, and edema. The mesenteric small intestine begins at the duodenojejunal junction and ends at the ileocecal valve. It is not included routinely as part of the upper GI series.

Colon radiography is used as an aid in the diagnosis of diverticulitis, cancer, colitis, polyps, and any form of obstruction (Figure 17-17) or bleeding. Barium or Hypaque are used as contrast media and are instilled through the rectal tube. The radiologist observes the barium through a fluoroscope as it flows into the large intestine. X-ray pictures are taken. If polyp formation is suspected, an air contrast medium may be used. A double contrast of air and barium may be indicated to derive the best resolution. Radiographic examinations of the large intestine are referred to as a *lower GI series.*

Endoscopy

A group of diagnostic devices, known as fiberoptic instruments or **endoscopes,** are used for direct visual examination of various internal body structures. Each of these instruments has a lighted mirror lens system attached to a flexible tube. The tube can be inserted into openings and tracts of the body not easily accessible as directly visualized by other means. The insertion can be both for diagnosis of pathological conditions and for therapy, such as the removal of foreign objects. Biopsies are submitted for cytological examination.

Gastroscopy allows visualization of the lumen of the upper gastrointestinal tract. The gastroscope is valuable in determining the cause of upper gastrointestinal bleeding, biopsy upper gastrointestinal lesions, confirm suspicious findings on an x-ray film, and diagnose hiatal hernia and esophagitis. Washings removed from the stomach are also helpful in determining if a gastric ulcer is benign or malignant.

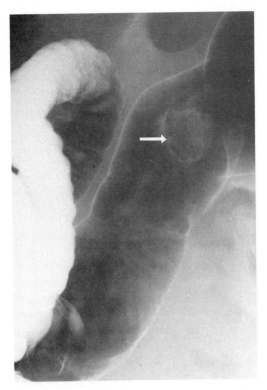

Figure 17-17 X ray of a barium enema demonstrating a sigmoid tumor (arrow). (Courtesy of Salem Hospital, Salem, Massachusetts.)

Proctoscopy or sigmoidoscopy involves the examination of a 12-inch area of the rectum and sigmoid with a proctoscope or sigmoidoscope (Figure 17-18). The main use of this instrument is the detection and diagnosis of cancers. Additional diseases of the rectum that may be diagnosed are hemorrhoids, fistulas, rectal polyps, and unexplained anemias due to active bleeding.

Colonoscopy is the examination of the large intestine. A colonoscope is inserted through the anus to the ileocecal valve. Air, which passes through an accessory channel of the colonoscope, is used to distend the intestinal walls. This technique is valuable in diagnosing inflammatory diseases such as colitis or in evaluating polypoid lesions (polyps and foreign bodies). Each specimen can be removed and examined.

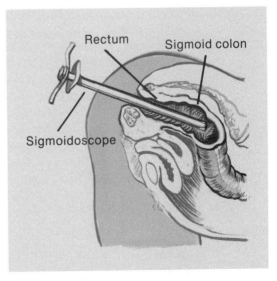

Figure 17-18 Use of sigmoidoscope to examine the colon for tumors and other disorders.

OUTLINE

I. Cellular requirements
  A. Water
  B. Energy
  C. Protein
  D. Enzymes
  E. Minerals
II. Foodstuffs
  A. Carbohydrates
    1. Monosaccharides—glucose
    2. Disaccharides—sucrose, lactose
    3. Polysaccharides—glycogen
  B. Fats
    1. Triglycerides
      a. Fatty acid molecules
      b. Glycerol unit
    2. Compound lipids
    3. Sterols
  C. Proteins
    1. Amino acids
      a. Manufactured in body
      b. Essential—taken in with food
    2. Polypeptides
    3. Simple versus conjugated proteins
  D. Vitamins
    1. Fat-soluble
      a. A, D, E, and K
    2. Water-soluble
      a. B complex and C
  E. Minerals
    1. Salts—chlorides, phosphates, and carbonates
    2. Calcium, phosphorus, and iron
    3. Trace elements—copper, manganese, and cobalt
III. Digestive organs and processes
  A. Mouth
    1. Salivary amylase—carbohydrates
      a. pH 6.8 to 7.0
    2. Mastication—teeth
      a. Incisors
      b. Canines
      c. Molars

3. Decalcification
4. Dental caries
5. Gingivitis
6. Herpes simplex type I
7. Cancer of tongue
  B. Esophagus
    1. Muscular tube
    2. Peristalsis
    3. Swallowing (deglutition)
    4. Esophageal varices and carcinoma
    5. Hiatus hernia
  C. Abdomen
    1. Peritoneum
    2. Mesentery
    3. Omentum
    4. Peritonitis
  D. Stomach
    1. Muscular layers
    2. Mucosa—mucous cells, chief cells, and parietal cells
    3. Pepsin, rennin, and hydrochloric acid—proteins
    4. Gastric juice—pH 1.0 to 4.0
    5. Chyme
    6. Enzyme activity—specificity
    7. Peptic ulcer
    8. Pyloric stenosis
    9. Gastritis
  E. Small intestine
    1. Duodenum, jejunum, and ileum
    2. Bile salts—fat emulsification
    3. Duodenal enzymes—assist pancreatic enzymes
      a. Enterokinase—proteins
      b. Maltase—carbohydrates
      c. Sucrase—carbohydrates
      d. Lactase—carbohydrates
      e. Lipase—fats
      f. Nuclease—nucleoproteins
    4. Villi
      a. Mucosal projections

b. Contain capillaries and lacteals
5. Duodenal ulcers and intestinal obstruction
6. Peptic ulcer
7. Diverticulum
8. Regional enteritis
9. Intussusception

F. Gallbladder
   1. Bile storage
   2. Gallstones
   3. Cholecystitis

G. Pancreas
   1. Pancreatic enzymes
      a. Pancreatic amylase—carbohydrates
      b. Pancreatic lipase—fats
      c. Trypsin—proteins
      d. Nuclease—nucleoproteins
   2. Pancreatitis

H. Large intestine
   1. Ascending, transverse, descending, sigmoid, and rectal colon
   2. Cecum and appendix
   3. Absorption of water
   4. Defecation of solid waste
   5. Diarrhea and constipation
   6. Hemorrhoids
   7. Colitis
   8. Tumors

I. Liver
   1. Largest gland in the body
      a. Sinusoids
      b. Hepatic portal system
      c. Phagocytic Kupffer cells
   2. Functions
      a. Produces plasma proteins, antibodies, heparin, and bile pigments
      b. Stores vitamins, minerals, and glycogen
      c. Converts fats, carbohydrates, and proteins to other metabolic forms

d. Removes old blood cells and toxins by phagocytosis, and waste products from amino acids
3. Hepatitis and cirrhosis
4. Jaundice
   a. Obstructive
   b. Hemolytic
   c. Toxic
5. Tumors

IV. Hormonal control of digestion
A. Stomach
   1. Gastrin
   2. Enterogastrone
B. Pancreas
   1. Secretin ⎫ Secreted
   2. Pancreozymin ⎭ in duodenum
C. Gallbladder
   1. Cholecystokinin (secreted in duodenum)
D. Duodenum
   1. Enterokinin (duocrinin)
   2. Villikinin

V. Nervous control of alimentary tract
A. Salivary reflexes—activated by chewing
   1. Unconditioned reflex—medulla oblongata
   2. Conditioned reflex—activated by smell, sight, or thought of food
B. Gastric glands—vagus nerve
C. Peristaltic contractions—vagus nerve
D. Intestinal movements
   1. Sympathetic nerves
   2. Defecation—complex reflex

VI. Mechanisms of metabolism
A. Calorimetry
   1. Carbohydrate and protein oxidation give off 4 cal of heat per gram
   2. Fat oxidation releases 9 cal/g
   3. Basal metabolic rate (BMR)
      a. T_3, T_4, and PBI

4. Metabolism
   a. Catabolism—breakdown
   b. Anabolism—buildup
B. Carbohydrate metabolism
   1. Insulin from pancreas
   2. End products absorbed by villi and transported to liver
      a. Glycogenesis
      b. Glycogenolysis
   3. Glycolysis
   4. Krebs citric acid cycle
   5. Electron transfer chain
C. Fat metabolism
   1. Fatty acids absorbed by villi
   2. Synthesized to form adipose tissue
   3. Found in form of cholesterol and triglycerides
   4. Lipolysis and lipogenesis supply 40 percent of body energy
   5. Ketone bodies—end products of fatty acid oxidation
D. Protein metabolism
   1. Amino acids absorbed by villi
   2. Used as building blocks of proteins
   3. Deamination—transamination
   4. Urea—major end product of protein metabolism
E. Basal metabolic rate
   1. Amount of energy used by body to maintain vital functions
   2. 3000 calorie normal adult diet—1500 to 1800 calories needed for vital functions

VII. Diagnostic procedures
   A. Liver
      1. Serum screening
         a. Alkaline phosphatase, LDH, SGPT, and SGOT
         b. Bilirubin
      2. Scanning
      3. Ultrasound
   B. Biliary tract disease
      1. Oral cholangiography
      2. Intravenous cholangiography
   C. Alimentary tract
      1. Radiography
      2. Fluoroscopy
      3. Endoscopy

STUDY QUESTIONS AND PROBLEMS

1. List the major minerals and vitamins and indicate their importance.
2. Trace the path of an indigestible object from the mouth through all parts of the alimentary canal to the outside, describing what happens on the way.
3. Distinguish between deciduous and permanent teeth. Describe the relation between their shape and their function.
4. How does peristalsis begin? How is it maintained throughout the intestinal tract?
5. pH has an important role in digestion. Explain.
6. Where is energy located in a chemical compound? What are exothermic and endothermic reactions?
7. What is meant by enzyme specificity? Does pH have anything to do with it?
8. Differentiate between a peptic and a duodenal ulcer. What problems does a perforated ulcer present?
9. Trace the three major foodstuffs consumed in a normal balanced meal through their digestion, and correlate the specific enzymes involved.
10. What constitutes a balanced diet?
11. What are the end products of digestion? Where are they absorbed and by which vessel?
12. What does the large intestine do? Explain the difference between diarrhea and constipation.
13. If a person has a colostomy, why must he be concerned with electrolyte and water balance?
14. What part does the pancreas play in digestion? What other role does the pancreas have in maintaining homeostasis?
15. What is the source of bile pigments?
16. What is infectious hepatitis? What other conditions may produce jaundice?
17. How is the liver involved in metabolism?
18. What is the relation between the hepatic portal system and digestion?
19. Why does hydrolysis play an important role in metabolism?
20. What does insulin do in carbohydrate metabolism?
21. Discuss ketogenesis and lipid metabolism. Why are cholesterol and triglyceride levels so closely watched?
22. What is a calorie? Are the PBI and T_3 significant tests for a BMR?
23. List at least two serum enzyme tests related to liver disease.
24. Differentiate between an oral and T-tube cholangiogram.
25. What is an endoscopy procedure? Give examples.

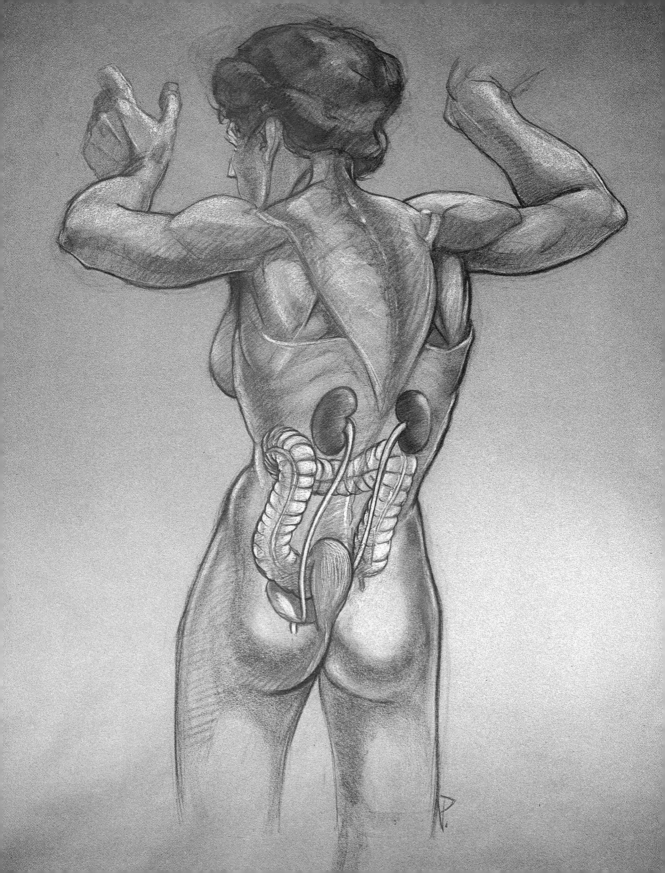

18

Excretion

LEARNING OBJECTIVES

- Label the anatomy of the urinary system

- Identify all urinary structures and relate their separate functions

- Define micturition and identify the factors involved in glomerular filtration rates

- Define diuresis and explain how it works

- Explain the role of ADH and aldosterone in renal tubular reabsorption

- Identify the active and passive processes occurring during tubular reabsorption

- Contrast the neural control of micturition in an infant with that in an adult

- Describe the phenomenon of thirst

- Identify the buffers in the blood that function to maintain an acid-base balance

- Explain how renal failure disturbs the body's homeostasis

- Define nephritis, glomerulonephritis, pyelonephritis, and urethritis

- Identify the major abnormal urinary constituents

- Explain the principle behind the creatinine and urea clearance tests

- Explain dialysis and identify the two major methods

- List different techniques used to diagnose abnormalities in the urinary tract

- Describe a renal echogram

- Explain the types of pyelography

- Explain a kidney scan

IMPORTANT TERMS

active transport	cystoscope	hemodialysis	nephron
antidiuretic	dehydration	hydronephrosis	retroperitoneal
axotemia	diuresis	hyperkalemia	urea
calculi	dysuria	incontinence	uremia
casts	excretion	micturition	urination

The active life of every cell is necessarily accompanied by the production of waste materials. These must be removed quickly if the activities of the tissues are not to be impaired. In the chemistry of the body, some reactions produce starting materials for others, but certain end products may accumulate and disturb the sequence of essential reactions if they are not removed. These materials are the excretory substances, and they are produced in different amounts by all cells.

The process of **excretion** consists of the separation and removal of substances harmful to the body. Excretion is carried out by the salivary glands, skin, lungs, liver, large intestine, and kidneys (Figure 18-1).

As discussed in Chapter 16, the lungs remove carbon dioxide from the blood, and other waste products of metabolism, such as ketone bodies (acetone), are removed by the respiratory system. The salivary glands excrete various salts, particularly sodium chloride. The sweat glands produce water, salts, amino acids, vitamins, and other materials that are carried from the body and deposited on the outer surface of the skin. Excretory functions of the liver include phagocytic action and the removal of bile pigments. The large intestine, following its principal function of water absorption, removes solid waste materials as feces.

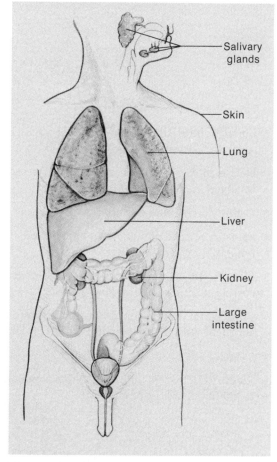

Figure 18-1 Organs of the excretory system.

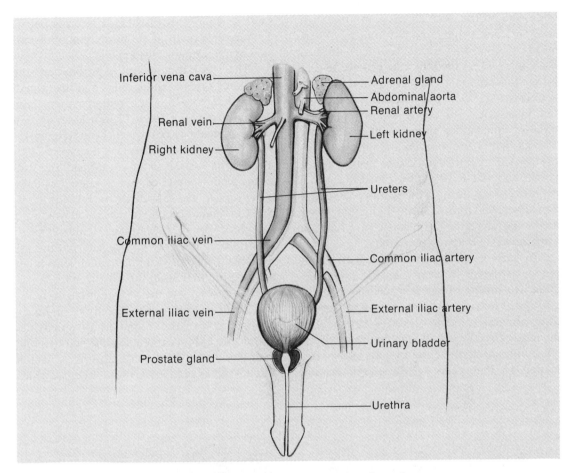

Figure 18-2 Urinary system with blood vessels.

URINARY SYSTEM

The two principal functions of the urinary system are the excretion of wastes and the regulation of the composition of blood. Blood composition must not be allowed to vary beyond tolerable limits, or the conditions in the tissues necessary for cellular life will be lost. Regulation of the composition of the blood involves not only the removal of harmful waste products but also the conservation of water and metabolites in the body.

The urinary system consists of two large kidneys that secrete urine; the ducts leading from them, the ureters; a large urinary reservoir, the bladder; and the tube from it to the surface of the body, the urethra (Figure 18-2).

Kidneys

The *kidneys* are deeply embedded in the abdominal cavity in the lumbar region (Figure 18-1). They are **retroperitoneal** (behind the peritoneum). Both are protected at the back by the last two pairs of ribs. They are surrounded by a layer of fat, the *perirenal* fat; in front they are covered by renal fascia, the *capsule*.

Each kidney is a dark red, bean-shaped organ about 11 cm (4.5 in.) long, 6 cm (2.5 in.)

wide, and a little over 2.5 cm (l in.) thick. An adult kidney weighs about 5 ounces. At the middle of the kidney's inner concave border is a deep depression called the *hilus* (Figure 18-3). It is here that the renal artery enters and the renal vein leaves the kidney to take blood to the inferior vena cava. The ureter also leaves at the hilus to connect the kidney with the bladder. Nerves, lymphatic vessels, and blood vessels control kidney function. Under resting conditions, about one-fourth of the blood that leaves the heart passes through the kidneys for regulation. The urine formed in the kidney is passed down the ureter to the bladder by peristaltic contractions of the muscular ureter.

In vertical section, the outer cortex of the kidney is darker in color than the inner medulla. The inner surface of the medulla is folded into projections, the *pyramids*, which empty their contents into the collecting space called the *renal pelvis* (Figure 18-3). Numerous collecting tubules bring the urine from its sites of formation in the cortex to the pyramids and calyces, where it is discharged into the pelvis. The ureter actually is no more than an extension of the funnel-shaped pelvis.

The renal tubules, or **nephrons,** are the functional units of the kidney, and the human kidney has more than one million of these. Each nephron tubule is about 12 mm long; its thin walls are made up of epithelial cells (Figure 18-4). It is open at one end, where it connects with a collecting tubule; at the closed end, the wall forms a cup-shaped concavity known as *Bowman's capsule.* The capsule drains into the proximal convoluted tubule, which is followed by a straighter portion, *Henle's loop.* Henle's loop enters the medulla of the kidney and continues as the distal convoluted tubule, which joins other similar tubules to form the collecting tubes.

Blood supply to the nephrons begins at the renal artery, which comes directly from the abdominal aorta. The renal artery subdivides within the kidney, and a small vessel (afferent arteriole) enters Bowman's capsule, where it forms a tuft of capillaries, the *glomerulus,* which

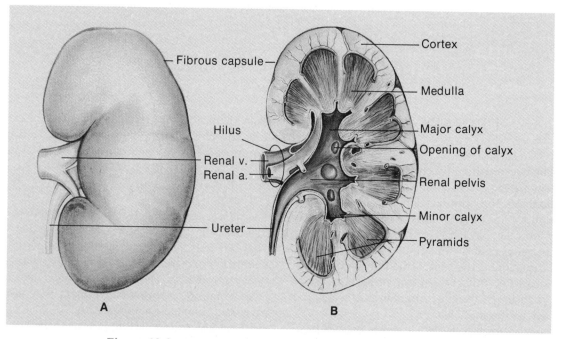

A   B

Figure 18-3 External **(A)** and internal **(B)** anatomy of the left kidney.

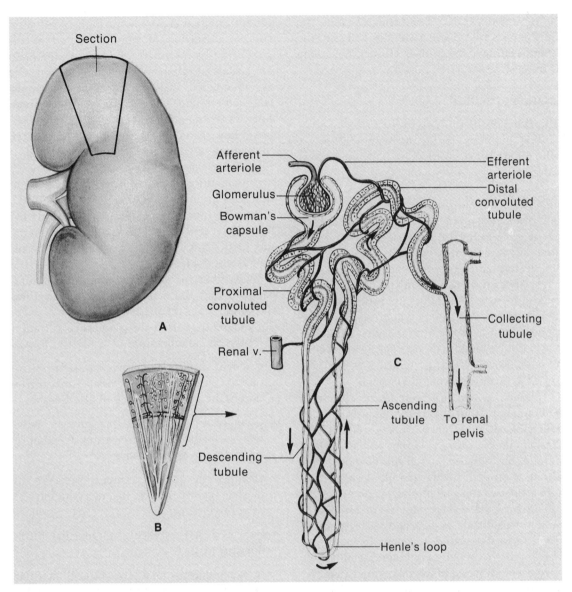

Figure 18-4 Kidney. **A,** External diagram of the left kidney. **B** and **C,** A section of the kidney has been enlarged to show the details of a nephron unit. Arrows indicate the direction of flow or urine through the nephron unit.

entirely fills the concavity of the capsule (Figure 18-4). The blood leaves the glomerulus through the efferent arteriole, which again subdivides into a network of capillaries that surround the proximal and distal tubules. These capillaries eventually unite as veins, which join to become the renal vein. The renal vein returns the now cleansed blood to the general circulation.

By movement of substances between the nephron and the capillaries of the tubules, the composition of the blood filtrate moving along in the tubules is changed, and it becomes a more concentrated solution. From the nephrons, the fluid moves to collecting tubules and into the ureter leading to the urinary bladder, where the urine is stored.

Urine Formation

The function of the nephron can best be described in terms of the formation of urine, which requires two processes: filtration and resorption.

FILTRATION The first process in the formation of urine is glomerular filtration. The walls of the glomerulus are permeable to water and other small molecules, but they are not permeable to formed blood elements (cells) and proteins. Since the blood entering the glomerulus is under pressure, some of the nonprotein constituents pass through into Bowman's capsule. The glomerular filtrate has a composition similar to plasma, since blood cells are not filtered. It contains the waste products, which it is the kidney's job to remove from the plasma.

Plasma, which has been selectively modified, emerges from the kidney as urine. In an average-sized man, approximately 600 mL of plasma flows through each kidney every minute, but only about 125 mL of glomerular filtrate is formed, owing to a glomerular pressure estimated to be 69 mm Hg. Glomerular filtration provides a large volume of essentially protein-free filtrate of plasma, which is largely resorbed in its passage through the nephron. The forces involved in glomerular filtration have been estimated in the following equation:

$$P_f = P_g - (P_{op} + P_c)$$
25 mm Hg 60 mm Hg 30 mm Hg 5 mm Hg

where P_f is the filtration pressure, P_g is the glomerular hydrostatic pressure, P_{op} is the colloid osmotic pressure, and P_c is the capsular hydrostatic pressure (Figure 18-5).

The most obvious factor likely to affect the glomerular filtration rate is the number of functioning glomeruli, which may be reduced by disease. The blood pressure is the driving force

for filtration; therefore with high blood pressure, the fraction of plasma filtered at the glomerulus is increased. The kidneys maintain a constant filtration rate over a wide range of systemic blood pressures. By appropriate vasoconstriction and vasodilation of the smooth muscles in the afferent and efferent arterioles, blood flow through the kidney may be adjusted to maintain a constant filtration rate, despite a fluctuating blood pressure.

As mentioned earlier, the glomerular filtration rate in an average-sized man is 125 mL/min or 7.5 liters (about 7.5 quarts)/hour or 180 liters daily. The normal urine volume is about 1 liter/day. Thus, 99 percent or more of the filtrate is resorbed. At the rate of 125 mL/min, the kidneys filter in one day an amount of fluid equal to four times the total body water and sixty-nine times the plasma volume.

Factors that affect the glomerular filtration rate include the following:

- Changes in glomerular hydrostatic pressure due to changes in blood pressure or constriction of the afferent or efferent arterioles

- Changes in hydrostatic pressure in Bowman's capsule due to ureteral obstruction or edema of the renal capsule

- Changes in colloid osmotic pressure of plasma proteins due to dehydration or hypoproteinemia

- Increased permeability of glomerular filtration due to disease

- A decrease in the total area of the glomerular capillary bed due to destruction of glomeruli from disease or from a partial nephrectomy

RESORPTION An equally important physiological process carried on by the nephron unit is resorption. The filtrate does contain such metabolic wastes as urea and creatinine, but most of the substances found in it (water, salts, glucose, amino acids) are extremely important to the body. If the water, for example, were not resorbed, the 180 liters that normally are resorbed

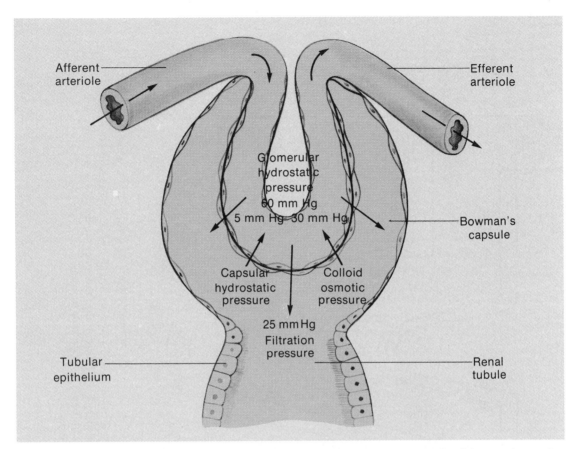

Figure 18-5 Urine formation: filtration. The glomerular membrane acta as a simple filter that allows passage of dissolved materials only because of a greater pressure on one side of the membrane than on the other.

would have to be replaced by drinking. The movement of water and ions, especially sodium (Na+), across different parts of the nephron tubules makes it possible for humans to form urine that is more concentrated than the blood and tissue fluids, thus conserving water.

Blood is filtered in Bowman's capsule. Blood proteins and blood cells do not pass into the kidney tubule, but everything else does. The filtrate entering the proximal convoluted tubule is isotonic (same osmotic pressure) with the blood. As the filtrate flows down toward Henle's loop, it becomes hypertonic because Na+ diffuses passively into the filtrate from the surrounding

tissue fluids (Figure 18-6) where the Na+ concentration is high. When the filtrate passes through the ascending loop, Na+ is actively pumped from the filtrate back into the surrounding tissue fluids. The filtrate becomes hypotonic, and the surrounding fluids develop a high Na+ concentration. The Na+ pumped back into the surrounding fluids can then diffuse into the filtrate flowing through the descending loop. In this way, Na+ is constantly recycled, and the loop of Henle is continuously surrounded by a salty fluid. As hypotonic filtrate flows out through the collecting tubule, water diffuses from the filtrate into the sur-

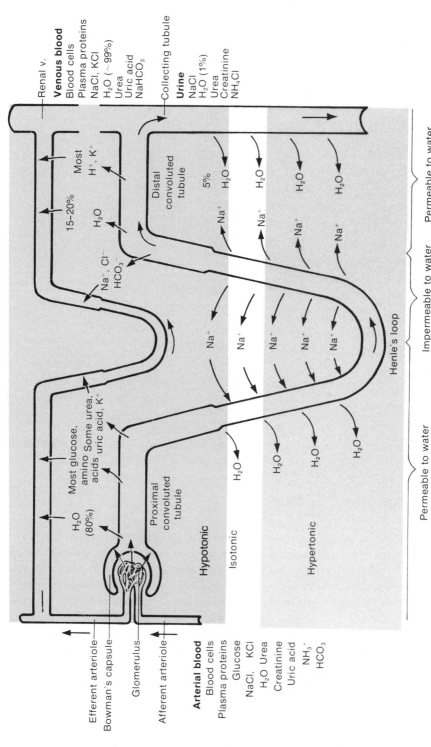

Figure 18-6 Urine formation: resorption. As the filtrate passes along the tubule, its concentrated and essential elements are reabsorbed by the capillary network. The entire mechanism depends on the recycling of Na^+, which causes the fluids in the tissues surrounding the renal tubules to be very salty and permeable to water.

for the urethra (Figure 18-8). The interior of the bladder is lined with highly elastic transitional epithelium. When the organ is full, the lining is smooth and stretched; when it is empty, the lining is a series of folds.

In the middle layer, a series of smooth muscle coats distends as urine collects and contracts to expel urine through the urethra. Urine is produced almost continuously and accumulates in the bladder until the increased pressure stimulates the organ's nervous receptors. When the muscles of the bladder contract, urine leaves by way of a single tube, the urethra. This process is called **micturition,** or **urination.**

Inflammation of the bladder, called *cystitis,* is often secondary to infection in some other part of the urinary tract or elsewhere in the body. Acute cystitis is commonly caused by gonorrhea, a type of venereal disease, or trauma to the bladder. Chronic cystitis may result from an enlarged prostate gland, pregnancy, or a stricture of the urethra that produces back pressure. Examination of the inner mucous membrane of the bladder is performed by means of a long tubular instrument, a cystoscope, which has a light at one end and an eyepiece at the other. Bladder tumors (neoplasms) may necessitate open-bladder surgery (cystostomy) or removal of the bladder (cystectomy).

Urethra

The *urethra* is a membranous tube that passes from the anterior part of the urinary bladder to the outside of the body. In the female it is only 4 cm (1.5 in.) long (see Figure 19-4). In the male it is about 20 cm (8 in.) long and serves to conduct both urine and sperm to the outside (see Figure 19-2).

Two sphincters are located in the urethra (Figure 18-8). The *internal sphincter* is formed of circular smooth muscle, and guards the exit from the bladder. It is controlled by the autonomic nervous system. The *external sphincter* is circular striated muscle and is under the voluntary control of the central nervous system.

Congenital anomalies may involve the external opening of the urethra. The opening may be too small, or the urethra may be narrowed to such an extent that back pressure of urine may cause serious damage to the bladder if the defect is not corrected. In adults, a stricture or closure of the urethra also can prevent normal micturition. The stricture may be caused by muscle spasms or by urethritis, an inflammation of the mucous membrane. This condition is more common in the male, and dilation of the inner tract may be required to permit urination.

If urine is retained in the bladder for a prolonged period, a back pressure is established, causing the kidneys to malfunction. Insertion of a rubber tube, a catheter, will temporarily relieve the problem. Persistent **dysuria** (difficult urination) may require continued catheterization until the cause of the stricture is found and treated.

MICTURITION

The movement of urine collected in the pelvis of the kidney down the ureters to the bladder is the result of both the force of gravity and peristaltic contractions in the muscle layers of the ureter. By stretching the bladder, a considerable amount of fluid (about 300 mL) can be accommodated without a significant rise in pressure. As the volume increases, the tension rises until, finally, stretch receptors are excited. These initiate an impulse that creates the sensation of distention and the desire to urinate. Sensory signals are conducted to the sacral segments of the spinal cord through the pelvic nerves and then back again to the bladder through the parasympathetic fibers in these same nerves (Figure 18-9).

In infants, as the sensory impulses increase in number and frequency, the motor impulses to the bladder, via the pelvic nerves, cause a reflex contraction of the bladder and a relaxation of the internal sphincter. The external sphincter is controlled by the pudendal nerve

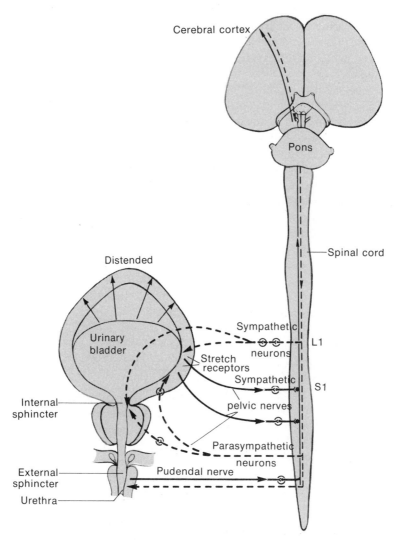

Figure 18-9 Micturition. Distention of the urinary bladder from accumulation of urine stimulates the stretch receptors in the bladder wall. Afferent fibers of the pelvic nerve conduct this impulse to the brain, which institutes a reflex mechanism. The brain controls the storage and expulsion of urine in the bladder by smooth muscle contraction or relaxation of the internal and external sphincters.

(motor). The bladder wall contains many stretch receptors, the afferent nerves of which enter the spinal cord and eventually synapse with these parasympathetic and motor neurons, stimulating the former to contract and inhibiting the latter, causing a relaxation.

In adults, higher brain centers control and inhibit the reflex mechanism so that micturition can be delayed or induced voluntarily. By learning to control these pathways, children achieve a capacity to inhibit their spinal reflexes.

The loss of voluntary control over micturition is called **incontinence.** Incontinence is normal in young children, since the infant voids whenever the bladder is sufficiently distended to arouse a reflex stimulus. The term nocturnal enuresis is used to describe the involuntary discharge of urine during sleep.

The normal daily output of urine ranges from 1000 to 2000 mL. Any suppression of urine flow is referred to as *oliguria,* as in urine retention in the kidney. Any increase in urine flow is called *polyuria,* as in diabetes insipidus. Complete suppression of urination is *anuria.* Inflammatory conditions that prevent normal blood flow to the glomerular filtering beds are a common cause of anuria. Any stricture or blockage of the ureters will also result in anuria.

URINE

Macroscopic Examination

Urine formed by the processes described in this chapter is a hypertonic solution of inorganic salts, nitrogenous waste products, and a few gases (Table 18-1). A good index of the degree of concentration of the dissolved materials is the specific gravity. The normal range, using a hydrometer, is between 1.015 and 1.025 (with reference to water as 1.000). An adult excretes an average of 1500 mL of urine every day. Higher and lower specific gravities may occur, depending on fluid intake and the quantity of dissolved substances in the urine. An elevated specific gravity generally indicates urine retention and an increased concentration of materials. Conversely, a low specific gravity is usually due to polyuria. When liquids are consumed at a greater than normal rate, urine forms more rapidly, and its specific gravity may approach that of water, falling to about 1.005. In severe diabetes mellitus, however, the high sugar content of the urine will keep its specific gravity around 1.040, even though urine output is very high.

The color of urine is generally pale to dark yellow (straw yellow), and the fluid is clear. Bile pigments cause the color, which can vary from a brown-yellow to olive-green, depending on the pigment concentration. Red hues, due to hemoglobin in the red blood cells, indicate that a degree of hemolysis has occurred. Urine containing fat droplets or pus may have a milky appearance.

The odor of urine is clinically insignificant. Normal urine is slightly aromatic. A smell of ammonia is usually due to bacterial action after voiding and can be very unpleasant. Sweet-smelling urine indicates a highly concentrated ketone level, especially acetone. Diabetic ketosis is a common cause.

Urine normally varies in pH from 5.0 to 8.0 and usually is slightly acidic. Reportedly, those who eat a high-protein (animal) diet have a urine pH of 5.0 to 6.0. Vegetarians have a slightly alkaline (pH 7.0 to 8.0) urine. Urinary acidosis is indicative of such problems as diabetic ketosis, starvation, and diarrhea. It also occurs in emphysema, in which carbon dioxide retention is high. Alkalosis, on the other hand, results from most urinary infections and high blood oxygen levels, as in hyperventilation.

An important clinical test of urine is that for protein. This is derived from plasma proteins and usually comes from the kidney. Normally no protein should be in the urine. Gross bleeding from the lower urinary tract can add large quantities of protein to the urine (protein-

Table 18-1   Composition of Urine

Constituent	Grams/Day
Water	1500.0
Organic substances	
Urea	30.0
Creatinine	1.2
Uric acid	0.6
Ammonia	0.8
Inorganic substances	
Sodium	5.0
Chloride	6.0
Calcium	0.2
Potassium	2.2
Phosphates	1.5
Sulfates	1.6
Magnesium	0.2

uria). Glomerular filtration problems, renal disease (nephritis), and inadequate resorption of extracellular protein by the lymphatic system all produce traces of albumin in the urine (albuminuria).

The glomerulus acts as an effective barrier to the filtration of large molecular weight (greater than about 60,000) serum protein; only an extremely small amount of albumin is lost in the glomerular ultrafiltrate in normal states. Low molecular weight proteins easily pass through the glomerular capillary walls but are largely catabolized by tubular cells so that they rarely appear in the urine in significant quantities. Based upon this brief outline of the handling of proteins by the normal kidney, it is possible to have some idea of the site of defect causing proteinuria. There are four major types of proteinuria:

- *glomerular*—the result of an abnormality of glomerular filtration

- *tubular*—resulting from a defect in functioning of tubule cells

- *overflow*—caused by the loss of low and medium molecular weight proteins present in the serum, usually in large quantities

- *nephrogenic*—due to the increased loss of proteins usually produced by the kidneys

Normally, simple sugar (glucose) is completely resorbed by the convoluted tubules. In diabetes mellitus, blood sugar is not broken down in the body cells because of a deficiency of insulin and therefore is excreted in the urine (glycosuria). As described earlier, the Clinitest is a simple, rapid, and accurate technique that can be used to detect sugar in the urine.

The presence of ketone bodies in the urine (ketonuria) is especially significant in diabetes. If the test for glucose is positive, a follow-up should include determination of the presence or absence of acetone. Uncontrolled diarrhea, starvation, and an altered carbohydrate metabolism that uses protein and fat as the energy source in place of carbohydrates can cause this condition.

It also may occur in children after prolonged vomiting.

Blood (hematuria) and pus (pyuria) appear in the urine during kidney or bladder infections. Their presence is of sufficient importance to warrant further testing.

Nitrogenous waste products, which include urea, uric acid, and creatinine, are classified as organic compounds (see Table 18-1) because they are formed by the breakdown of cells in the body and from food proteins. In other words, they originated from living organisms. Inorganic substances include the chlorides, phosphates, sulfates, and so forth that make up the body's salt and mineral content. These do not originate in living organisms.

Human chorionic gonadotropin (HCG), a hormone produced by the placenta during pregnancy, can be detected in the urine 4 to 6 weeks after conception.

Microscopic Examination

Cells

Cells in the urine may indicate disease, depending on the type and number present. It is therefore important that anyone doing microscopic examinations be constantly aware of the significance of particular types of cells.

Urine sediment does not normally contain more than two to five red blood cells per high power field, although the number can vary with the manner of preparation. Increased numbers of red blood cells suggest inflammatory disease, acute glomerulonephritis, pyelonephritis, hypertension, renal infarction, the use of anticoagulants, or bleeding due to trauma, stones, or tumors in the urinary tract. The appearance of red blood cells depends on the pH and specific gravity of the urine as well as on the stain used.

When examining urine for white blood cells, a fresh sample is preferable because per field white blood cells range from five to eight or more. If more than fifty are found, acute inflammation of the genitourinary tract is suggested. White blood cells accompanied by proteinuria are likely manifestations of pyelonephritis; if little protein is present, the cells may be indicative of lower urinary tract infection. Neutrophils are the most commonly seen white blood cells, although lymphocytes, monocytes, and histocytes may also be found. Monocytes generally suggest tissue damage with severe inflammation.

Epithelial cells often appear in the urine sediment. Squamous epithelial cells are particularly common in women, usually arising from vaginal contamination. Transitional epithelial cells may also be found in normal urine, but a large number may indicate disease of the bladder or renal pelvis or use of a catheter. Renal tubular epithelial cells are the most clinically significant, implying possible renal disease such as acute tubular necrosis, glomerulonephritis, acute infection, and renal toxicity.

Malignant cells are rare in the urine, and their identification is difficult because their appearance varies widely. Generally the nucleus is much larger than normal and contains abnormal chromatin and several large nucleoli. Since malignant cells readily accept Papanicolaou stain, its use may aid in their identification. Presence of malignant cells in the urine may indicate tumors in the renal pelvis, renal parenchyma, ureter, or bladder. Usually malignant cells are accompanied by hematuria.

Casts

Casts are cylindrical rods, molds of kidney tubules. They can include various kinds of cells and materials (Figure 18-10):

1. *Epithelial cell casts* are formed by shedding tubular cells. Since the renal epithelium is constantly sloughing and regenerating new cells, occasionally epithelial casts appear in normal urine. Occurrence of many epithelial casts, however, is indicative of acute tubular necrosis.

2. *Hyaline casts* are clear, colorless casts formed when protein within the tubules precipitates and gels. Occasional hyaline casts are observed in normal urine and are increased in number in febrile illness, in congestive heart failure, and following strenuous exercise. They indicate possible damage to the glomerular capillary membrane permitting leakage of proteins.

3. *Granular casts* result from the disintegration of the cellular material in white and epithelial blood cells into coarse and then fine granular particles. Granular casts are invariably accompanied by proteinuria in patients with renal disease. The presence of numerous broad granular casts indicates advanced glomerulonephritis, pyelonephritis, or malignant nephrosclerosis.

4. *Waxy casts* are thought to evolve from pre-existing cellular and granular casts. They have the highest refractive index of all urine casts and are easily recognized by having broken off ends, irregular margins, and a smooth waxy-appearing surface. Many waxy casts usually indicate a relatively long renal transit time and therefore depressed renal function.

5. *Red blood cell* or *erythrocytic casts* are easily recognized in fresh urine. They have a high refractive index and have red blood cells on the surfaces of the cast matrix. They are invariably indicators of intrinsic renal disease and ordinarily pinpoint the glomerulus as the site of injury.

6. *White blood cell* or *leukocyte casts* found in the urine indicate renal pathology. They are easily recognized because of their high refractive index and presence of nucleated white blood cells. White cell casts mostly originate when the kidney is inflamed (pyelonephritis). The leukocytes are attracted to the area of injury and become incorporated into casts where they are found in the urine.

7. *Fatty casts* are casts in which either free fat or oval fat bodies have become incorporated into the cast matrix. Glomerular, tubular,

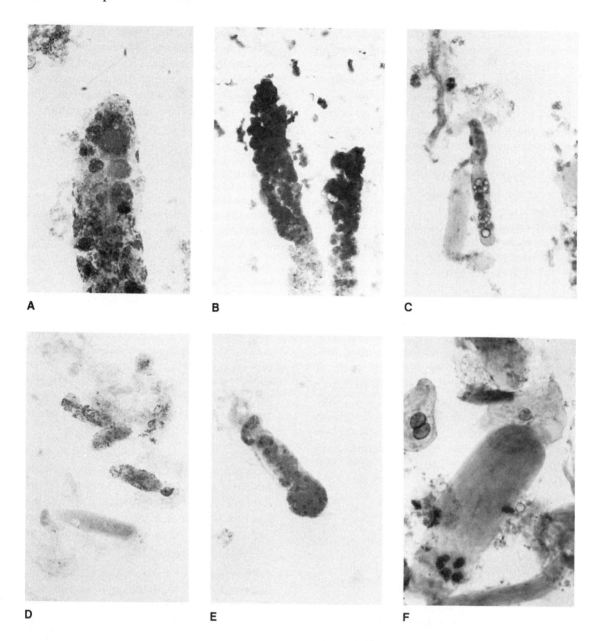

Figure 18-10 Various casts of kidney tubules. **A,** Epithelial cast with rows of renal tubular cells seen in diabetic nephropathy. **B,** Hyaline cast with neutrophils and superficial squamous cell in chronic renal failure. **C,** Finely granular casts and a waxy cast in chronic glomerulonephritis. **D,** Fatty and hyaline casts seen in lipoid nephtrosis. **E,** Red-blood-cell casts in acute poststreptococcal glomerulonephritis. **F,** White-blood-cell cast in acute pylonephritis. (From *An Atlas of Urinary Sediment* by W. Jao, R. Padleckas, and M. Swerdlow. Reprinted by permission of Michael Reese Hospital and Medical Center and Abbott Laboratories.)

and interstitial diseases, as well as certain generalized illnesses such as diabetes mellitus, are known to cause the nephrotic syndrome. Urinary fat (lipid) is present in a wide variety of renal diseases but is most commonly associated with the nephrotic syndrome.

DISORDERS OF THE URINARY SYSTEM

Uremia

Uremia is a clinical condition resulting from severe loss of renal excretory function. This is a toxemia of the blood produced by the retention of urinary constituents. Although nitrogenous substances such as urea are a good index to the severity of renal dysfunction, there is little evidence to justify the assumption that urea, creatinine, or uric acid contribute to clinical symptoms. The most common causes of uremia are the acute and chronic renal diseases. Symptoms of visual disorders, nausea, vomiting, headaches, and anuria may lead to convulsions and coma.

Clearance tests may be applied to any substance present in the blood and excreted in the urine. These tests make it possible to gauge the progress of renal damage. Creatinine and urea are filtered through the glomerulus and, to some degree, are subject to tubular resorption; therefore, both creatinine and urea clearance tests are used as guides to the glomerular filtration rate. They are sensitive guides to kidney function. In early kidney disease, a fall in the clearance rate may occur before the serum creatinine level or the blood urea nitrogen (BUN) rises above the normal range. It must be remembered, however, that although these two clearance tests can signify anomalies earlier than other tests do, they still yield positive results only after the disease is advanced. They cannot diagnose kidney disease but merely determine its progress.

Urea is the nontoxic chief end product of protein metabolism. **Azotemia,** elevated urea concentrations in the blood, found by the BUN test signify inadequate kidney function.

Creatinine and uric acid are nonprotein nitrogenous compounds found in low concentrations in the blood. In kidney dysfunction and uremia, creatinine is retained in the blood, and plasma creatinine levels may increase markedly, indicating early renal disease.

Urinary Tract Infections

Glomerulonephritis

Nephritis means inflammation of the kidney. *Glomerulonephritis* designates that the inflammation is specifically related to the glomerulus of the kidney. Glomerulonephritis is an allergic disease caused by an antigen-antibody reaction.

Acute glomerulonephritis is most common in children 3 to 10 years of age and in adults over the age of 50 years. The common cause is an antecedent infection of the pharynx (strep throat) and tonsils with group A β-hemolytic streptococci approximately 1 to 4 weeks before hypertension, edema, and hematuria (blood in urine) occur. The symptoms are chills and fever, loss of appetite, and general malaise. Urinalysis shows hematuria, albuminuria (albumin in urine), and a moderate number of urinary casts (large groups of coagulated proteins or blood cells). Following the initial streptococcal infection, antibodies are formed against the bacterial antigens, and this antigen-antibody complex is deposited on the glomerular membrane (Figure 18-11). The antigen-antibody complexes become trapped in the glomeruli, blocking them and causing the inflammatory response. If a renal biopsy is taken early in the disease, the glomeruli are large and contain numerous neutrophils. Blood flow to the nephron is reduced due to the congested glomeruli causing many glomeruli to degenerate reducing urine formation. The recovery rate is approximately 95 percent in children and a little lower in

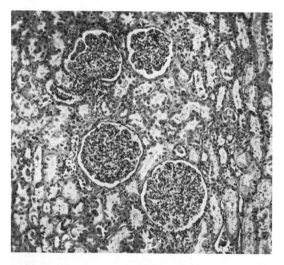

Figure 18-11 Acute diffuse glomerulonephritis. Note the large size and cellularity of glomeruli. (From *Pathology*, 5th Ed., by W. A. Anderson. St. Louis: C. V. Mosby Company, reprinted by permission.)

adults. Numerous drugs may also cause immune glomerulonephritis.

Pyelonephritis

Infections of the urinary tract are common. In most parts of the tract, an infection can clear up and not leave any residual damage. An infection that reaches the kidney, however, causes permanent damage, with destruction of some of the complex kidney tissue.

Acute *pyelonephritis* is a suppurative inflammation of the kidney caused by pyogenic (pus-forming) bacteria, such as *Escherichia coli*, streptococci, or staphylococci. The abscess formed destroys interstitial tissue of the kidney, leaving small fibrous scars that alter the adjacent kidney tubules. If the abscess ruptures, pus can enter the renal pelvis (pyuria).

The symptoms of acute pyelonephritis are chills, fever, and weakness. The attacks are brief but can persist for years. Microscopically, leukocyte casts, numerous pus cells, and bacteria are indicative of an upper urinary tract infection.

Repeated infections may destroy a large amount of kidney tissue and leave a distorted kidney. This is called *chronic pyelonephritis*, and it is a cause of renal failure and hypertension. The shrunken kidney exhibits extensive fibrous tissue and lesions that are usually not reversible. Renal failure and uremia develop over a period of time. The treatment for pyelonephritis is antibiotic therapy.

Pyelitis

Pyelitis is an inflammation of the renal pelvis. It is caused by *Escherichia coli* or other pyogenic bacteria that are in the bloodstream or the urinary bladder. Microorganisms can enter the urethra from the outside and travel to the bladder. The infection may then spread up the ureter to the renal pelvis. Painful urination (dysuria) is experienced. A urinalysis shows numerous pus cells and bacteria. Early diagnosis and treatment with antibiotics and sulfa drugs may prevent spread of infection into the kidney tissue.

Cystitis

Inflammation of the bladder is called *cystitis*. Under normal conditions, the bladder lining offers a natural resistance to infection. However, a reduced urine flow or introduction of bacteria generally leads to cystitis. The bacteria are similar to those involved in pyelonephritis with the addition of *Proteus vulgaris* and *Trichomonas vaginalis*. Acute cystitis is associated with frequent and painful micturition. Bacteria, pus, and blood cells are present in the urine. A major difference in diagnosis between cystitis and pyelonephritis is the absence of casts in cystitis. Infectious organisms invade the urinary tract by way of the urethra or the bloodstream.

A major danger in cystitis is that the peristaltic contractions of the walls of the ureters, which normally force the urine downward, may be partially reversed. If this happens, the infection can ascend to the kidneys and cause pyelonephritis.

Urethritis

Urethritis is inflammation of the urethra. Specific urethritis is caused by *Neisseria gonorrhoeae*, also called gonococcal urethritis. Nonspecific urethritis is caused by microorganisms other than *N. gonorrhoeae* and by other factors, such as traumatization due to catheterization, vigorous sexual intercourse, or chemical agents such as alcoholic beverages or therapeutic drugs. More will be said about urethritis in the next chapter on diseases of the reproductive system.

Neoplasms of the Urinary Tract

Renal Cell Carcinoma

A renal cell carcinoma, also called hypernephroma, arises from renal tubular epithelium (Figure 18-12). The clinical symptoms may be painless hematuria, a mass in the flank, fever, weight loss, and anemia. The mass may be detected by intravenous pyelogram (IVP). Renal cell carcinoma tends to spread along the veins, and typically it causes secondary tumors (metastases) in the lungs, liver, bones, bone marrow, and brain. Surgical removal is the indicated treatment, followed by chemotherapy if metastasis has occurred.

Wilms' Tumor

Wilms' tumor (or nephroblastoma) is a congenital malignant neoplasm. It is detected in young children up to the age of 2 to 3 years. Clinically Wilms' tumor is usually detected by the mother as an asymptomatic mass (Figure 18-13). There may be a fever due to necrosis or hemorrhage in the tumor. The tumor grows rapidly and may metastasize to the lungs, liver, and lymph nodes. Surgical removal and chemotherapy are the prescribed treatment. Although prognosis for this cancer is poor, a cure may be achieved if

Figure 18-12 Renal cell carcinoma (hypernephroma) of the kidney. (From *Pathology*, 5th Ed., by W. A. Anderson. St. Louis: C. V. Mosby Company. Reprinted by permission of C. V. Mosby Company.)

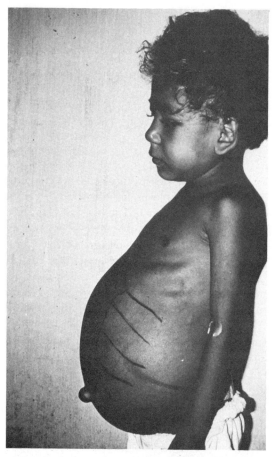

Figure 18-13 Child with Wilm's renal tumor. (From *A Survey of Human Diseases* by D. T. Purtilo. Menlo Park: Addison-Wesley Publishing Company. Reprinted by permission.)

metastases have not occurred before nephrectomy.

Obstruction of the Urinary System

Obstruction of the urinary system may occur anywhere from the renal tubules to the urethral meatus. Tubular obstruction may be caused by luminal blockage due to necrotic epithelial debris or by various casts formed either in the renal pelvis or in the tubule itself.

Kidney Stones

Urolithiasis is the formation of **calculi** (stones) within the urinary tract. They most often begin in the renal calyces and pelvis (Figure 18-14), sometimes later being passed into the ureter or bladder. Some calculi develop in the lower urinary tract.

Crystallization of minerals in the urine form stones. They are a mixture of proteins, calcium, oxalate, and other salts. Stones containing calcium are visible on x-ray examination. Other stones, such as those composed of urates or cystine, are not visible on x-ray film. Small stones may be passed spontaneously in the urine, but larger stones may require surgical removal (ureterolithotomy).

Certain metabolic diseases (gout and cystinosis) cause increased secretion of urates and cystine, leading to stone formation. In addition, stones are caused by hyperparathyroidism and severe bone diseases, which result in increased calcium in the urine.

Clinically the patient is asymptomatic. The stone may be discovered during a routine x-ray examination of the abdomen. Passage of a small stone from the renal pelvis into the ureter produces severe pain (colic). This intense pain is felt in the flank and ureteral areas. If the stone fails to dislodge from the ureter, the stone may have to be removed surgically or by passing a catheter (cystoscopy) through the urethra, bladder, and up into the ureter. An impacted stone may lead to a backup of urine into the pelvis of the kidney. This condition of dilated ureter and/or kidney pelvis is called **hydronephrosis,** or *hydroureterosis* (Figure 18-15). Hematuria is generally present, and if an infection develops due to the blockage, pyuria may be detected.

Renal Failure

Renal failure is the abnormal biochemical state that exists when the kidneys do not function properly. The possible causes include congenital abnormality (polycystic kidneys), glomeru-

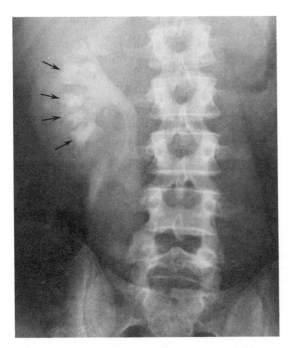

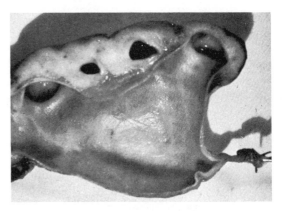

Figure 18-15 Hydronephrosis caused by obstruction of the upper end of the ureter. (From *Pathology*, 5th Ed., by W. A. Anderson. St. Louis: C. V. Mosby Company. Reprinted by permission.)

Figure 18-14 X ray of urolithiasis of the right kidney. Note the presence of numerous calculi (arrows) in the upper pelvis of the kidney. (Courtesy of Salem Hospital, Salem, Massachusetts.)

lonephritis, pyelonephritis, and hypertension. The kidneys are unable to excrete urea and creatinine from the blood. An increase in the urea nitrogen and creatinine levels in the blood develops into a toxicity that affects the whole body—uremia.

Acute renal failure is a term applied to a state of sudden cessation of kidney function. Among the causes are toxic agents, trauma, infectious diseases, electrolyte and water depletion, and major surgery. Renal tubular necrosis is the life-threatening development. It is characterized by oliguria (30 to 300 mL/day), anorexia, nausea, and lethargy. An excess of potassium (**hyperkalemia**) may cause muscle weakness and can result in total cessation of the heart (cardiac arrest). The prognosis is good if the patient's blood electrolytes can be kept balanced and kid-

ney dialysis performed over a 2-week period. This will allow the renal epithelium to regenerate and restore normal renal function.

Chronic renal failure results from prolonged progressive kidney disease. The urine output is usually large, with low specific gravity. The serum potassium is decreased or even depleted. Chronic renal failure is detected by elevated serum BUN and creatinine levels. The retention of the nitrogenous wastes (azotemia) results in uremia and neuromuscular and cardiovascular disturbances. As with acute renal failure, survival depends on hemodialysis and a controlled electrolyte balance. Renal transplantation is a suggested procedure, but this involves problems of rejection, autoimmunity, and tissue compatibility. The prognosis in chronic renal failure is poor due to the long duration of development of the condition.

Hemodialysis

The artificial kidney, or **hemodialysis,** is a continuous-flow dialysis system (Figure 18-16) of large surface area. A stream of blood from an artery spreads out in a film over a thin cello-

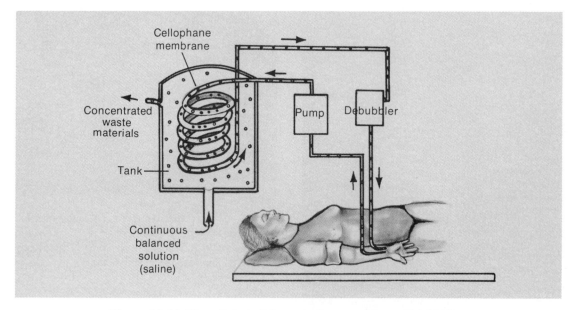

Figure 18-16 Hemodialysis. Schematic diagram of the artificial kidney.

phane membrane in contact with a balanced salt solution. Exchange of all diffusible components of the two fluids then occurs across the artificial semipermeable membrane. Blood is returned to the body through a venous catheter, and the dialyzing medium is rapidly renewed from a large reservoir.

Any substance diffusible in higher concentration in the blood also diffuses into the dialysis fluid; therefore the artificial kidney extracts toxic waste products from the blood. The reverse procedure also works: diffusible substances, for example, bicarbonate ions in acidosis, can be added to the blood.

The rate of dialysis depends on the instrument used, the surface area of the dialyzing membrane, and the concentration of waste materials in the blood. A patient may be connected to the dialysis machine for 6 to 8 hours at a time or for longer periods. The artificial kidney is not a substitute for the patient's own kidney, but it does afford a temporary therapeutic device in the continuing effort to save the lives of patients whose kidneys are diseased or destroyed.

Peritoneal dialysis is also based on the principle of diffusion of substances across a semi-

permeable membrane. The surface area of the peritoneum acts as the semipermeable membrane. A catheter is inserted into the peritoneal cavity, and 2 L of a sterile dialysis fluid are allowed to run into the peritoneal cavity at intervals (10 to 20 minutes). Equilibrium between the dialysis fluid and the highly vascular peritoneal membrane takes place (30 to 45 minutes), and the fluid is allowed to drain by gravity into a closed collecting bottle.

Peritoneal dialysis is used in renal failure to remove toxic substances and body wastes that are normally excreted by healthy kidneys. The procedure may be intermittent or continuous but is not indicated in patients with peritonitis. The advantage of this technique is its simplicity. The disadvantage is that it takes five to six times longer and is often painful.

DIAGNOSTIC PROCEDURES

Cystoscopy

A **cystoscope** is a long, hollow, lighted tube used to view the inside of the bladder and ure-

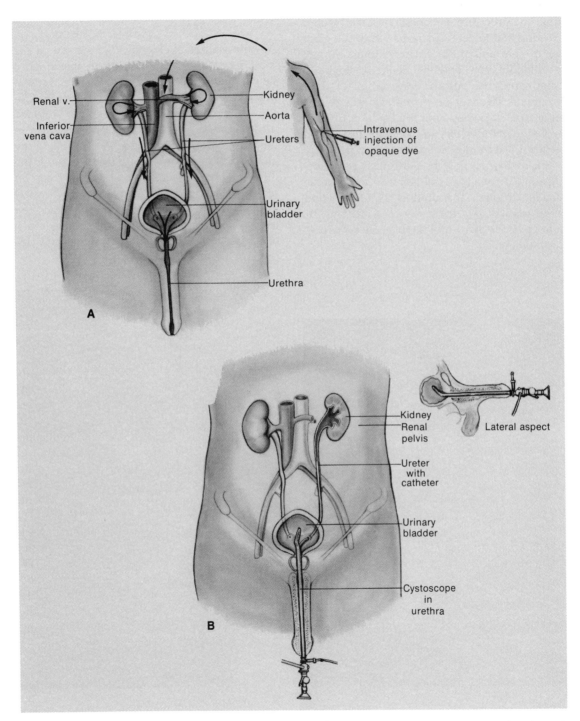

Figure 18-17 Pyelogram. **A,** An opaque dye is injected intravenously to determine the function, filling, and emptying of the upper urinary tract. **B,** Retrograde pyelogram is a direct inspection of the lower urinary tract using a cystoscope. A catheter is inserted to check the ureters and renal pelvis for any pathologic condition or obstruction.

thra. The procedure is done to observe bladder function and to detect deformation of the bladder, ureters, and prostate gland. Tumors, stones, or inflammation may be detected. Such an examination may also be done to obtain tissue specimens (biopsy) or to remove stones and other foreign bodies from the bladder. Bladder stones can be crushed, and bladder tumors can be destroyed by electric sparks (fulgurated).

A cystogram may be performed by injecting a dye through the catheter inserted into the bladder through the urethra. An x-ray film taken immediately thereafter or while the patient voids the dye outlines the bladder and urethral lumen.

Intravenous Pyelography

An intravenous pyelogram (IVP) is one of the most frequently ordered tests in instances of suspected renal disease or urinary tract dysfunction. A radiopaque iodine contrast medium substance (Hypaque or Diodrast) is injected intravenously, concentrating the substance in the urine. It is cleared from the blood by glomerular filtration and affords a view of the renal parenchyma and collecting system on multiple x-ray films (Figure 18-17A). Because the test visualizes the entire urinary tract, the test is helpful in diagnosis of renal masses, cysts, ureteral obstruction, trauma, tumors, and calculi strictures.

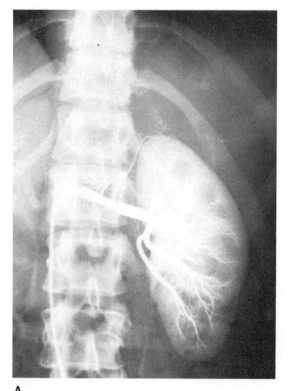

A

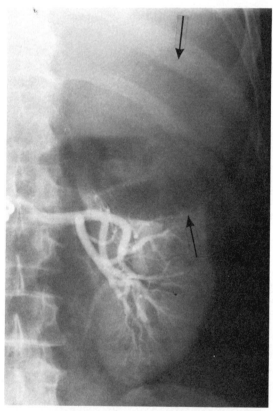

B

Figure 18-18 Renal arteriogram. **A,** This arteriogram shows normal vascularity in the right kidney. **B,** Injection material clearly outlines a mass (arrows) in the upper half of the right kidney. (Courtesy of Salem Hospital, Salem, Massachusetts.)

Retrograde pyelography is performed in conjunction with a cystoscopy by passing a catheter from the urethra to the urinary bladder, then to the left or right ureter, and injecting a contrast medium (Figure 18-17B). This test is helpful in the diagnosis of ureteral obstruction.

Kidney Echogram

Kidney echogram is a noninvasive test performed to differentiate renal masses. It is usually done following an IVP to denote the exact location of a renal mass. This technique is valuable in detecting renal or perirenal masses, cysts, hydronephrosis, and obstructions of the ureters.

Renal Angiography

Renal angiography (arteriography) provides visualization of the entire arterial, capillary, and venous systems. This test is performed by introducing a catheter into the renal artery or the aorta proximal to the origin of the renal arteries, and injecting contrast medium while rapid x-ray filming is performed (Figure 18-18). Renal angiography is helpful in diagnosing renal artery stenosis, renal masses, trauma, venous thrombosis, or urinary obstruction.

Kidney Scan

A radioactive renogram (radiorenography) is a procedure in which a radioactive material (^{99m}Tc) is injected intravenously. It is used to determine anatomical outlines and renal plasma flow and to detect renal masses. Scanning will demonstrate the size, shape, and position of the kidneys as well as the distribution of the radioisotope in the kidneys (Figure 18-19). The iodine-sensitive patient who cannot tolerate an IVP can be evaluated by radiorenography.

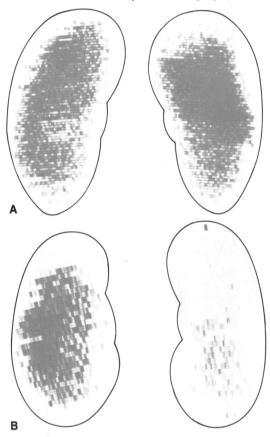

Figure 18-19 Renal scans. **A,** Normal scan, posterior aspect. **B,** Abnormal scan using ^{99m}Tc, which accumulates in renal tubular cells and indicates an abnormally large kidney in the presence of polycystic disease.

OUTLINE

A. Functions
1. Excretion of wastes
2. Regulation of composition of blood
B. Kidneys
1. Abdominal—lumbar region
2. Bean-shaped
3. Anatomy
   a. Cortex
   b. Medulla
   c. Pyramids
   d. Hilus
   e. Renal pelvis
4. Nephron units—functional units
C. Urine formation
1. Filtration
   a. Walls of glomerulus are semi-permeable
   b. Glomerular filtration force
   c. Factors affecting glomerular filtration rate
      (1) Changes in hydrostatic pressure in glomerulus or Bowman's capsule
      (2) Changes in colloid osmotic pressure of plasma proteins
      (3) Increased permeability of glomerular filtration
      (4) Decrease in total area of glomerular capillary bed
2. Resorption
   a. Water
   b. Sodium control
   c. Urea and creatinine—principal waste products
   d. Final urine more concentrated than initial glomerular filtrate
D. Regulatory mechanism
1. Hormonal control
   a. Posterior pituitary—antidiuretic hormone (ADH)
   b. Adrenal cortex—aldosterone
2. Electrolyte regulation
   a. Sodium transport
   b. Electropositive charge in blood
   c. Chloride and bicarbonate (negative) attraction
3. Acid-base balance

a. Buffers
   (1) Carbonates
   (2) Phosphates
b. Acid-base derangements
   (1) Respiratory acidosis
   (2) Respiratory alkalosis
   (3) Metabolic acidosis
   (4) Metabolic alkalosis
E. Ureters—rhythmical muscle contractions propel urine
F. Urinary bladder
1. Muscular reservoir
2. Middle layer of smooth muscle contracts to expel urine
3. Openings for ureters and urethra
4. Cystitis, cystotomy, and cystectomy
G. Urethra
1. Size difference in male and female
2. Internal and external sphincters
3. Urethritis
4. Strictures
III. Micturition
A. Gravity
B. Neurochemical control
1. Sensory receptors—stretch
2. Motor impulses—pelvic nerves
3. Reflex mechanism in infants
4. Incontinence
C. Oliguria, polyuria, and anuria
IV. Urine
A. Macroscopic examination
1. Specific gravity
2. Color
3. Clarity
4. Odor
5. pH
6. Protein
7. Glucose
8. Ketone bodies
9. Blood or pus
10. Gonadotropic hormones
B. Microscopic examination
1. Blood cells
2. Bacteria
3. Casts
4. Pus

V. Clinical considerations
  A. Uremia
    1. Blood retention of urinary constituents
    2. May be present in acute or chronic kidney disease
  B. Clearance tests
    1. Creatinine
    2. Urea (BUN)
VI. Urinary tract infections
  A. Glomerulonephritis
  B. Pyelonephritis
  C. Pyelitis
  D. Cystitis
  E. Urethritis
VII. Neoplasms of the urinary tract
  A. Renal cell carcinoma
  B. Wilms' tumor
VIII. Obstruction of the urinary system
  A. Kidney stones
IX. Renal failure
  A. Acute—sudden cessation of kidney
  B. Chronic—prolonged disease
  C. Hemodialysis
    1. Artificial kidney
    2. Peritoneal dialysis
X. Diagnostic procedures
  A. Cystoscopy
    1. View bladder and urethra
  B. Intravenous pyelography (IVP)
    1. Injected dye—x-ray examination after filtration
    2. Retrograde—use of dye during a cystoscopy
  C. Echogram
    1. Noninvasive x-ray technique
  D. Renal angiography
    1. Blood test
    2. Dye used to detect urinary tract disorders
  E. Kidney scan
    1. Radioisotope outlines renal flow

STUDY QUESTIONS AND PROBLEMS

1. In what way might secretion and excretion be the same, and how are they different?
2. What is the kidney's role in homeostasis?
3. What do we mean when we say that the kidneys are not in the abdominal cavity?
4. Differentiate between the cortex and medulla of the kidney.
5. What is the significance of peristaltic ureteral contractions?
6. How do the female and male urethras differ in structure and function?
7. Describe the anatomy of a nephron unit.
8. What substances are filtered by the glomerulus?
9. Is all resorption passive? Describe the physiology of resorption.
10. List the abnormal contents of urine and correlate them with specific diseases.
11. What is urea? How is it formed? Is all of it excreted from the blood into urine?
12. Describe the neurochemical control of micturition.
13. Describe the hormonal control of the kidney. What is diabetes insipidus?
14. How does sodium affect the acid-base balance of the body?
15. What is nocturnal enuresis?
16. What is incontinence?
17. How does the kidney aid in maintaining the body's pH?
18. What do casts in the urine normally indicate?
19. What does the specific gravity of urine have to do with excretion?
20. Define renal failure and list a few tests used to diagnose it.
21. What is renal colic?
22. What are calculi, and where do they come from?
23. What is meant by hemodialysis?
24. What is the difference between an intravenous and retrograde pyelogram?
25. Differentiate between an ultrasonogram and a kidney scan.

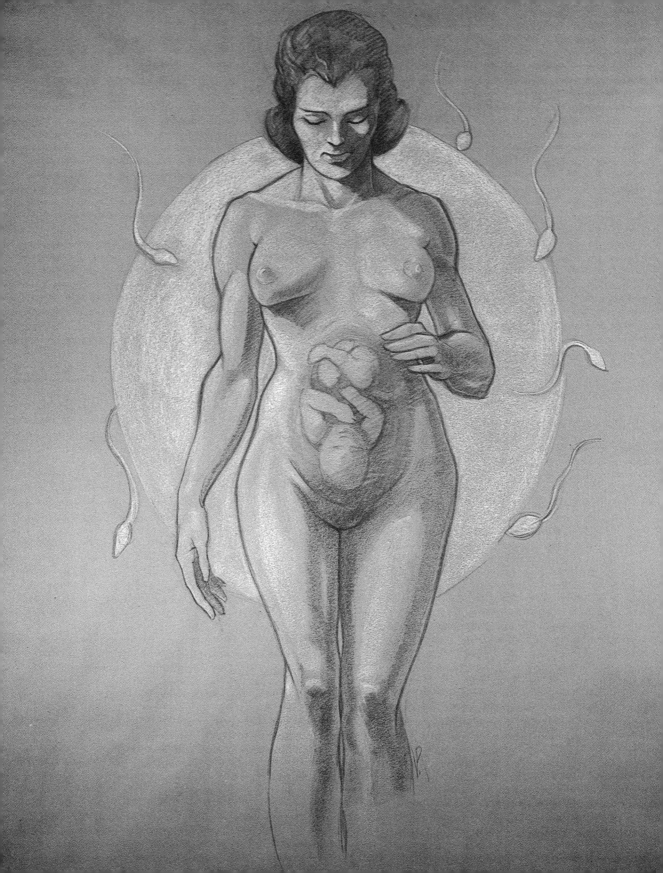

19

Reproduction, Growth, and Development

LEARNING OBJECTIVES

- Contrast mitosis and meiosis in terms of tissue, location, and number of chromosomes

- Define spermatogenesis, and explain the role of the pituitary gland

- Identify the structures of the male and female reproductive systems

- Contrast spermatogenesis and oogenesis

- List the constituents of semen

- Define erection, impotence, and ejaculation

- Name the primary and secondary sex characteristics and the effect of sterilization on them

- Name the ovarian and pituitary hormones that regulate menstruation

- Define myometrium, endometrium, and endometriosis

- Describe the changes that occur in the uterus during menstruation

- Define puberty, menopause, and menarche in relation to hormonal production and cessation

- Describe the glandular structure of the breast and the hormonal influences of lactation

- Explain the suckling phenomenon

- State the duration of sperm and ovum viability after emission of each

- Describe fertilization in the oviduct

- Identify the stages of early development of the embryo in the oviduct and uterus

- Explain the purpose of HCG

- Identify the symptoms of a possible breast tumor and how it can be diagnosed

- Name blood and radiation tests used to diagnose abnormalities of the reproductive system

IMPORTANT TERMS

abortion	cryptorchidism	gametes	menopause	ovum
autosome	dysmenorrhea	gastrulation	menorrhagia	parturition
blastula	ectopic	hermaphroditism	menorrhalgia	placenta
chromosome	ejaculation	hypermenorrhea	menstruation	semen
cleavage	embryo	lactation	metrorrhagia	somatic
coitus	endometriosis	leukorrhea	oocyte	sperm
colostrum	eunuch	mammography	oogenesis	spermatogenesis
colposcopy	fertilization	meiosis	orgasm	vasectomy
copulation	fetus	menarche	ovulation	zygote

The simplest form of reproduction is a cell splitting in half to yield two daughter cells, both identical to the original parent cell (binary fission). The cells of every animal species contain a definite and characteristic number of **chromosomes,** gene-containing filamentous structures in cell nuclei. For example, a roundworm, the ascaris, has only two chromosomes in each cell; a moth has 224. In humans—both male and female—there are forty-six chromosomes in each cell. It is essential to remember, however, that these are double sets of chromosomes: twenty-two pairs called **autosomes** (chromosomes without sex traits) and two sex chromosomes. In the female the sex chromosomes are identical and are designated by XX; the male sex chromosomes are not identical and are designated by XY.

The **somatic** (body) cells of the human body are those primarily concerned with growth and differentiation. These cells form the main structural tissues of the body—bones, muscles, nerves, and the like. They are *diploid* because each somatic cell contains double sets of chromosomes (twenty-three sets, or a total of forty-six). Through mitosis new cells are produced with exactly the same number of chromosomes as the parent cell (see Chapter 2). Hence, whenever a somatic cell divides, the reproduced cells have forty-six chromosomes, two of each type. In this sense, reproduction refers to the growth or replacement of tissues.

A sexual type of reproduction also occurs, whereby two germinal, or sex, cells join, allowing for the continuation of the species. This form of reproduction is unlike mitosis because the union of two sex cells results in a new organism not identical to its parents.

In human sexual reproduction, two cells (the egg and sperm) unite by a process called **fertilization** to produce the first cell, or **zygote,** of the new individual. Since the chromosome number must remain constant within a species, the new zygote, although formed from two separate cells, must contain a total of no more than forty-six chromosomes. It follows, therefore, that a different kind of cell division takes place prior to fertilization. This other special process of cell division is called **meiosis** (Figure 19-1).

The entire reproductive process is regulated by a series of chemical substances from the brain and pituitary gland, which influence the gonads and order the successive events of egg and sperm development, transport, fertilization, implantation, and gestation. In both females and males this remarkable relay of molecular messages begins in specialized nerve cells in the brain. Sensory stimuli from the external environment and/or hormonal stimuli from the bloodstream activate these neurons, causing them to release small neurotransmitter molecules that reach neurosecretory cells in the hypothalamus at the base of the brain (Figure 19-2). As an appropriate molecular message is

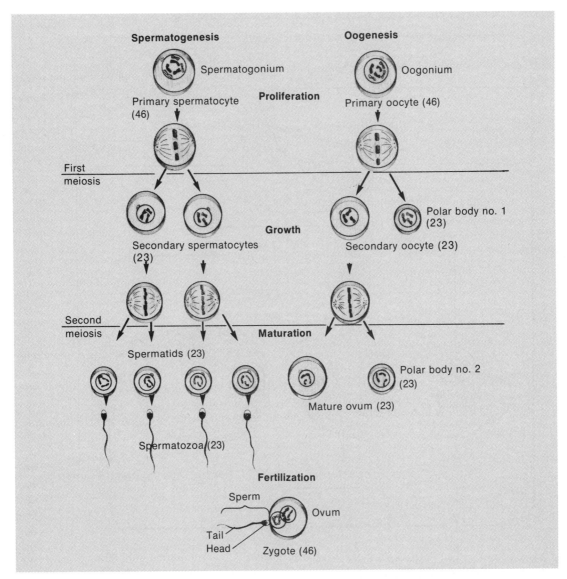

Figure 19-1 Meiosis (gametogenesis). *Left*, The various stages of spermatogenesis that give rise to four viable, mature sperm, each having 23 chromosomes. *Right*, Oogenesis, the production of a single, viable ovum (egg) and the two polar bodies from each oogonium. The union of sperm and egg (fertilization) produces a zygote. The numbers in parentheses indicate the number of chromosomes.

received, the hypothalamic cells discharge their stored supply of a gonadotropin-releasing factor. Aggregates of releasing-factor molecules move from the hypothalamic cells into a short local system of small capillaries and veins that carry them only a few centimeters to the anterior lobe of the pituitary. The releasing factor causes the pituitary to discharge its stored supply of gonad-influencing hormones—luteinizing hormone (LH), interstitial cell-stimulating

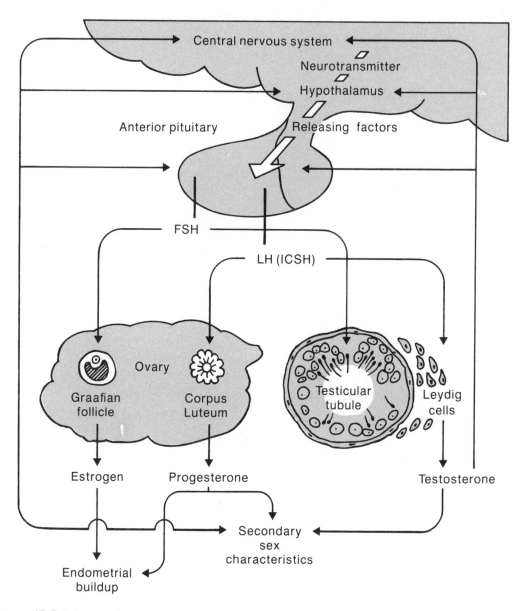

Figure 19-2 Releasing factors involved in the reproductive process. The central nervous system, prompted by external or internal stimuli, causes the hypothalamus to secrete a releasing factor that stimulates the anterior pituitary gland. The pituitary then releases the gonadotropic hormones FSH and LH, which stimulate specific structures in the gonads to secrete the steroid hormones—either estrogen and progesterone or testosterone. The gonadal steroids affect reproductive organs and feed back to the hypothalamus and perhaps to other structures to stimulate and/or inhibit their activity.

hormone (ICSH), and follicle-stimulating hormone (FSH)—which enter the bloodstream and are carried to the sex glands (see Chapter 10).

MEIOSIS

Meiosis is restricted to the germinal cells, since these are the only cells concerned with fertilization. Meiosis reduces the chromosomal number in the cell by half *(haploid cells)* so that, when the egg and sperm unite in fertilization, the normal diploid number of chromosomes is restored. The sexual haploid cells of the species produced by meiosis are known as **gametes.** Meiosis is also known as *gametogenesis* (see Figure 19-1). The production of the male sex gamete, called **sperm,** is called **spermatogenesis; oogenesis** is the production of the female sex gamete, the egg, or **ovum.** Spermatogenesis begins at puberty and continues until old age; oogenesis, however, is a cyclical process that begins at puberty and terminates at **menopause,** the period of life when menstruation normally ceases, usually between the ages of 35 and 50 years in most American women.

First Meiotic Division

The two meiotic divisions, illustrated in Figure 19-1, occur in a series of stages similar to those of mitosis: interphase, prophase, metaphase, anaphase, and telophase. The chromosomes in prophase are long and thin, and each pair of like chromosomes moves together and unites longitudinally in a process known as *synapsis.* The synaptic pairs become shorter and thicker; at this point it becomes evident that they are doubled. During metaphase, the paired chromosomes (tetrads) arrange themselves so that like pairs line up at the midline of the dividing cell. It is this maneuver that distinguishes meiosis from mitosis. In mitosis like chromosomes line up and separate, resulting in one identical chromosome going to each respective daughter cell. In the metaphase of meiosis, like chromosome pairs line up and separate, but the result is a reduction in the number of chromosomes in each cell. The rest of the first meiotic division is mechanical, producing two daughter cells that together have the same number of chromosomes as the parent cell.

Second Meiotic Division

The second meiotic division follows a mechanical sequence identical to that of the first. Maturation and division of the two intermediate-stage daughter cells result in four haploid cells or gametes. The chromosomes do not synapse because no like chromosomes are available (see Figure 19-1). Each double-stranded chromosome moves into the metaphase stage, and the new single-stranded chromosomes separate. The second meiotic division is essentially a mitotic division.

In summary, the first meiotic division produces two haploid cells, containing double-stranded chromosomes. Each of these cells divides in the second meiotic division, producing a total of four new haploid cells, containing single-stranded chromosomes.

It is evident now that the chromosomes are not all functionally equal. Chromosomes in adult human cells do occur in pairs with similar genetic function, but the pairs constitute twenty-three distinct and unique sets of genetic information. Thus the necessity for a second remarkable aspect of meiosis is obvious. Not only does meiosis halve the chromosome number, it does it in such a way that each mature germ cell receives one member of each pair of chromosomes. The zygote, therefore, receives a member of each pair of chromosomes from each parent. Sexual reproduction is advantageous in that it provides for genetic variability—a mixture that can produce countless variations in the offspring.

Spermatogenesis and Oogenesis

At the time of reproduction, meiosis produces gametes (sperm and eggs), which then unite in fertilization to produce a zygote. The zygote then grows mitotically to produce the new diploid individual. The sperm and eggs are, therefore, the only haploid stage in the life cycle of the individual.

In males, sperm cells (spermatozoa) are produced by the germinal epithelium of the testes. When one of the epithelial cells undergoes meiosis, the four haploid cells that result differentiate into swimming sperm cells. This is spermatogenesis.

In the female, eggs are produced in the follicles of the ovaries by a comparable process called oogenesis. When a cell in the ovary undergoes meiosis, however, the haploid cells are unequal in size. The first meiotic division produces one relatively large cell, the primary **oocyte** (immature ovum or egg cell), and a tiny dead one, the first polar body (Figure 19-1). The second meiotic division of the living, larger cell produces a secondary oocyte (the egg) and a smaller dead secondary polar body. Thus, when an ovarian cell undergoes meiosis, only one viable mature egg, or ovum, is produced. The polar bodies are not functional and ultimately are absorbed into the metabolic system as food.

MALE REPRODUCTIVE PROCESSES AND ORGANS

Male Reproductive Organs

The organs involved in the production and ejaculation of human sperm are illustrated in Figure 19-3. Sperm are produced in the testes, which are contained in the scrotum outside the abdominal cavity. Spermatogenesis occurs best at temperatures slightly lower than normal body temperature, and the scrotum is suspended outside the body where the temperature is a few degrees lower. The coiled seminiferous tubules make up the mass of each testis (Figure 19-4). The epithelial lining of their walls contains continuously proliferating cells

called *spermatogonia*. These develop to form primary spermatocytes, which undergo meiotic division to yield two secondary spermatocytes each (Figure 19-4). These divide again, mitotically, to yield spermatids and motile sperm. The newly formed sperm pass along the vas deferens into the epididymis, where they remain for a short time. If not ejaculated, the sperm soon degenerate and are absorbed in these tubules. Some sperm are stored in the ampulla, an enlargement of the excretory duct (vas deferens) of the testis. Adjacent to the ampulla are two seminal vesicles that secrete a nutritive fluid for the sperm. From the ampulla, the duct passes through the prostate gland into the urethra, a tube that passes out of the abdominal cavity down the middle of the penis (Figure 19-3).

Interspersed between the seminiferous tubules of the testes are a number of interstitial cells, called *Leydig cells*, which produce and secrete the male hormone, testosterone. Testosterone passes directly into the bloodstream to control the process of spermatogenesis and the development and activity of the secondary sex organs. The hormone also promotes skeletal and muscular growth during adolescence and is necessary for the development of secondary sex characteristics such as deepening of the voice and growth of pubic hair.

Another cell of importance in the testis is the Sertoli cell, which is located in the seminiferous tubule between the spermatogonia (Figure 19-4). It is primarily responsible for supplying the developing germinal cells with any necessary nutritional support. Sertoli cells are commonly called "nurse cells."

The *prostate gland* is fairly large, about 4 to 5 cm in diameter, and almost completely surrounds the urethra (see Figure 19-3). In prostatitis, the gland may become so large that it constricts the urethra and prevents urination. This problem is common in old age and usually results in the surgical removal of the prostate gland (prostatectomy). Just below the prostate gland is Cowper's gland. The two glands together secrete material that mixes with the sperm and provides it with nourishment, a

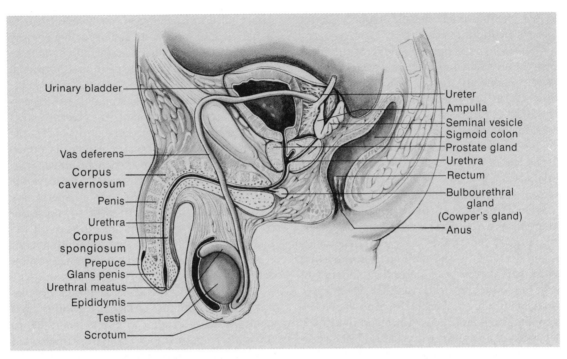

Figure 19-3 Midsagittal section of the male reproductive system.

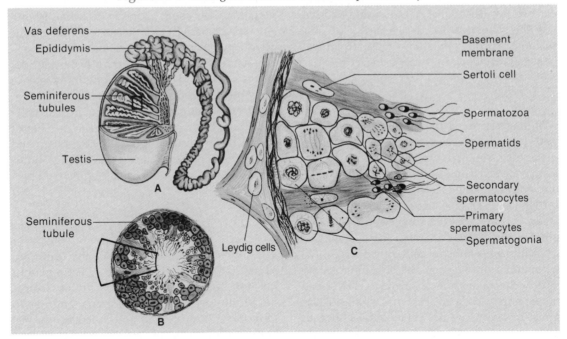

Figure 19-4 Testes. **A,** Diagram showing the duct system of the human testes. **B,** Cross section of a seminiferous tubule of the testes. **C,** Enlarged diagram of a seminiferous tubule, showing the development of spermatogonia.

proper alkaline medium, and support for the sperm's swimming motions.

The *penis* is composed chiefly of three columns of erectile tissue (see Figure 19-3). Two of these columns are the paired *corpora cavernosa*, which form the bulk of the penile structure. The third is called *corpus spongiosum*, and it is considerably smaller, running parallel and ventral to the other two. It also contains the urethra. At the end of the penis, the corpus spongiosum enlarges and extends toward the dorsal surface to form a conical cap, the *glans penis*, which covers the blunt ends of the corpora cavernosa. Anatomically, the penis differs from its female counterpart, the clitoris, only by virtue of its larger size and by the location of the urethra within the corpus spongiosum. A network of large endothelium-lined vascular areas separated by fibrous trabeculae occupy the three corpora. These are filled with blood or empty, depending on whether the penis is erect or flaccid. The organ is covered by skin that is thin and freely movable over the body but is firmly attached over the glans. A fold of skin overlapping the glans is called the *prepuce*. The penis functions as the copulation organ, a pathway for semen, and a pathway for urine.

Sperm

A human sperm is microscopic in size (0.06 mm). It is a specialized, motile germ cell that carries the male chromatin material and consists of an oval head, short neck, and body. The tail is long and capable of violent swimming actions (see Figure 19-1). This vigorous activity is facilitated by closely packed rows of mitochondria found in the neck.

A limited number of collecting ducts funnel the sperm coming from the seminiferous tubules to the epididymis, a long tube convoluted into a compact body adjacent to the testes. These immature sperm are not yet able to fertilize an egg or even to move under their own power. As they pass from the head of the epididymis through its slender body to its distended tail, they achieve motility and a degree of maturity.

The final critical changes that enable them to penetrate and fertilize an egg are achieved only in the female reproductive tract, and even then only if the tract maintains the proper hormonal balance.

The process of development from spermatogonium to spermatozoan (sperm) takes about 3 days (72 hours). Some of the sperm die and are disposed of by white blood cells; others enter the urethra in a steady stream and are carried away in the urine. The remainder leave the male tract at **ejaculation,** when sperm are forced rapidly into the urethra by muscular contractions. These sperm are mixed with the fluid secretions of the prostate and seminal vesicles, whose ducts lead into the terminal portion of the vas deferens or into the urethra.

The material expelled in a normal ejaculation is about 3 mL of viscous milky fluid called **semen.** Each ejaculation contains about 200 to 300 million sperm. Only a single spermatozoan is needed to fertilize an ovum. However, it is less likely for fertilization to occur when semen contains a low count of sperm (less than 60 million sperm per ejaculum). The need for a large number of sperm for fertilization may be due to the fact that the ovum is surrounded by several layers of cells. Many sperm containing special enzymes are needed to break a path to the ovum itself, even though only one sperm may actually penetrate. More important than the number of sperm in fertilization are the morphology and motility of sperm. Some sperm exhibit abnormal structure, and these usually show poor motility.

Hormonal Function

The interstitial cells (Leydig) of the testes are stimulated by the interstitial cell-stimulating hormone (ICSH) of the anterior pituitary to produce testosterone, a male hormone (androgen). Little or no testosterone is secreted before puberty, which normally occurs between the ages of 12 and 16 years. Evidence indicates that ICSH is stored in the pituitary until puberty, but the hormone is not actually released until then. A second gonadotropic hormone, follicle-stimulating hormone (FSH), is also secreted by the anterior pituitary. FSH assists ICSH in stimulating spermatogenesis in the testes.

If the testes are absent because of castration, congenital defects, or hyposecretion of gonadotropins at the time of puberty, testosterone is not produced, spermatogenesis cannot occur, and the secondary sex characteristics of a man do not develop. A male individual without testes is called a **eunuch.** Loss of the testes after puberty causes sterility, but there is only some recession of the secondary sex characteristics.

CLINICAL CONSIDERATIONS AND DISORDERS OF THE MALE REPRODUCTIVE SYSTEM

Benign Prostatic Hyperplasia

The prostate is a gland that surrounds the urethra just below the bladder. It makes a watery fluid that is released with the spermatozoa in sexual intercourse. In men over 50 years of age, the gland often enlarges. Sometimes the enlargement obstructs the flow of urine from the bladder, ureter (hydroureter), or kidney (hydronephros) causing associated infection. This condition is called a *benign enlargement* (hyperplasia). Surgical removal (prostatectomy) may be indicated.

Prostatitis

Prostatitis, inflammation of the prostate, may be acute or chronic. However, acute prostatitis commonly develops into a chronic state. Bacteria may reach the prostate from the bloodstream or from the urethra. Urethritis is commonly associated with prostatitis caused by infection by the gonococcus bacterium in a male with gonorrhea or by *Escherichia coli*, which is common in urinary tract infections.

Perineal pain, fever, and dysuria (pain while urinating) are symptoms of prostatitis. Pus may often be seen dripping from the penis, and a culture and Gram's stain will identify the gonococcus. Penicillin is the treatment of choice.

Carcinoma of the Prostate

Carcinoma of the prostate is most commonly found in men over 50 years of age. The tumor usually develops at the back of the gland, causing obstruction of urine flow. The presence of the tumor may be noticed after metastasis within bones with resultant pain. Other symptoms include urinary infection, excessive urination at night (nocturia), and perineal pain.

Growth of the tumor is increased by androgens and inhibited by estrogens. The prostatic tissue is rich in acid phosphatase. When cancer has extended beyond the prostate to the surrounding tissues, the serum acid phosphatase level is increased and provides a good index of growth and extension of the tumor.

Rectal examination usually indicates a hard prostatic mass in contrast to the enlarged soft prostate when it has hypertrophied.

Cryptorchidism

The testes are located in the scrotal sac where the temperature averages 2.2°C less than in the peritoneal cavity. Any rise in temperature is harmful to spermatogenesis because it arrests maturation of the germinal epithelium.

In the embryo, the testes develop within the body cavity, descending before birth through the inguinal ring to their location in the scrotum. Sometimes one or both testes do not descend, a condition called undescended testicle, or **cryptorchidism.** An undescended testis will develop its hormones but will not produce sperm at puberty, presumably because of the high temperature within the abdomen. In some patients, hormonal treatment may stimulate spermatogenesis after the testis has been moved surgically to its normal position.

Orchitis

Inflammation of the testis (*orchitis*) may be caused by injury. The most common cause, however, is mumps after puberty. Swelling of the testes and severe pain usually develop within 2 weeks after acquiring mumps of the parotid gland. Usually the inflammation subsides, but there may be a residual fibrosis in the testis that could cause sterility.

Epididymitis

The epididymis is a common site for infection by gonococcus or chlamydial organisms, which cause inflammation. Symptoms of *epididymitis* include severe pain in the testes, swelling, and tenderness in the scrotum. Abscesses sometimes form, and scar tissue develops, which can cause sterility.

Hydrocele

A *hydrocele* is a collection of fluid in the epididymis. Acute cases are usually due to a spread of gonorrhea or to tubercular infections. Chronic hydrocele can be caused by a low-grade infection. Chemotherapy and surgical repair of the water sac are the recommended treatments. Hematocele is the name given to a hemorrhage into a hydrocele following trauma.

Phimosis

Phimosis is a congenital condition in which the prepuce (foreskin) is too tight around the head of the penis and cannot be drawn back. This situation normally may be remedied by removing all or part of the foreskin, a process called *circumcision.*

Vasectomy

Vasectomy is a surgical procedure, resulting in sterilization of a male. A portion of the vas deferens (Figure 19-3) from both testes is removed to prevent sperm from reaching the urethra. The sperm are absorbed into the vascular system, and the individual retains his ability to produce hormones as well as to have sexual intercourse.

Hermaphroditism

Hermaphroditism is a blending of male and female attributes. In *true hermaphroditism*, both male and female sex organs are present in one person. There are no known cases in which this condition has resulted in fertility, either as a man or woman. In *false hermaphroditism*, the genital glands of one sex are evident, but the secondary sex characteristics of the opposite sex predominate.

FEMALE REPRODUCTIVE PROCESSES AND ORGANS

Female Reproductive Organs

The female reproductive organs are illustrated in Figure 19-5. The ovaries are the counterparts of the testes in the male. They are small bodies located in the pelvic part of the abdomen and are attached to the uterus by two layers of peritoneum called the *broad ligament* (Figure 19-5, *B*). Within the ovaries the female sex gametes, the ova, are formed by oogenesis. All the eggs to be produced—about 750,000 during the lifetime of a woman—are already present in the ovaries of the female child. At the beginning of puberty the eggs begin to mature, one at a time. Approximately every 28 days in a process called **ovulation** an egg is released from one ovary and passes down the *oviduct (fallopian tube)* toward the uterus. The egg lands on the endometrium of the uterus and, if not fertilized, is expelled with the endometrial lining through the cervix, down the vaginal canal, and to the outside of the body. This process is called **menstruation,** and, as mentioned, occurs about once every 28 days. This regularity is controlled by the pituitary gland.

The fallopian tubes are small and muscular

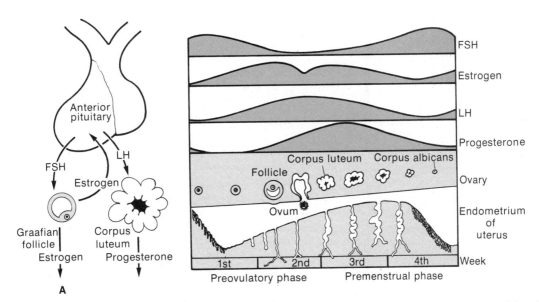

Figure 19-8 Relationships among the anterior pituitary, ovarian, and endometrial cycles. **A,** Menstrual cycle of a nonpregnant woman. The hormone levels in the blood are graphically illustrated to show their effect in the ovary and on the endometrium. Note the correlation of FSH and estrogen in the preovulatory phase, and of LH and progesterone in the premenstrual phase.

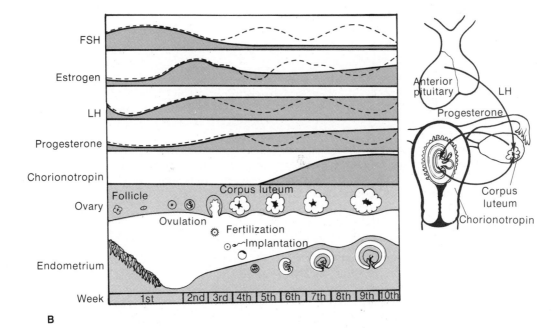

Figure 19-8 *Continued* **B,** Pregnancy cycle. The addition of chorionotropin from the embryo eliminates the premenstrual phase by keeping the corpus luteum alive and continually secreting progesterone to the endometrium. Abbreviations: LH, luteinizing hormone; FSH, follicle-stimulating hormone.

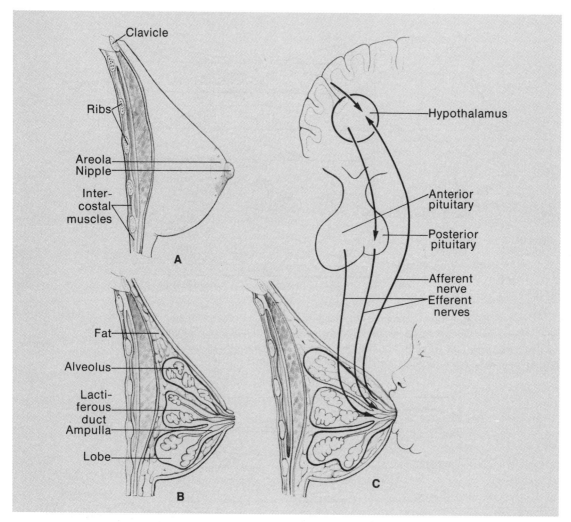

Figure 19-9 Mammary glands. **A** and **B**, Lateral aspect and sagittal section of the mammary glands, showing the external and internal anatomy of the milk (lactiferous) glands and duct system. **C**, Lactation—The child's suckling gives rise to afferent nerve impulses which, in turn, stimulate the mammary glands to secrete milk.

and their ducts. The zone around the nipple, the areola, and the nipple itself become darkly pigmented. After childbirth, the fall in blood levels of estrogen and progesterone triggers the release of prolactin (lactogenic hormone) from the anterior pituitary. Prolactin stimulates the acinar glands to secrete the milk produced by blood flowing through the glands.

Lactation starts 3 to 4 days after childbirth, and is maintained by prolactin. Afferent nerve fibers convey impulses established by the child's suckling (Figure 19-9). These impulses give rise to efferent nerve impulses to the posterior pituitary gland for the release of oxytocin, which is carried by the blood to the ancinar glands to maintain secretion of milk. If continued lactation is not in the mother's best interest, it can be suppressed by injections of estrogen.

Physiologically the advantages of breastfeeding are clear-cut in the first few days after childbirth. The first milk secreted is called **colostrum.** Its nutritional content is a little different from that of ordinary milk, but more importantly, it contains antibodies from the mother that can be absorbed by the infant to provide a short-term passive immunity against disease.

Menopause

Menopause is the period of time during which menstruation is irregular and finally ceases. This occurs usually between the ages of 49 and 51 years. It involves changes in the endocrine system, the first step being a decrease in LH supplied by the pituitary. As a result the corpus luteum is not developed, the ovum does not mature, and the endometrium of the uterus does not develop. Menstruation continues to occur, but the female is unable to conceive. After several months or years of anovulatory menstruation, FSH is no longer secreted, and menstruation ceases completely.

DISORDERS OF THE FEMALE REPRODUCTIVE SYSTEM

Each structure of the female reproductive system—vulva, vagina, uterus, ovaries, and mammary glands—reacts to disease in its own characteristic way. This section presents the clinical manifestations as they pertain to each part of the female reproductive system.

Vulval and Vaginal Disorders

Vulvovaginitis

The vulva and vagina are frequently involved in a disease process. The woman with vulvovaginitis may complain of external irritation or itching (pruritis). Bacterial infections are usually related to uncleanliness and scratching or to widespread skin infections, such as furunculosis. Fungus infections, often caused by *Candida albicans* (yeast), tend to occur in diabetic women under antibiotic therapy. The thick, cheesy exudate contains spores and hyphae.

Vulvovaginitis is frequently caused by the protozoan *Trichomonas vaginalis.* This organism, a flagellate, is detected in the copious, purulent vaginal discharge characteristic of the infection. When trichomonas infects the vagina, it causes a burning sensation upon urination (dysuria).

The major symptoms of vulvovaginitis are local swelling, itching, and a burning sensation produced by a white vaginal discharge known as **leukorrhea.** Diagnosis is achieved by examining a wet smear from the vagina. The infecting organism can be seen swimming in the fluid (if a protozoan), or spore sacs and hyphae can be identified (if a fungus). Antibiotics and antifungal chemotherapy are the indicated treatment.

Other inflammations of the vulva are syphilis, chancroid, Herpes simplex II, and lymphogranuloma venereum.

Bartholin's Cyst

Bartholin's glands are located in the deep tissues on each side of the vaginal opening. Obstruction of the excretory ducts by inflammatory scarring or epithelial growths results in a cystic formation. The cyst is usually unilateral, about 4 cm in diameter, and appears as a round mass in the labia minora. If the duct of the gland becomes occluded from an inflammation, pus collects in the gland and an abscess forms. Treatment is by incision and drainage of the infected material and administration of antibiotics. If the cyst recurs, surgical removal is indicated.

Neoplasms

Benign neoplasms (squamous papilloma, nevus, and lipoma), occur occasionally in the vulva. Microscopically a neoplasm is a cystic structure, usually filled with a papillary growth or fat covered with columnar epithelium.

The most common malignant neoplasm of the vulva and vagina is the epidermoid carcinoma, which usually affects women over 50 years of age. The neoplasm begins as a small nodule on the labia, gradually enlarges, and often ulcerates. It may infiltrate the external genitalia and spread to inguinal lymph nodes. The 5-year survival rate is about 30 percent. Histologically, the neoplasm consists of a squamous cell carcinoma.

Melanoma is the second most important malignant neoplasm of the female genitalia. It can arise either from a preexisting nevus or develop from the heavily pigmented skin of the area. Metastases occur to the inguinal lymph nodes.

The treatment of invasive cancer of the vulva and vagina is radiation therapy and/or radical surgery. The importance of early diagnosis is obvious.

Uterine Disorders

Cervicitis

Infections of the cervix, or *cervicitis,* are the most common gynecological disorders. These infections may be acute and of specific origin, caused by *Neisseria gonorrhoeae* or streptococci, or nonspecific, which is much more common. Sometimes nonspecific chronic cervicitis is a sequel to acute cervicitis of pyogenic origin, but it more commonly seems to be a result of laceration and low-grade infection following childbirth.

Cervicitis is characterized by leukorrhea, erosion and eversion of the cervix, and a mucoid exudate that is mucopurulent and may be blood-streaked. Symptoms include low back pain, difficult menstruation (**dysmenorrhea** or **menorrhalgia**), dysuria, and urinary frequency. Cervical polyps are frequently associated with chronic cervicitis.

Treatment with antibiotics and electrocautery can prevent the infection from developing into carcinoma of the cervix.

Carcinoma of the Cervix

Carcinoma of the cervix is the third most common malignancy in women. Cervical cancer occurs chiefly in women who have had many pregnancies, especially if those pregnancies occurred early in life. The neoplasm is rare in virgins and in the marital partners of males circumcised in infancy. It has been associated with type II herpes simplex virus. Antibodies against type II herpes simplex virus are present significantly more frequently in women with cervical cancer.

Cervical carcinoma appears first in the epithelial layers of the external opening of the cervix (Figure 19-10). Preinvasive cancer is a common diagnosis in women 30 to 40 years of age, whereas invasive carcinoma occurs about

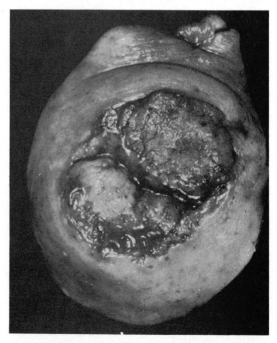

Figure 19-10 Carcinoma of the cervix. The cervix is markedly enlarged by an invasive tumor. (From *Basic Pathology* by S. L. Robbins. Philadelphia: W. B. Saunders, 1981. Reprinted by permission.)

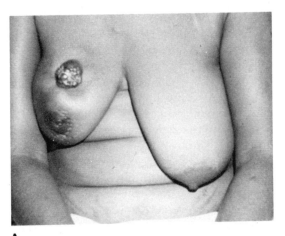

A

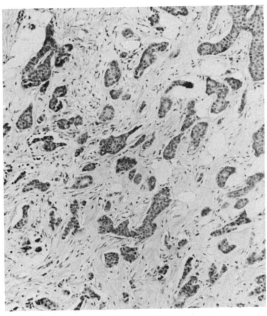

B

Figure 19-11 **A,** Carcinoma of the breast. **B,** Tumor cells invading the connective tissue. (From *A Survey of Human Diseases* by D. T. Purtilo. Menlo Park, Calif.: Addison-Wesley Publishing Company, 1978. Reprinted by permission.)

Heredity factors have to be considered the foremost possibility according to the data, which states that the risk of developing breast cancer is two to three times higher in women whose mothers or sisters have had this disease. A study of patients from families with a high incidence of breast cancer shows that breast cancer tends to occur at a younger age and is more likely to be bilateral than in women without such a family history.

There is a great deal of evidence indicating that prolonged and excessive estrogen stimulation leads to the development of breast cancer. The incidence of breast cancer is high in women who experience an early puberty and late menopause, thus extending the cyclic stimulation of the breasts by ovarian estrogen.

Screening for breast cancer, using self-examination and mammography, has prevented the advancement of the disease and greatly improved the survival rate in thousands of patients. Once detected, the lump undergoes further examination. A tissue biopsy (frozen section) may be performed while the patient is on the operating table. Microscopically, cancers

of the breast are composed of cells in a more or less glandular arrangement and surrounded by connective tissue. When the tissue is cut open, it feels gritty due to the presence of streaks of calcified tissue and the dense fibrosis of the tumor.

Based on the pathologist's diagnosis, the surgeon may remove just part of the tumor or the entire mammary gland and lymph nodes of the axilla. This procedure is called a *modified mastectomy.*

Other prescribed treatments involve the use of radioisotope implants, radiation therapy, and chemotherapy. Patients with estrogen-dependent tumors and known metastases are often treated with hormones (estrogen-receptor assay).

Survival rates for breast cancer patients are based on the size and type of tumor, its degree of infiltration, its location (unilateral or bilateral), and condition of the patient at the time of discovery (women who develop cancer during pregnancy have a poorer survival rate). Overall the 5-year survival rate is approximately 55 percent. Early detection and treatment greatly improve the chances of survival.

COPULATION, FERTILIZATION, AND EARLY DEVELOPMENT

Copulation

During sexual intercourse, or **copulation,** sperm is ejaculated in semen from the penis into the vagina. Ejaculation is impossible without erection of the penis, which is controlled by autonomic impulses from the thalamus and cerebral cortex. Psychological or physical stimulation results in the dilation of the arteries and blood sinuses of the penis so that they become engorged with blood and an erection occurs. The friction of the copulatory act stimulates the nerve endings of the glans penis, and these impulses activate a reflex mechanism that terminates in ejaculation itself. During this act, the vas deferens, seminal vesicles, and prostate gland contract, propelling the semen and sperm into the vagina. This process of ejaculation and the feeling of intense pleasure accompanying it are termed an **orgasm.** The sperm cells, being extremely motile, swim up the vagina toward the uterus and eventually enter the fallopian tubes. It is said that the sperm can swim from the vagina to the upper end of the tubes in less than an hour.

In women sexual excitement may be accompanied by nipple erection, increase in breast size, and a diffuse red "flush" spreading over the upper part of her body. The normally flattened labia minora and clitoris become engorged and protrude through the labia majora. Within thirty seconds of sexual stimulation (either physical or psychological), the vagina becomes lubricated by an exudate from the Bartholin glands. These physical changes intensify with anticipated entrance of the penis, and orgasm occurs as a spastic contraction of the vagina and uterus.

Fertilization

The microscopic sperm cells are fragile and susceptible to damage and destruction before they encounter the egg. To preserve the sperm for at least 48 to 72 hours, the seminal fluid contains energy sources, pH buffers, and other chemicals. Of the 300 million sperm in each ejaculation, several thousand are necessary for fertilization. These high numbers may sound ridiculous, but it must be remembered that millions are destroyed during the act of copulation; millions are destroyed by the normal acidity of the vagina; and the remaining number of sperm is halved because both fallopian tubes receive sperm, but only one contains an egg. Sperm within the female decompose within 72 hours of ejaculation.

When the minute ovum ruptures from the ovarian follicle, it is surrounded by numerous cells, called the *corona radiata* or *protective zone.* Sperm arriving near the egg secrete an enzyme, *hyaluronidase,* necessary to break down an organic acid barrier, hyaluronic acid, secreted by the cells of the corona radiata. Individual sperm contain the enzyme, but it takes thousands of sperm to secrete enough enzyme so that one sperm can successfully enter the egg and accomplish fertilization. Once this sperm has entered the egg, its tail is absorbed and its nucleus joins the nucleus of the egg to complete the transfer of genetic information.

When the process of fertilization is complete, the fertilized egg, or zygote, is restored to the diploid (forty-six chromosomes) number. The zygote, therefore, contains a new arrangement of genes on the chromosomes—an equal contribution from each parent. If the ovum was fertilized by a sperm bearing a Y chromosome, the result will be XY, a male; if the sperm contained an X chromosome, the offspring will be XX, or female.

The site of fertilization in the tubes is important. Normal embryonic growth and development will occur if fertilization occurs one-third of the way down the fallopian tube from the ovary (Figure 19-12). The zygote needs the following few days of development to produce the chorionic sac, which, in turn, secretes chorionotropic hormone, the principal hormone needed to maintain the corpus luteum for

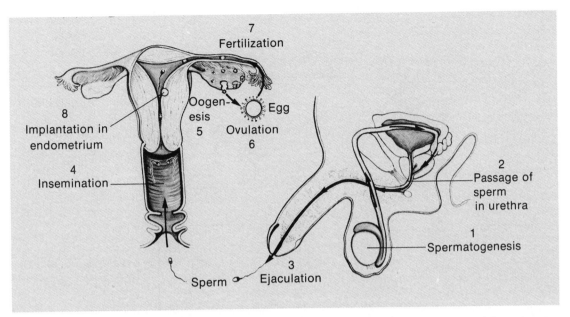

Figure 19-12 Conception. Schematic diagram showing the events leading up to fertilization of the egg and eventual implantation in the uterus. The arrows indicate the direction of flow of the swimming sperm; the numbers indicate the sequence of events.

a 3-month period. The corpus luteum is then able to secrete enough progesterone to maintain the endometrium awaiting the new embryo. Failure of the zygote to develop the chorionic sac in sufficient time will result in normal menstruation and loss of the fertilized egg.

Sometimes the zygote becomes lodged in the fallopian tube, and partial development occurs there. Sometimes sperm swim the length of the tube and fertilize an egg ejected from the ovary and dropped into the pelvic cavity. In either case, the pregnancy is said to be ectopic. In the first instance, surgical removal of the fallopian tube and ovary is necessary. Degeneration of the zygote occurs in the latter case.

More than one egg may be fertilized if more than one is present. This is typical in animals such as dogs and cats, in whom ovulation releases many eggs. When a human female releases eggs from both ovaries during the same cycle, the conception of twins is possible. The twins in this case would be fraternal, meaning that each child grows separately from its own

placenta. Identical twins, triplets, and so forth occur when the zygote divides, and more than one embryo develops from the same original egg.

Family planning sometimes involves the use of hormones (FSH-derivative) to produce many eggs during a cycle—the so-called fertility drugs—for couples having difficulty conceiving a child. The intent is to narrow the odds for conception. Multiple births have occurred following use of these drugs.

Early Development

The zygote passes through a series of distinct developmental stages. **Cleavage** is the first step after fertilization. This involves a process of cell division (mitosis) without any significant growth. The one-celled zygote divides and passes through stages described as two-cell, four-cell, eight-cell, and so forth (Figure 19-13). Since the cells do not grow very much between

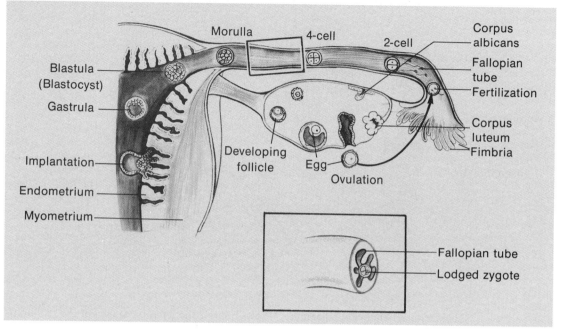

Figure 19-13 Implantation of the fertilized ovum. Approximate stages of the fertilized egg during development and implantation in the endometrium. The inset shows a section of the fallopian tube, with partial development and eventual surgical removal of an ectopic pregnancy.

divisions, cleavage makes them smaller and smaller. At the end of the cleavage stage, the zygote consists of a ball of many small cells, each bearing a complete set of genes from the original zygote.

By the time cleavage has produced several hundred cells, the zygote has become a ball-shaped structure consisting of a single layer of cells surrounding a fluid-filled cavity and is called a **blastula** (Figure 19-13). The blastula (blastocyst) is commonly referred to as the "job-assignment" stage. The time element is now becoming important. Blastula formation took approximately 5 days, and the zygote is now located just inside the horns of the uterus, ready for implantation. The corpus luteum, meanwhile, is about halfway through its 10- to 12-day lifespan in the ovary, and the quantity of pituitary hormones has already begun to diminish.

In the next stage, **gastrulation,** the cells of the zygote begin to grow and differentiate into cells clearly different from one another. At this point there is an inner layer of cells, the *endoderm,* and an outside layer, the *ectoderm.* Gastrulation represents the tissue formation of the new embryo, which becomes implanted in the endometrium 7 days after fertilization (Figure 19-13).

During gastrulation, one side of the hollow wall of cells becomes indented, and this indentation continues until a cup-shaped structure is formed. The resulting cavity now becomes the inner layer, and the previous inner layer becomes a new middle layer of cells, called the *mesoderm.* As the embryo continues to differentiate and develop, the mesoderm will produce bones, muscles, the heart and circulatory system, and the excretory and reproductive systems (Figure 19-14). The endoderm

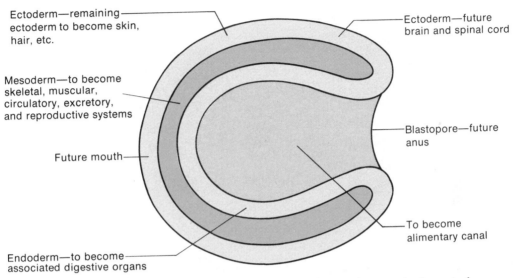

Figure 19-14 Early differentiation of the various tissue layers as seen in the gastrula stage of the human embryo.

differentiates into the intestinal tract, stomach, pancreas, liver, and lungs. The ectoderm produces cells that will become the skin, brain, and nervous system.

IMPLANTATION THROUGH BIRTH

Once the developing mass of cells has reached the uterus, it generally takes 4 to 5 more days before it becomes firmly implanted in the uterine lining. Implantation begins when the blastula digests some of the endometrium. The zygote is now called an **embryo.** The embryo, in its gastrula form, sinks into the soft lining, and later becomes firmly attached. When implantation is complete, the cells of the embryo and the endometrium form a structure called the **placenta,** which enables the mother and **fetus** (term for developing human organism after the second month of pregnancy) to exchange materials throughout gestation (Figure 19-15). The lining of the uterus thickens, and fingerlike projections, the villi, develop. These function in the exchange of products of metabolism between the mother and the fetus. Loops of blood capillaries are connected to the fetus through the umbilical cord. These capillary loops are bathed in tiny pools (sinuses) of blood supplied by the mother. It is important to remember that the maternal and fetal circulations are separated by thin membranes and do not mix. The exchange of nutrients and waste products occurs by diffusion through these membranes.

Fetal Membranes

The developing embryo forms two principal extraembryonic membranes, the *amnion* and *chorion* (Figure 19-15). The amnion is a thin protective layer that directly covers the embryo and is filled with amniotic fluid. The amniotic sac is very small at first but expands during the embryo's development until it eventually fills the entire extraembryonic cavity.

The chorion is a thicker membrane that invades the endometrium with villi and forms the

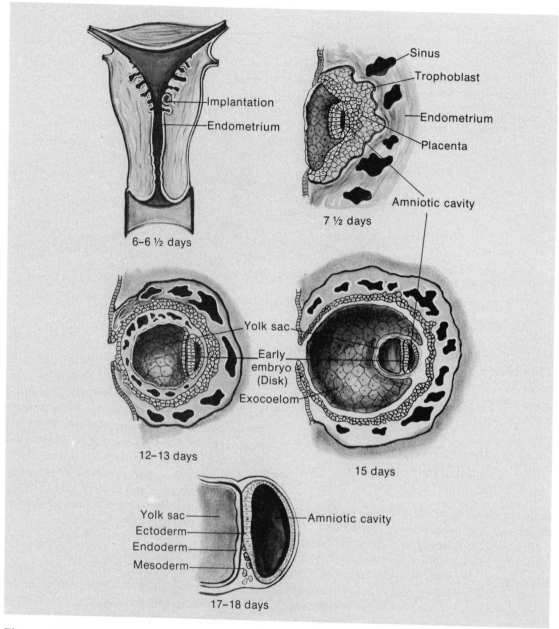

Figure 19-15 Implantation. Schematic representation of the development of an embryo up to three weeks. The three primary embryonic germ layers are indicated.

embryonic portion of the placenta. The chorion produces the chorionotropic hormone that will maintain the corpus luteum for 3 months. The placenta requires 2 months to develop suffi-ciently to produce adequate levels of progester-one; therefore, the first 2 months of endometrial stimulation must be supplied by the corpus lu-teum. During the third month of its existence, the corpus luteum works in conjunction with the placenta to maintain the progesterone level. As an outer membrane, the chorion is also responsible for the interchange of nutrients and waste products between maternal and fetal circulations.

The human ovum does not contain the large quantity of yolk present in the eggs of many lower animals. A *yolk sac* develops but is evident only during early embryonic existence (Figure 19-16). The yolk sac eventually becomes incorporated into the umbilical cord, where it degenerates. A second rudimentary sac, the *allantois,* also degenerates as the embryo grows

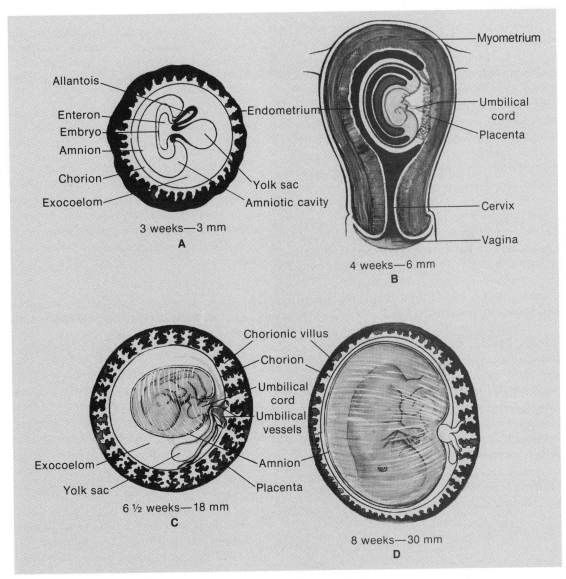

Allantois

Enteron

Embryo

Amnion

Chorion

Exocoelom

Endometrium

Yolk sac

Amniotic cavity

3 weeks—3 mm

A

Myometrium

Umbilical cord

Placenta

Cervix

Vagina

4 weeks—6 mm

B

Chorionic villus

Chorion

Umbilical cord

Umbilical vessels

Exocoelom

Yolk sac

Amnion

Placenta

6 ½ weeks—18 mm

C

8 weeks—30 mm

D

Figure 19-16 Fetal membranes and early human development. **A,** Schematic representation of a three-week-old embryo enclosed within its fetal membranes. **B,** Coronal section of a pregnant uterus. **C** and **D,** Development of the human embryo through its sixth to eighth weeks of growth.

and reportedly functions as an excretory organ during its short existence (Figure 19-16).

Fetal Alcohol Syndrome

During the past few years a pattern of malformation in infants born to women with severe chronic alcoholism has been identified and termed *fetal alcohol syndrome (FAS).*

The alcohol that first reaches the stomach is quickly absorbed. The liver has only a limited capacity to detoxify and metabolize any agent. Under normal conditions, it is able to reduce alcohol from the stomach into innocuous acetaldehyde and acetic acid. These two substances subsequently break down into carbon dioxide and water. The liver can handle about 10 mL/hour or 4 ounces of whiskey (2½ pints of beer) over a 5- to 6-hour period of time. The trouble begins when the liver receives more alcohol than it can handle. It simply lets the excess pass into the general circulation and picks out that which it can detoxify at its leisurely pace. If the amount is such that the transfer of acetaldehyde exceeds the body's capacity, the acetaldehyde concentration increases, and that is what causes a hangover or is passed through the maternal-fetal barrier to the growing fetus.

Alcohol has variable effects on fetal development. The syndrome is associated with craniofacial anomalies (microcephalia, strabismus, and cleft palate), musculoskeletal anomalies (small nails and diaphragmatic anomalies), growth deficiencies, and central nervous system anomalies (hypotonia and retardation). These conditions vary in severity, but all are permanent. It is a critical disorder to be recognized and prevented. Alcoholic women should know the risk of alcohol consumption giving rise to a serious problem in the developing fetus.

Course of Human Development

The cells inside the blastocyst form a plate, the *embryonic disk* (see Figure 19-15). This disk eventually undergoes gastrulation between 8

and 20 days after fertilization, becoming an embryo. By the time a human embryo is 1 month old, it is about ¼ inch long, and the developing features of the brain, eyes, heart, stomach, and liver can be distinguished. Its heart beats regularly and its weight is many thousand times what it was at the moment of conception.

A convenient way to follow the development of the embryo is to indicate its course of development chronologically, rather than by structure. The following sequence begins a week after fertilization and terminates at birth:

- *1 week* Beginning of implantation

- *2 weeks* Implantation complete; placenta begins to form; fetal membranes functional

- *3 weeks* Embryo is 3 mm (⅛ in.) long; growth centers (somites) appear; mesodermal tissue develops into major filtering organs (liver, spleen, lymph nodes); heart begins development; a two-lobed pancreas develops from the intestine; an ear pit starts development internally; the embryo demonstrates a forward and tail end

- *4 weeks* Embryo is 6 mm (¼ in.) long; heart has fully developed into a four-chamber pump; amniotic and chorionic sacs are clearly visible; eye cup forms externally

- *5 weeks* Embryo is 10 mm (⅖ in.) long; placenta is 1 inch in diameter; forebrain and midbrain differentiate into their respective areas; arm and leg buds appear, the arm bud developing faster; bronchial tubes begin development; alimentary tract differentiates into stomach and intestine—one loop of the enormously long intestine pushes out into the umbilical cord; head seems to elongate; eye cup appears on the neck

- *6 weeks* Embryo is 16 mm (⅔ in.) long; a dark retina and eyelids appear; ear centers now visible as the eye cup begins to assume its normal position; esophagus elongates, differentiating into pharynx and pharyngeal pouches; embryo has a cartilage form for a skeleton; external genital features appear but not enough to determine sex; a large heart and liver are prominent

- *7 weeks* Embryo is 25 mm (1 in.) long; fingers and toes are evident, but no joints are present; eyes are now in a forward position; a diaphragm separates the pleural and peritoneal cavities; the two-lobed pancreas fuses to form one; external ear structures appear; a shimmering spinal cord can be seen bordering the embryo's back

- *8 weeks* Embryo is 3 cm (1⅕ in.) long; now called a fetus because the last functional system, the bony skeleton, has begun to form from cartilage; muscles become evident in the contracting arms and legs; the heart is very active at this point; the fetus is now a recognizable human being that will stir the amniotic fluid with its weak kicking

- *9 weeks* Fetus is 3.7 cm (1½ in.) long; a prominent liver makes up about 10 percent of the body weight; the eyes close and will remain closed until delivery; skull formation is evident

- *10 weeks* Fetus is 5 cm (2 in.) long; bone formation has markedly increased (almost double in 2 weeks); the extruding loop of the intestine is now pulled back in from the umbilical cord; the male and female sex organs are formed but are not evident for another 5 to 6 weeks

- *11 weeks* Fetus is 6.5 cm (2½ in.) long; now weighs approximately 3/4 ounce (weight of an envelope), yet has been functional for quite a few weeks

- *12 weeks* Fetus is 7.5 cm (3 in.) long; nail beds develop on fingers and toes

- *13 to 14 weeks* Fetus is 10 cm (4 in.) long; ears have now moved into proper position with final formation of head shape

- *16 weeks (4 months)* Fetus is 15 cm (6 in.) long; feet and legs start to become chubby; skin hair begins to develop; fetus is now becoming larger and quite active—the mother should

feel the first signs of life at this point; respiratory bronchioles are developing

- *17 weeks* Fetus is 20 cm (8 in.) long; solid trunk formation as final systems are fully developed

- *18 weeks* Fetus is 22.5 cm (9 in.) long; sex organs are evident at this point and can be identified; the first hair becomes evident

- *20 weeks (5 months)* Fetus is 28 cm (11 in.) long; fingernails are fully developed; muscles are very active; umbilical cord measures about 50 cm (20 in.)

- *6 months* Fetus is 32.5 cm (13 in.) long; hair is evident over the body; oil (sebum) is secreted to protect the skin and hair from drying; fetus is quite active (sleeping and waking as if it has been born); remaining functional structure of the respiratory tract is developed—alveolar sacs are present, and the lungs are somewhat functional despite their airless quarters

- *7 to 9 months* Fetus is 35 to 50 cm (14 to 20 in.) long, averaging 7½ pounds at birth; this is the time for growth and putting on weight as all the systems needed for life are developed and ready. Chances for survival increase if fetus is born after the seventh month

Birth

Birth (**parturition**) is preceded by considerable hormonal changes in the mother, particularly secretion of a decreased proportion of progesterone compared to that of estrogen. These hormonal changes act to increase the contractility and irritability of the uterus, thus facilitating delivery (Figure 19-17).

Approximately 280 days after fertilization, the uterine hormone levels suddenly drop, and the birth process starts. The baby cannot pass out of the uterus into the vagina unless the cervix is first dilated. Contractions of the uterus force the baby's head against the cervix, providing the main factor in cervical dilation. During this process the amniotic sac is also forced into the cervix, contributing to its dilation. This stage usually requires several hours; the strong uterine contractions that accompany the initial stages of childbirth are commonly called labor pains.

Other sources of discomfort during and after childbirth are the rips or tears that can occur in the vagina and other parts of the birth canal. The vagina will stretch many times its size to allow passage of the fetus. However, if the baby's head is larger in circumference than the vagina's distention, the vaginal wall may tear. The area most likely to tear is in the lower, more delicate region leading to the anus, the perineum. When tears occur in the anal sphincter, contamination and complications may arise. To prevent this, the obstetrician will make a small cut in the vaginal wall (**episiotomy**), angling out and down away from the anus to allow expansion in that direction, away from the anus (Figure 19-18).

At the end of the first stage of the birth process, the amniotic sac usually ruptures, allowing the fetus, head first as a rule, to pass through the cervical canal and the vagina to the outside (Figure 19-17). During the third and last stage, the umbilical cord is cut, freeing the fetus from the mother. Finally, the placenta (afterbirth) is expelled.

At birth, the newborn baby is no longer protected by its mother's womb. It must now breathe, eat, grow, and rely on its own homeostatic mechanisms to maintain its life.

Abnormalities of Birth

Placenta previa is an abnormal condition in which the placenta is attached to or near the internal opening of the cervix. Normally it is

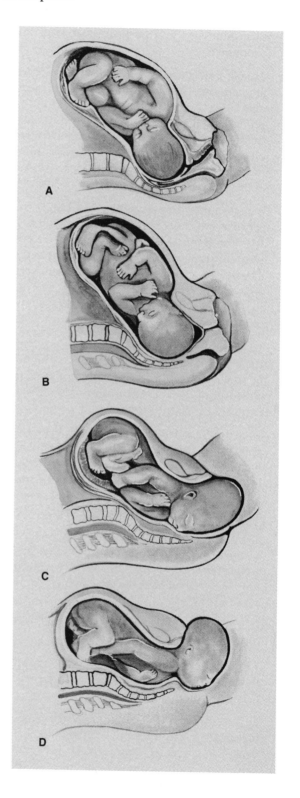

Figure 19-17 Birth. **A,** At childbirth (parturition), the cervix dilates, and the fetal membranes rupture. **B,** Rhythmic uterine muscle contractions move the fetus through the dilated cervix into the birth canal. **C,** The head now rotates so that it faces the pubic symphysis. **D,** The final phases of birth are well underway.

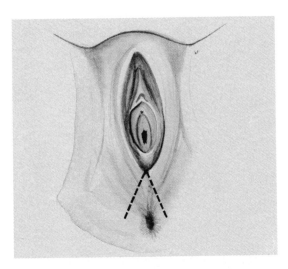

Figure 19-18 Episiotomy.

attached to the upper area of the uterus, and a low attachment can cause an early separation from the uterine wall, resulting in hemorrhage as well as danger to the continued development of the fetus. Surgical removal of the fetus through an incision made in the abdominal wall and into the uterus (cesarean section) is generally the indicated treatment in a case of placenta previa.

An *abortion* is the loss of the embryo or fetus during the first 20 weeks of pregnancy, regardless of the cause. A spontaneous abortion is the expulsion of the embryo due to natural causes, such as infection or disorders of the reproductive tract. Abortions are also induced by artificial means. A *miscarriage* is termination of a pregnancy by natural causes, occurring between the twentieth and twenty-eighth week of pregnancy.

Instances of fetal death from causes other than abortion or miscarriage have prompted a new terminology to distinguish this phenomenon. *Early fetal death* occurs when the fetus or embryo is less than 20 weeks old; *intermediate fetal death* when the fetus is more than 20 weeks but less than 28 weeks old; and *late fetal death* when the fetus is more than 28 weeks old. A

premature infant is a baby who weighs less than 5 1/2 pounds or whose gestation has lasted less than 37 weeks instead of the normal 40 weeks.

CLINICAL TESTS INVOLVING THE REPRODUCTIVE SYSTEM

Cancer of the prostate gland is most often diagnosed when the patient seeks medical advice because of urethral obstruction. The obstruction involves the prostate gland, urethra, and bone in the area, resulting in pain as well as dysuria. Phosphatase is an enzyme produced by the prostate, and elevated levels of acid phosphatase in the blood serum suggest cancer of the prostate. Since the acid phosphatase is not absorbed by the blood until the cancer has extended into other tissues, this is not a good test for early diagnosis.

Aschheim-Zondek (A-Z) Test

The A-Z test is based on the knowledge that chorionotropic hormone is secreted by the growing embryo after the chorionic sac has been developed. Sufficient levels of the hormone must be present for the test to be meaningful; therefore, it should be taken only if pregnancy is already suspected to be in the fifth or sixth week. If the test is given in the first few weeks of pregnancy, the result may prove to be false negative.

The test is done on an early morning urine specimen because it is more concentrated. If chorionotropic hormone is present in the urine, it will produce a reaction either on a slide test (immediate) or after the urine is injected into an animal (24-hour reaction observed on the ovaries of the animal).

The A-Z test can also be used to diagnose a disorder of the male reproductive system. Unfortunately, in this instance a positive test indicates terminal cancer of the testes or prostate gland.

Mammography

Mammography is a soft tissue x-ray examination of the breast. Its primary purpose is to discover cancers that escape detection by other methods. Cancers less than 1 cm in size cannot be regularly detected by routine examination. Since most breast cancer develops in the breast for many years before it grows to detectable size, it is imperative that detection and treatment be performed in the preclinical stage.

Mammography diagnosis of breast cancer is based on gross characteristics. A low-energy x-ray beam is used to delineate the breast structures. Benign lesions push breast tissue aside as they expand, while malignant lesions invade the surrounding breast tissues (Figure 19-19). The x-ray criteria for diagnosing lesions of the breast are 85 percent accurate in identifying cancers and 10 percent in giving false positive diagnosis.

Papanicolaou (Pap) Smear

The Pap smear is used principally for diagnosis of precancerous and cancerous conditions of the genital tract, which includes the vagina, cervix, and endometrium. Exfoliated cells in body tissues and fluid are studied to count the cells, determine the type of cells present, and detect and diagnose malignant and premalignant conditions. All specimens are examined for surface modifications, appearance and staining properties, functional adaptations, nuclear structure, and inclusions. Deviations from the normal are noted.

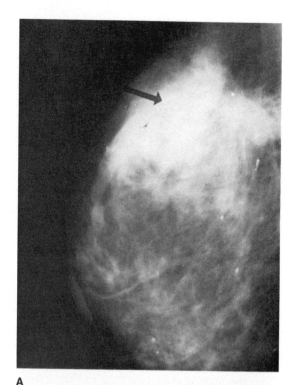

A

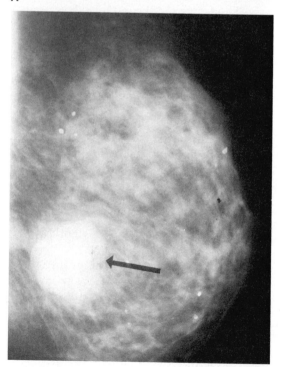

B

Figure 19-19 Mammography. **A,** Invasive carcinoma (arrow) diagnosed as a soft nodule in the left breast. **B,** Clinical evaluation of a lump in the right breast was diagnosed as a fibroadenoma (arrow). (Courtesy of Salem Hospital, Salem, Massachusetts.)

Vaginal cytological examination allows the study of the condition of the tissues. Scrapings taken directly from the cervix provide material for highly accurate interpretation. Vaginal fluid and smear give reliable results as well, ranging in accuracy up to 95 percent. The advantage of the vaginal smear is that it may reveal malignant cells not only from the cervix but also from the endometrium, ovaries, and even other abdominal organs.

OUTLINE

I. Introduction
   A. Cell division types
      1. Binary fission
      2. Mitosis
   B. Types of tissue cells
      1. Somatic (body)
      2. Germinal (sex)
II. Meiosis (gametogenesis: spermatogenesis—production of male sperm; oogenesis—production of female egg)
   A. First meiotic division—stages similar to those of mitosis
      1. Long homologous chromosomes unite longitudinally (synapsis)
      2. Chromosome pairs line up at equatorial plate—metaphase
      3. Resulting daughter cells together have same number of chromosomes as the parent cell
   B. Second meiotic division—basically a mitotic sequence, ending in four haploid cells or gametes
      1. Each daughter cell of first division divides into two daughter cells, each containing single-stranded chromosomes
      2. Four haploid daughter cells result for every diploid parent
   C. Spermatogenesis and oogenesis in mammals
      1. Four viable sperm are produced from every germinal epithelial cell in testes
      2. A single viable egg and two polar bodies are produced from every germinal cell in ovaries

III. Reproduction in the human male
   A. Gross anatomy
      1. Testes
      2. Vas deferens
      3. Seminal vesicles
      4. Prostate gland
      5. Cowper's gland
      6. Penis
      7. Epididymis
   B. Microscopic structure
      1. Seminiferous tubules
      2. Spermatogonia in germinal epithelium—undergo spermatogenesis
      3. Associated cells
         a. Leydig—interstitial cells between the seminiferous tubules; produce and secrete testosterone
         b. Sertoli—dispersed between germinal spermatogonia to nourish them
   C. Hormonal function
      1. Anterior pituitary—secretes ICSH, which stimulates Leydig cells in testes
      2. Testes—Leydig cells, in turn, secrete testosterone to promote spermatogenesis
      3. Abnormalities
         a. Hyposecretion before puberty produces sterility and lack of male development (eunuchism)
         b. Hyposecretion after puberty results in sterility and some regression of virility
      4. Sperm
         a. Microscopic—0.06 mm (60μ)

b. Specialized for motility—tail
c. Contains many mitochondria in its body and neck to facilitate vigorous activity

IV. Clinical considerations of the male reproductive system
   A. Benign prostatic hyperplasia
   B. Prostatitis
   C. Carcinoma of the prostate
   D. Cryptorchidism
   E. Orchitis
   F. Epididymitis
   G. Hydrocele
   H. Phimosis
   I. Vasectomy
   J. Hermaphroditism

V. Reproduction in the human female
   A. Reproductive organs
     1. Ovaries—egg production
     2. Oviduct—fallopian tubes
     3. Uterus—endometrial lining
     4. Vagina—Bartholin glands
     5. External genitalia—vulva, labia majora and minora, clitoris

VI. Hormonal function
   A. Menstruation
     1. Preovulatory phase (proliferation)
       a. FSH stimulates follicle development
       b. Follicle secretes estrogen to promote oogenesis
       c. LH, from anterior pituitary, and remaining FSH promote ovulation
     2. Premenstrual phase (secretory)
       a. LH promotes formation of corpus luteum from preexisting follicle
       b. Corpus luteum secretes progesterone
       c. Corpus luteum dies after 12 to 14 days, and menstruation occurs
     3. Accessory effects
       a. Estrogen promotes development of endometrium in uterus

b. Progesterone maintains endometrial development
c. Degeneration of corpus luteum and decrease in progesterone level promote endometrial repression and ultimate menstruation

     4. Hormonal effects
       a. Menopause at 40 to 50 years is a gradual regression of hormonal production, especially of estrogen, necessitating physiological and psychological adjustments

   B. Mammary glands and lactation
     1. Gross anatomy
       a. Many lobes composed of fat and glandular tissue
       b. Lactiferous ducts—convey milk to nipple
     2. Estrogen production at puberty stimulates ducts to grow and branch
     3. Progesterone stimulates specialized acinar glands to produce milk
     4. During pregnancy, estrogen and progesterone promote acinar gland growth
     5. After childbirth, fall in estrogen and progesterone levels stimulates anterior pituitary to secrete prolactin, which promotes milk production (lactation)
     6. Child's suckling action stimulates posterior pituitary to secrete oxytocin, which maintains lactation

   C. Menopause

VII. Diseases of the female reproductive system
   A. Vulva and vagina
     1. Vulvovaginitis
     2. Bartholin's cyst
     3. Neoplasm
   B. Uterus
     1. Cervicitis
     2. Carcinoma of cervix

a. Stages of cancer
3. Carcinoma of endometrium
4. Leiomyoma
C. Fallopian tubes and ovaries
  1. Salpingitis
  2. Endometriosis
  3. Neoplasms of the ovaries
D. Breast
  1. Mastitis
  2. Fibrocystic disease
  3. Fibroadenoma
  4. Carcinoma—radical or simple mastectomy
VIII. Copulation, fertilization, and early development
A. Copulation
  1. Erection
  2. Ejaculation
  3. Autonomic control
B. Fertilization
  1. Sperm
    a. Swim in seminal fluid; nourished by semen
    b. Life expectancy only 48 to 72 hours
    c. Numbers in millions
    d. Each sperm head secretes enzyme hyaluronidase
    e. Thousands of sperm needed to supply enough enzyme to penetrate egg barrier
  2. Egg
    a. Corona radiata, a protective shield of cells that secretes hyaluronic acid
  3. Fertilization occurs when one sperm penetrates the shield and its nucleus joins with the egg nucleus to form a zygote
  4. Location of fertilization
    a. Should be one-third down oviduct for normal development
    b. Abnormal fertilizations are ectopic

5. Multi-egg production may result in multi-births
  a. Fraternal twins—more than one egg
  b. Identical twins—same egg divides in two
C. Early development (first 5 to 7 days)
  1. Cleavage—division of zygote into 2, 4, 8 cells, etc.
  2. Blastula—hollow ball: single-layered zygote
  3. Gastrulation
    a. One side of hollow ball becomes indented
    b. Development of endoderm, mesoderm, and ectoderm—three cell layers
IX. Implantation through birth
A. Implantation
  1. 4- to 5-day period following blastula formation
  2. Blastula digests endometrium
  3. Placenta formation
    a. Finger-like villi
    b. Exchange of nutrients and wastes
    c. Blood sinus pools
B. Fetal membranes
  1. Amnion—covers and protects fetus
  2. Chorion lines endometrial cavity and forms embryonic portion of placenta
    a. Secretes chorionotropic hormone, which maintains corpus luteum for 3 months
  3. Yolk sac—rudimentary food source
  4. Allantois—rudimentary excretory organ
C. Fetal alcohol syndrome
  1. Craniofacial anomalies
  2. Musculoskeletal anomalies
  3. Growth deficiencies

D. Course of human development
   1. First month
      a. Embryo develops systems essential for life
      b. Embryo grows to about 6 mm (¼ in.) long and rapidly multiplies its weight
   2. Second month
      a. Nervous system differentiates
      b. Arm and leg buds appear; fingers and toes develop
      c. Embryonic skeleton formed of cartilage
      d. Alimentary tract differentiates into stomach and intestine
      e. Diaphragm separates pleural and peritoneal cavities
      f. Eye and ear structures first appear
   3. Third month
      a. Embryo now a fetus
      b. Bony skeleton develops
      c. Fetus now over 5 cm (2 in.) long and weighs ¾ ounce
      d. Liver is prominent
      e. Eyes close
   4. Fourth month
      a. Rapid growth to 10–15 cm (4 to 6 in.)
      b. Skin, nails, and facial characteristics develop
      c. Fetus very active—first signs of life
   5. Fifth month
      a. Fetus over 20 cm (8 in.) long
      b. Fetus very active—gaining weight
   6. Sixth month
      a. Fetus 32.5 cm (13 in.) long
         b. Hair and sebum evident
         c. Fetus sleeps and wakens
   7. Seventh to ninth months
      a. Survival possible if birth occurs during this period
      b. Fetus 35–50 cm (14 to 20 in.) long; average weight 7½ pounds
      c. Time of growth and development of systems begun before seventh month
   8. Birth (parturition)
      a. Hormone level decreases
      b. Fetal head pushes into cervix, causing dilation
      c. Uterus contracts—childbirth
      d. Episiotomy performed, if needed
      e. Afterbirth delivered
E. Abnormalities of birth
   1. Placenta previa
   2. Abortion
   3. Miscarriage
   4. Fetal death: early, intermediate, or late
   5. Prematurity
X. Clinical Tests Involving Reproductive System
   A. Acid phosphatase—blood test to detect prostatic cancer
   B. Aschheim-Zondek (A-Z) test urine test
      1. In women—positive result indicates pregnancy
      2. In men—positive result indicates terminal cancer
   C. Mammography
   D. Papanicolaou (Pap) smear

STUDY QUESTIONS AND PROBLEMS

1. Describe gametogenesis of the sperm and the egg. How do the two processes differ?
2. What part do the anterior pituitary hormones play in the reproductive cycle?
3. What are primary and secondary sex characteristics?
4. How do Leydig cell secretions affect spermatogenesis?
5. What is the importance of the male prostate gland to the life expectancy of sperm?
6. If diseased, how does the prostate gland interfere with the genitourinary tract?
7. Describe the menstrual cycle.
8. Discuss the effect of estrogen on the ovum, endometrium, secondary sex characteristics, mammary gland development, and lactation.
9. What stage must be initiated for the corpus luteum to remain functional?
10. Discuss ovulation. How does the birth control pill affect it?
11. Briefly describe the periods of pregnancy on a developmental basis.
12. What is lactation? How does the baby's suckling affect it?
13. How would you "dry out" the mother if she were not going to breast-feed?
14. What is ejaculation? Describe its reflex action and autonomic control mechanism.
15. Where does fertilization normally occur? Why is it important that it occur at this location?
16. List the three embryonic germ layers and their derivatives.
17. What is cleavage?
18. What is an ectopic pregnancy?
19. What is chorionotropic hormone? Where does it come from?
20. Describe implantation and the eventual blood supply connection between the fetus and the mother.
21. Explain the function of the fetal membranes.

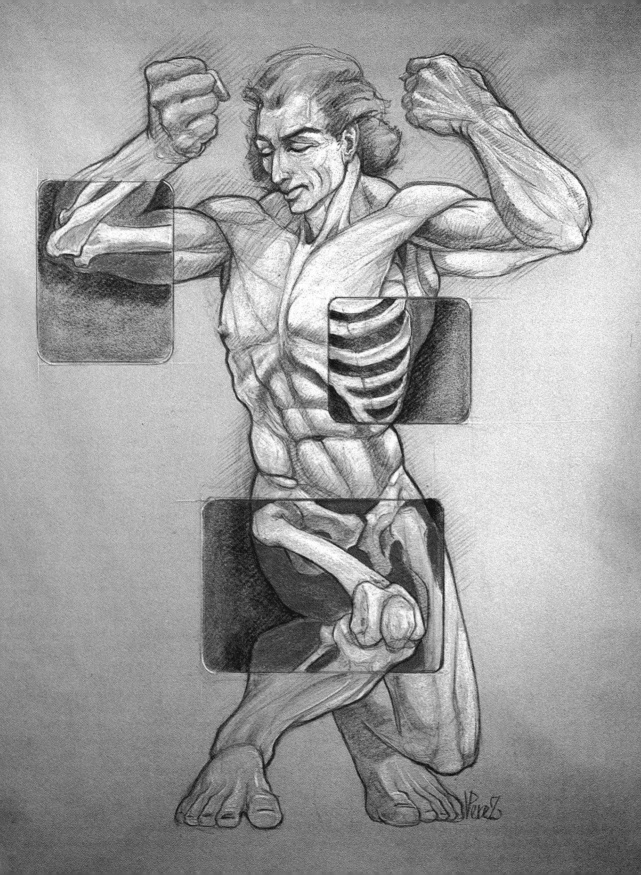

20

Radiological Health

LEARNING OBJECTIVES

- Describe the nature of radiation

- Describe the electromagnetic spectrum

- Differentiate between ionizing radiation and radiation sickness

- Describe the two types of particulate radiation

- Differentiate between particulate and photoelectric

- Explain radioactivity

- Differentiate between a natural and artificial isotope

- List methods used to detect radiation

- Describe the protective measures used to defend against radiation

- List the cellular, tissue, and systemic effects of radiation exposure

- Define radiosensitivity and compare it to radioresistance

- Describe the half-life of a radioactive isotope

- List the methods used to diagnose abnormalities of the body

- Define radiopharmaceutical and give examples of its use in nuclear medicine

- Describe how radiation is used in the treatment of cancer

- Compare computerized tomography (CT) with computerized fluoroscopy

- Describe how ultrasonography is used

- Explain why nuclear magnetic resonance is safer to use in diagnosis than are radiation techniques

IMPORTANT TERMS

absorbed dose
angiography
bolus
chemotherapy
computerized
   fluoroscopy
computerized
   tomography (CT)

curie
dose rate
dosimeter
dysfunction
electromagnetic
gamma ray
half-life
oncology

particles
particulate
   radiation
photon
quantum
radiation
radiation therapy
radioactive
radiography

radioisotope
radioresistance
radiosensitivity
REM
scan
ultrasonography

Radiological technology provides a practical application to many biological sciences. It has been introduced here, and not earlier in this text, because a broad and basic understanding of the structure and function of tissues, skeletal anatomy, and basic control mechanisms is necessary if the significance of radiography and radiation therapy is to be appreciated.

NATURE OF RADIATION

You may have wondered why certain electrical instruments are harmless, whereas others are extremely dangerous. Everyone knows that radio and television waves exist, and no one is afraid to stand in front of a radio or a television set. However, too long an exposure to the radiation of an x-ray machine can cause serious injury. Electrical waves are used for the operation of radio, television, and x-ray machines—so why should there be such a difference?

Radiation is the emission and projection of energy. You are exposed daily to a wide variety of radiations, whether you are aware of it or not. Some of these can be felt, such as the radiant energy through which heat is transferred. Others can be seen as light, and this phenomenon can be further subdivided into all the colors of the visible light spectrum. All of these radia-tions are within the electromagnetic spectrum and are spoken of in terms of waves of energy.

The electromagnetic spectrum is a grouping of all energy waves (**photons**) according to their wavelengths (Figure 20-1). The amount of energy of a photon is determined by its size (wavelength) and its frequency. The quantum theory, $E + 1/\lambda$ where λ = Wavelength, sums up the entire energy concept by stating that the amount of energy emitted by a photon is inversely proportional to its wavelength (λ). This means the shorter the wavelength the greater the power it imparts.

Notice in Figure 20-1 that the entire scale can be divided almost in half. The left side of the spectrum includes electrical power, heat, television, and the infrared rays, all of which are visible. The right side contains the dangerous waves, such as ultraviolet rays, x-rays, and cosmic radiation. The reason for this division is simple. In the first chapter of this text, we saw that compounds are not just thrown together. They are composed of elements held together by an energy bond. If this bond is broken, the compounds become ionized, and this process may cause injury if the compound is water.

It takes approximately 13 ev (electron volts) to ionize a molecule of H_2O:

$$H_2O \rightarrow H^+ + OH^-$$

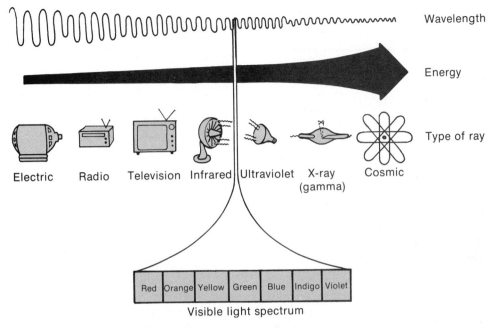

Figure 20-1 The electromagnetic spectrum.

This alone is not significant, but the fact that the free hydroxyl radicals can combine to form hydrogen peroxide, a cellular poison, is significant:

$$2OH^- \rightarrow H_2O_2$$

The body is prepared to adapt to a normal amount of hydrogen peroxide production, but it cannot accommodate massive quantities produced during a short period. This occurs in radiation sickness. A person exposed to a large quantity of high-energy radiation over a short period produces large quantities of hydrogen peroxide, and the resultant cellular disruption may cause permanent physical defects or death.

Let's return to the initial question: why are we not afraid to stand in front of a radio while we are afraid to stand in front of an x-ray machine? As shown in Figure 20-1, the energy spectrum is set up so that waves are arranged in order of decreasing size (Figure 20-1). As the energy level increases, the quantity of radiation needed to destroy body tissues appears just after

the visible light range—in the ultraviolet range. Therefore, all energy waves beginning with the ultraviolet range are ionizing and dangerous. All energy waves below the ultraviolet range are nonionizing and are not considered dangerous. Thus, to answer the question, a radio or television wave is not small enough to be an ionizing wave, but x rays must be ionizing to penetrate the body when used in radiographic examinations.

TYPES OF RADIATION

Ionizing radiations may be classified according to either their origin or their physical properties. They fall into two general categories: those that have a mass, or weight (called **particles**), and those that are energy only (referred to as **electromagnetic**).

The following experiment was performed to study the radiations produced by a radioactive element. A piece of radium was placed at

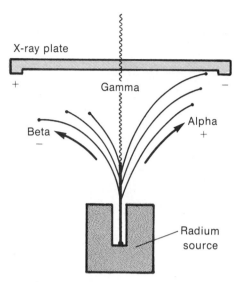

Figure 20-2 Electrical properties of radioactive emissions. The alpha particles are attracted to the negative field; therefore, they are positively charged. Conversely, the beta particles are negatively charged. The gamma emissions are not affected by either electrical field and are neutrally charged.

the bottom of a thick lead well. The escaping radiation was allowed to fall on a photographic plate (Figure 20-2). When these radiations were passed through a strong electrical field, three different areas showed up on the photographic plate. These were called alpha, beta, and gamma rays:

Alpha rays are attracted to the negative field, which indicates that they are positively charged. These rays are particles and are relatively harmless because they do not penetrate the skin.

Beta rays are attracted toward the positive field, which indicates they are negatively charged particles. They have only a slight penetrating power and will penetrate the skin or an organ to a depth of only a few millimeters. Most of the radioisotopes used in medicine emit beta rays.

Gamma rays are not affected by an electrical field because they have no charge. Strictly speaking, they are not particles but rather a

form of electromagnetic radiation similar to x radiation. Gamma rays penetrate tissue readily and will cause cellular damage as they travel through the body.

A fourth type of radiation involves the emission of a neutron from the nucleus of an atom. Such changes are called *nuclear reactions* and result in the formation of either an isotope or an entirely new element.

Particulate Radiation

Any subatomic particle in motion is capable of causing ionization. Consequently, electrons, protons, neutrons, and even rare nuclear fragments can all be classified as *particulate ionizing radiation* if they are in motion and possess sufficient kinetic energy. At rest, ionization cannot occur. There are two principal types of particulate radiation—alpha particles and beta particles—both of which are associated with radioactive decay.

Once emitted from a radioactive atom, the alpha particle travels with enormous velocity through matter. However, because of its great mass and charge, it easily transfers this kinetic energy to orbital electrons of target atoms. Ionization frequently accompanies alpha radiation. The average alpha particle possesses 4 to 7 mev (million electron volts) of kinetic energy and ionizes approximately 40,000 atoms for every centimeter of travel through air. This ionization rate is called *specific ionization*. Specific ionization is usually specified in ion pairs per centimeter of air or per micron of water.

Because of this rapid transfer of energy, the energy of an alpha particle is quickly dissipated and thus has a very short range through matter. In air, alpha particles can travel about 5 cm, while in soft tissue the range may approach 100 μ. Consequently, alpha radiation from an external source is nearly harmless, since the radiation energy is deposited in the superficial layers of the skin. As an internal source of radiation, just the opposite is true. If an alpha-emitting radioisotope is deposited in the body, it can severely irradiate the local tissue. When an alpha particle finally loses all of its kinetic energy, it comes to rest, attracts two free electrons, and becomes an atom of helium gas.

Beta particles differ from alpha particles in both mass and charge. They are light particles with atomic mass number of 0, carry one unit of negative charge, and are emitted from the nucleus of radioactive atoms. The only difference between electrons and beta particles is that beta particles originate in the nuclei of radioactive atoms while electrons exist in shells outside the nuclei of all atoms. Once emitted from a radioisotope, beta particles traverse air with a specific ionization of several hundred ion pairs per centimeter. Their range is longer than that of an alpha particle. Depending on its energy, a beta particle may traverse 10 to 100 cm in air and about 1 to 2 cm in soft tissue. Once a beta particle has transferred all of its kinetic energy, it comes to rest and combines with an atom deficient in electrons.

Electromagnetic Radiation

The ancient Greeks recognized the unique nature of light. It was not one of their four basic essences but was given entirely separate status. They called an atom of light a photon. Today many types of electromagnetic radiation in addition to visible light are recognized, but the term photon is still used. A photon is the smallest quantity of any type of electromagnetic radiation, just as an atom is the smallest quantity of an element. A photon may be pictured as a small bundle of energy, sometimes called a **quantum,** traveling through space at the speed of light. We speak of x-ray photons, light photons, and other types of electromagnetic radiation as *photon radiation.*

X rays and gamma rays are forms of electromagnetic ionizing radiation and are often called photons. They have no mass and no charge, travel at the speed of light ($c = 3 \times 10^8$ m/sec), and may be considered as energy disturbances in space.

Unlike radio emissions or visible light, ionizing electromagnetic radiation is usually char-

acterized by the energy contained in a photon. When an x-ray machine is operated at 80 kvp (kilovolt peak), the x rays it produces contain energies varying from 0 to 80 kev (kilo-electron volt). An x-ray photon contains considerably more energy than a visible-light photon or a photon of a radio broadcast. The frequency and wavelength of x radiation are much higher and shorter, respectively, than those of other types of electromagnetic radiation.

The distinction is sometimes made between x-ray and gamma-ray photons that gamma-ray photons have higher energy. In the early days of radiology this was true because of the limited capacity of the available x-ray machines. Today, with large particle accelerators available, it is possible to produce x rays with energies considerably higher than those of gamma-ray emissions. Consequently, the distinction by energy is not appropriate. The only difference between x rays and gamma rays is their origin. X rays are emitted from the electron cloud of an atom that has been artificially stimulated; gamma rays, on the other hand, come from inside the nucleus of a radioactive atom. X rays are produced in electrical machines, whereas gamma rays are emitted spontaneously from radioactive material. Nevertheless, given an x ray and a gamma ray of equal energy, one could not tell them apart.

This situation is analogous to the difference between beta particles and electrons. These particles are the same except that beta particles come from the nucleus while electrons come from outside the nucleus.

NATURAL AND ARTIFICIAL RADIOACTIVITY

Some elements, such as uranium and radium, are naturally radioactive; that is, they emit energy continually. Natural **radioactivity** may be defined as the spontaneous change of one element into another. Most of the heavier elements are naturally radioactive. Nonradioactive elements, generally the lighter ones, can be made

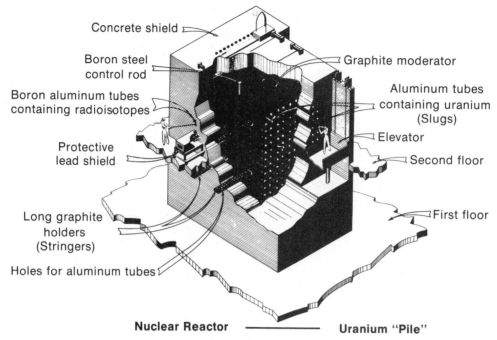

Nuclear Reactor ——————— **Uranium "Pile"**

Figure 20-3 A nuclear reactor where isotopes are formed: an atomic "pile." (Courtesy of United States Atomic Energy Commission.)

radioactive if the element is bombarded with protons, neutrons, electrons, or alpha particles. The radioactive substance thus produced in a nuclear reactor is artificially radioactive, and is called a **radioisotope** (Figure 20-3). Radioisotopes have the same chemical properties as non-radioactive isotopes of the same element because chemical properties are based on electrons only, and isotopes of an element have the same electron structures (Figure 20-4).

Radioisotopes are used in medicine in the diagnosis and treatment of disease. The main criterion in choosing radioisotopes is the dosage required. If the dose is small but large enough to

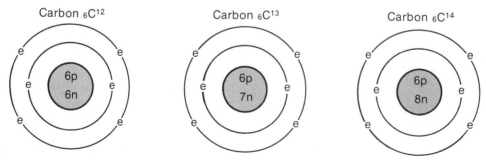

Figure 20-4 Isotopes of carbon. The nucleus contains six protons (p) and six, seven, or eight neutrons (n), whereas the outer energy rings or shells contain six electrons (e).

be detected, the radioisotope can be used to diagnose abnormalities.

Iodine-131 (^{131}I) is an isotope used in the diagnosis and treatment of thyroid disorders. The thyroid gland requires iodine to function normally and, in turn, converts the available iodine into thyroxine, which acts to control the metabolic rate. If a patient is suspected of having a metabolic problem, ^{131}I is given in the form of liquid sodium iodide. The patient is then placed in front of a specialized Geiger counter (Figure 20-5), and the amount of radiation is measured. If the thyroid is functioning normally, the gland should take up about 12 percent of the radioactive iodine within a few hours. If less than the normal amount of ^{131}I has

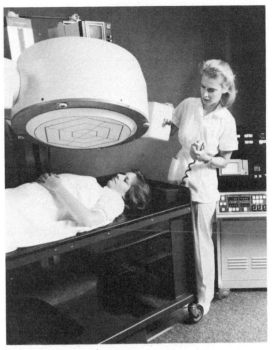

Figure 20-6 Radioisotope camera. A machine used to detect localized radiation and produce a print (scan) of the distribution of the isotope. (Courtesy of Technicare, Cleveland Ohio.)

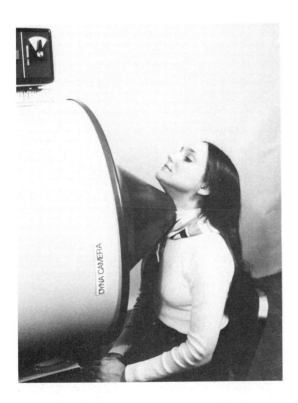

Figure 20-5 Thyroid uptake collimator. A machine that measures the amount of ^{131}I absorbed by the thyroid gland. (Courtesy of Picker International, Northford, Connecticut.)

been taken up, the patient may have a hypothyroid condition. Conversely, if the amount of ^{131}I taken up by the thyroid is greater than normal, the patient may have a hyperthyroid condition.

Other radiation counters involve the use of a scanner to detect the presence of localized radiation and to produce a picture, called a **scan,** of the area involved (Figure 20-6). Proper interpretation of a scan can reveal, for example, whether the thyroid is normal or abnormal. A patient is given a selected radioisotope that will accumulate in the body area suspected of having a problem. The scanner moves back and forth across the site and detects the radiation present.

More detailed information on the type and use of various radiological instrumentation will be discussed later in this chapter.

DETECTION AND MEASUREMENTS OF RADIATION

How does one know whether something is radioactive? This is a very important question—particularly in the protection of medical personnel. **Dosimeters** are devices that measure radiation exposure. One device used for this purpose is the Geiger counter, which consists of a glass tube containing a gas at low pressure (Figure 20-7). When the counter is exposed to a radioactive substance, the energy given off causes a current to flow through a small wire inside the glass tube. This wire carries the current to an amplifier, which gives off a clicking sound and records the intensity of energy emitted by the radioactive source. It does not tell you what the source is, just that it is there.

X-ray technicians and other workers continually exposed to radiation usually are required to wear a badge or a "pencil" device that registers the total amount of radiation (the absorbed dose) to which they have been exposed (Figure 20-7). The badge can be developed like an x-ray film; the pencil can be held up to the light and read directly. In either case, if the device shows that a dangerous level of exposure (the REM) has been reached, the person must leave the area until his body has recovered from the accumulated radiation. Depending on the extent of exposure, this period may be as long as several weeks because the physiological effects of radiation are cumulative.

RADIATION PROTECTION

The x-ray technologist can minimize radiation exposure of patients and radiological personnel in many ways. Most do not require sophisticated equipment or especially rigorous training but simply require a conscientious attitude in the performance of assigned duties.

Type of Radiation

In nuclear medicine three types of emissions are of primary concern: alpha particles, beta particles, and gamma rays. In addition, x rays result from several phenomena of interaction with matter, but these are generally very weak and of no particular concern as far as radiation protection is concerned. One fact to be considered is that of external emission versus internal emission—whether the radiation comes from outside the body and penetrates the epidermis into the body or whether the emitters are already inside the body, having been introduced via ingestion, inhalation, or intravenous injection. Gamma rays and x rays are able to penetrate the epidermis of the skin and, therefore, present the same hazard whether they are external or internal emitters. Such is not the case with alpha and beta emitters. Alpha and beta particles ordinarily cannot penetrate the outer layers of the skin. As external emitters they do not usually constitute a serious problem with radiation protection. If they are used as internal emitters, however, the problem of alpha and beta radiation damage becomes severe.

Penetration Power

As stated above, alpha and beta particles are not generally regarded as radiation hazards as external emitters. A 1 mev alpha particle has a range in tissue of 0.0006 cm and a 5 mev alpha particle has a range of 0.0037 cm. An alpha particle would require an energy of 7.5 mev to penetrate human skin. Under ordinary circumstances, a piece of paper will stop an alpha particle.

The penetration power of beta particles is about 100 times that of alpha particles. About 2.5 cm of wood or 6 mm of aluminum is required to stop a beta particle. The range in tissue for a 1 mev beta particle is 0.42 cm and for a 5 mev beta particle, 2.2 cm. Although an external beta emitter is generally considered not to be of consequence as far as radiation protection is concerned, a beta particle can penetrate from a few millimeters to 1 cm beneath the skin. There

Figure 20-7 Radiation detection devices. A, Ionization type. B, Geiger, Mueller type. C, Ionizing self-reader "dosimeter" (pencil) worn by all personnel who are continually exposed to radiation of any type. D, Health physics radiation measuring system. (Courtesy of Victoreen Instrument Division, Cleveland, Ohio.)

the particle rapidly decelerates as a result of its interaction with tissue.

Gamma rays have extremely high penetration power and can create radiation hazards as either external or internal emitters. The gamma ray cannot be stopped by paper or small amounts of aluminum or lead. In general, protection is spoken of in terms of inches of lead and feet of concrete. In contrast to total absorption of an alpha or beta particle, only 3 percent of the gamma-ray energy is absorbed in 1 cm of tissue. The rest is either absorbed in a much larger volume of tissue or travels through until completely out of the body.

Ionization

Ionization in tissue is considered the most important biological interaction of radiation. Almost all of the damage to tissue is a result of this phenomenon. The ability to ionize varies tremendously among alpha, beta, and gamma emissions. A term mentioned earlier that is used to describe this phenomenon is *specific ionization*, which is defined as the number of ion pairs produced per unit of path. The specific ionization for a 1 mev alpha particle in air is 60,000 ion pairs per centimeter in air. This is greatly increased over the specific ionization for a 1 mev beta particle, which is 45 ion pairs per centimeter in air. The specific ionization of gamma photons is reduced even more.

There seems to be an important correlation between the ionization ability and the penetration power of the various emissions. The alpha particles are very weak in their penetration power, but their ionization ability reaches tremendous proportions. If alpha particles bypass the epidermis, they present a tremendous problem in radiation biology. This is also true of beta emitters. Although beta emitters do not have as great an ionization ability, they have increased penetration power. However, their penetration power is not such that they can penetrate far beyond the skin. Based on their ionization ability, beta emitters can cause tremendous biological damage, provided the protective layer of the skin is bypassed. It is for this reason that phosphorus 32 (^{32}P), a pure beta emitter, is used therapeutically in cases of leukemia and polycythemia. The rationale for its use is based solely on its ability for localized destruction of tissue function. Since most radiopharmaceuticals are beta-gamma emitters, the beta component represents the largest contribution of radiation dose as an internal emitter. In some cases, 90 to 95 percent of the dose is from the beta component.

Although alpha and beta particles cannot penetrate the skin, there must still be protection against them. Because of their ability to ionize tissue once they have bypassed the skin, they present more of a radiation hazard than gamma photons, although the damage will be localized.

A bombardment of cosmic radiation from outside the earth's atmosphere occurs constantly (Figure 20-1). These cosmic rays produce so much energy that no amount of shielding can protect us from them. Cosmic radiation is known to be the cause of many biological mutations that occur in nature.

Distance

Distance constitutes one of the best methods of radiation protection and is a method used routinely. It is not only an effective means of radiation protection, but in many instances it is the least expensive. As one moves away from the source of radiation, one naturally expects to receive less radiation. One might think that as the distance is doubled from the source at a given position, the radiation to the person would be reduced by one-half; however, the radiation is reduced by one-fourth. This is known as the *inverse square* law, which states that the amount of radiation at a given distance from a source is inversely proportional to the square of the distance. By doubling the distance, the dose is one-fourth the original; by halving the distance, the dose is four times the original.

The inverse square principle explains the suggested use of long-handled tongs and remote control handling devices with application

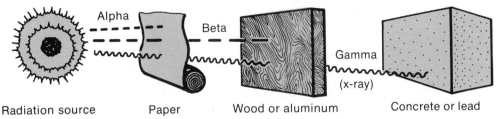

Radiation source   Paper   Wood or aluminum   Concrete or lead

Figure 20-8 Radiation shielding. Note the increased density of the material necessary to protect the user against alpha, beta, and gamma (x-ray) radiation types.

to large quantities of radiation. It also plays a role in the calculation of visiting time to a recently treated patient. For example, a patient's visitors may be allowed to stay only 10 minutes at the bedside. But if they were not allowed within 6 feet of the patient, their stay could be prolonged based on inverse square relationships.

Shielding

Shielding is also a practical method of radiation protection. The use of shielding materials such as lead sheets and lead bricks is nothing new to even the most inexperienced radiation worker. This shield is simply a body of material used to prevent or reduce the passage of radiation. In the case of alpha and beta radiation, very little shielding is required to absorb the emissions completely. An alpha particle is stopped by a sheet of paper, and a beta particle is stopped by an inch of wood. However, feet of concrete or inches of lead are necessary to absorb gamma radiation (Figure 20-8). The general practice is to use enough shielding for complete absorption of alpha and beta particles. This is not true, however, with gamma or x radiation. With these two types of emissions, shielding is used to reduce the amount of radiation.

Because gamma and x rays can be stopped by lead, aprons of this metal are worn by technicians who must be in the room during x-ray procedures (Figure 20-9). Normally, however, the technician either stands behind a lead-lined screen with a window in it or leaves the room.

Time

Time can be used as a method of radiation protection. Obviously the longer an individual is exposed to a field of radiation the greater the total exposure. Common sense dictates that time should be used as a control of radiation exposure. In diagnostic applications of nuclear medicine, time is not as important a protection device as in therapy.

BIOLOGICAL EFFECTS OF RADIATION

Alpha and beta rays are relatively harmless because of their slight penetrating power. Gamma rays, however, can cause ionization of the atoms inside the cells, leading to the production of substances harmful to the cell's metabolism. Such radiation also may affect the genes, causing a cell to grow abnormally, as in cancer, or to die.

The physiological effects of radiation are cumulative, and they can shorten life. Radiation of the whole body can be lethal if enough radiation (measured in roentgen units, or r) is absorbed. Such a dose is called LD_{50}—the amount of radiation necessary to kill 50 percent of a population. Table 20-1 lists the levels of radiation necessary to kill 50 percent of the population of certain types of organisms.

Of all the physical agents, ionizing radiation causes the most damage because it injures not only tissues, organs, and cells, but the mole-

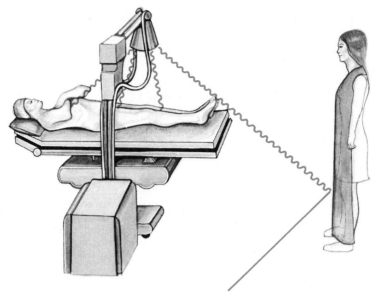

Figure 20-9 Technician shielded by means of a lead apron while patient receives x-ray therapy.

cules and atoms of protoplasm. Exposure to radiation causes a change, either a loss or gain, in electrons in an atom. The free electron may become attached to some other atom, which then loses its electrical neutrality and becomes charged, or ionized. Alterations in the affected molecule can manifest itself in many ways.

Cellular Effects

No living cell is completely resistant to radiation. Cell damage may vary from alteration of a single molecule to death of the cell. There is a significant difference between the radiosensitivity of the nucleus and the cytoplasm of the cell. Nuclear structures, particularly the chromosomes (DNA) at time of division, can undergo dramatic changes as a result of radiation. The cytoplasm (RNA), however, does not appear to be radiosensitive under the same radiation conditions.

One of the most obvious effects on the cell from radiation is growth suppression. Cells decrease in number following radiation because mitosis has been disrupted. Cells with greater mitotic activity are more sensitive to radiation. The cell actually loses its ability to divide successfully.

Chromosomal aberrations are a demonstrable effect of radiation (Figure 20-10). Radiation causes chromosomes to adhere, clump together,

Table 20-1   LD$_{50}$ of Various Organisms*

Subject	LD$_{50}$ In Roentgens
Guinea pig	250
Dog	325
Pig	375
Human	450
Monkey	600
Rat	675
Bacteria	100,000
Viruses	1,000,000

*Notice that the microscopic organisms have a decidedly greater resistance to radiation than do the larger animals.

and break, resulting in higher mortality in those people who have been exposed to varying quantities of radiation. DNA organization of the chromosomes may be locked, duplicated, inverted, or moved to a new position on the same chromosomes, which is the basis of the lethal effects of radiation on dividing cells.

Medical utilization of **radiation therapy** (radiotherapy) utilizes the principle that ionizing radiation is cumulative in effect and destroys the cells that have a greater mitotic index. Small doses of radiation directed at a malignant tumor that has a greater mitotic activity than the surrounding normal tissue will destroy the

tumor while allowing the normal tissues to recover from the injury.

Tissue Damage

There are many examples of local tissue damage immediately following radiation exposure. In fact, if the dose is high enough, any local tissue will respond. Examples of local tissues affected immediately are gonads, bone marrow, nervous tissue, and skin.

The response of skin is a good example of an acute radiation effect on local tissue. Erythema (reddening of the skin) was perhaps the

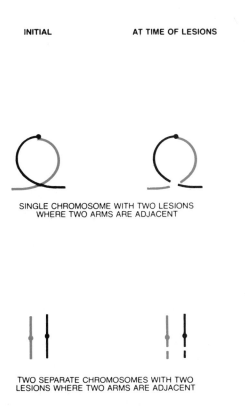

INITIAL

AT TIME OF LESIONS

SINGLE CHROMOSOME WITH TWO LESIONS WHERE TWO ARMS ARE ADJACENT

TWO SEPARATE CHROMOSOMES WITH TWO LESIONS WHERE TWO ARMS ARE ADJACENT

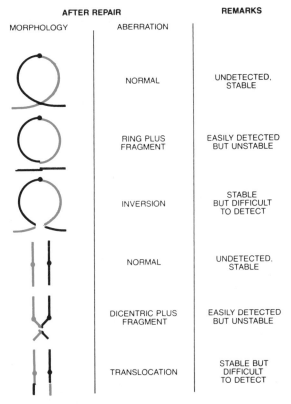

AFTER REPAIR

MORPHOLOGY	ABERRATION	REMARKS
	NORMAL	UNDETECTED, STABLE
	RING PLUS FRAGMENT	EASILY DETECTED BUT UNSTABLE
	INVERSION	STABLE BUT DIFFICULT TO DETECT
	NORMAL	UNDETECTED, STABLE
	DICENTRIC PLUS FRAGMENT	EASILY DETECTED BUT UNSTABLE
	TRANSLOCATION	STABLE BUT DIFFICULT TO DETECT

Figure 20-10 Damage to chromosomes by radiation. This damage is often repaired by the body but is sometimes misrepaired, resulting in chromosomal abnormalities. The situation is depicted for a single chromosome (*top*) with a radiation-induced lesion in each arm and for two separate chromosomes (*bottom*), each with a lesion in one arm. If the two breaks are

not too close in space and time, the repair mechanism restores the chromosome to its normal condition. If such aberrations are close in space and time, abnormalities can arise from various kinds of misrepair. (From "The Biological Effects of Low-Level Ionizing Radiation" by Arthur C. Upton. Copyright © 1982 by Scientific American, Inc. All rights reserved.)

first observed biological response to radiation exposure. Many of the early x-ray pioneers, including Roentgen, suffered x-ray–induced skin burns. A hazard to the patient during the early years of radiology was x-ray–induced erythema. During those years, x-ray tube potentials were so low that it was usually necessary to position the tube very close to the patient's skin, and not uncommonly 10 to 15 minutes of exposure were required to obtain a suitable radiograph. Often the patient would return several days later suffering from a severe x-ray burn.

Before the advent of cobalt machines, the limitation of radiation therapy was determined by the tolerance of the patient's skin. The object of x-ray therapy is to deposit x-ray energy in the tumor while sparing the surrounding normal tissue. Since the x rays must pass through the skin to reach the tumor, the skin was sometimes necessarily subjected to higher radiation doses than the tumor. The resultant skin damage, erythema followed by moist desquamation (peeling), often required that the course of therapy be interrupted.

Systemic Effects

The effects of radiation vary considerably from organ to organ and system to system. The following discussion points out some of the major effects of overexposure to radiation.

METABOLISM   When cells are irradiated, interactions of the free radical (OH^-) and even many of the genetic changes will alter the metabolic structure of the cell. If the change is a small one, and the structure of the metabolite is not radically changed, they are not prevented from being incorporated into larger, more complex units. But once incorporated, they must conform perfectly to the cell's rigid requirements. An enzyme that is "wrongly" synthesized may not be able to function at all with the intended substrate or may function at a lower or insignificant rate than is proper for the system. If the enzyme is responsible for the crucial synthesis of nucleic acids, their transcription problem can occur with changes in cellular identity.

HEMOPOIETIC TISSUE   Organs that produce blood cells (bone marrow, spleen, lymph nodes) are extremely responsive to radiation or are **radiosensitive,** as are the blood cells. This radiosensitivity of the precursors is demonstrated within hours following irradiation. Leukocytes, including lymphocytes, are the most sensitive and disappear from circulation first. Erythrocytes disappear next, followed by the blood platelets. Lymphocytopenia (decrease in lymphocyte count) may be a radiosensitivity indicator. If the number of lymphocytes with their 2-day life span decreases to a level of 100 to 200/mm^3 within 24 hours, the radiation dose is probably lethal (the count should be 2000/mm^3). If the count remains above 1000/mm^3, the prognosis for recovery is good. Hemorrhage may be an indicator of the blood platelet level and effects of radiation on the capillaries. A decrease in platelets (thrombocytopenia) from radiation can lead to spontaneous hemorrhage with resultant circulatory collapse.

REPRODUCTIVE ORGANS   The gonads (ovaries in females and testes in males) are highly radiosensitive. Total body radiation produces an immediate suppression of mitosis and disintegration of the germinal epithelium, resulting in mutations and possible sterility. The principal effect on the testes is altered spermatogenesis. Production of testicular lesions on the germinal epithelium prevents further development of spermatagonia and resultant loss of sperm motility (see Chapter 19). At moderate exposure to radiation, however, the testes have shown a remarkable ability to regenerate their germinal epithelium and the subsequent development of sperm. Although radiation to the testes depresses or prevents spermatogenesis, it does not affect hormonal secretion; therefore potency is not affected.

The ovary is affected temporarily by a smaller dose (200 r) than the testis (300 r) and is permanently affected by a dose of 300 r compared to 1000 r in the testis.

LYMPHATIC SYSTEM   Characteristics of poor lymphocyte production have been discussed in

The development of bone-scanning agents has been one of the most intensive areas of radiopharmaceutical research in the late 1960s and early 1970s. This work has produced a number of agents that have been reported to be effective bone-seeking materials. The most promising radiopharmaceutical reported has been technetium polyphosphate, which employs ^{99m}Tc and has shown exceptional properties as a bone-scanning agent.

The great improvement in bone scanning brought about by the use of ^{99m}Tc-polyphosphate can be seen in Figure 20-17. The definition in the scintiphotos is good enough to be able to distinguish ribs and vertebral bodies. Pathological conditions show up as hot spots—areas that have accumulated more activity than normal bone.

Whole Body

Gallium-67 (^{67}Ga) imaging, ultrasonography, and computerized tomography represent recently developed, highly sensitive modalities for detecting abdominal abscesses. Most studies involving large series of patients demonstrate that each of these procedures has approximately 90 percent sensitivity (patients with a positive study) for intra-abdominal abscesses. Gallium-67 has proven of greatest clinical use in demonstrating the presence and extent of certain malignancies such as bronchogenic carcinoma, Hodgkin's disease, and lymphoma.

Radiation Therapy

Oncology is the study and treatment of tumors, which are made up of clusters of abnormal cells. Radiation oncology, therefore, is the radiation treatment of such growths, although abnormal cell growth may be treated by surgery, **chemotherapy** (treatment using chemicals), radiation, or any combination of these modalities.

Radiation therapy involves the emission of radiant energy in varying degrees of intensity, discharging chemical ions (chemically bound group of atoms) through the use of x rays, gamma rays, and electrons. Radiation oncology

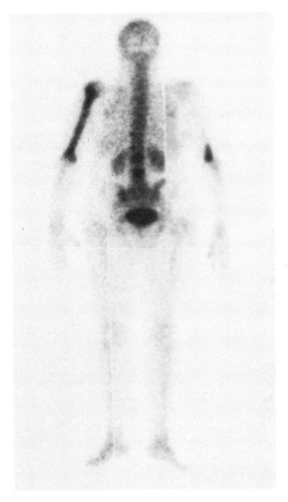

Figure 20-17 Bone scan. Posterior scan after an injection of ^{99m}Tc. Note marked increase in uptake (hot spot) in left humerus. (Courtesy of Salem Hospital, Salem, Massachusetts.)

provides methods and techniques for destroying abnormal cells while preserving the surrounding normal tissues.

Although radiation therapy may be used as the only method of treatment for malignant disease, a more common approach is to use radiation therapy in conjunction with surgery, chemotherapy, or a combination of the two. Some patients may be treated by only surgery or chemotherapy; however, approximately 60 percent of all diagnosed cancer patients will be treated with radiation. The choice of treatment involves consideration of a number of variables such as the patient's overall physical and emotional condition, the histological type of the disease, and the extent and anatomical position of the tumor. If a tumor is small and its margins well defined, only a surgical approach may be prescribed. On the other hand, if the disease is systemic, a chemotherapeutic approach may be chosen. Most tumors, however, exhibit degrees of size, invasion, and spread and require varia-

tions in the treatment approach that in all likelihood include radiation therapy as an adjunct to or in conjunction with surgery or chemotherapy. These limitations determine the goal of treatment—cure or as an adjunct to surgery.

The biological effectiveness of ionizing radiation in living tissue is dependent partially on the amount of radiation that is deposited within the tissue and partially on the condition of the biological system. The terms used to describe this relationship are *linear energy transfer (LET)* and *relative biological effectiveness (RBE)*.

One of the first laws of radiation biology,

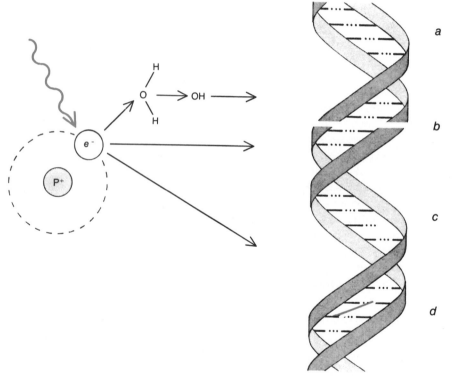

Figure 20-18 Damage to DNA. This damage is the most critical effect of low-level radiation. The effect can be direct or indirect. At the left a hydrogen atom, consisting of one proton and one electron, is ionized. (The electron is dislodged when the atom absorbs a photon of radiation.) In the indirect effect (*top arrow*) a secondary electron interacts with a molecule of water (H_2O) to give rise to a free radical (OH), which does the damage to the DNA. In the direct effect (*lower arrows*) the electron itself interacts with the DNA. The schematic representation of DNA at the right shows a normal segment of the molecule (**A**) and three of the many types of damage that can be caused by either direct or indirect effects of ionization: a double-strand break in the DNA double helix (**B**), the deletion of a base (**C**) and the chemical cross-linking of the two DNA strands (**D**). (From "The Biological Effects of Low-Level Ionizing Radiation" by Arthur C. Upton. Copyright © 1982 by Scientific American, Inc. All rights reserved.)

Table 20-3   Relative Radiosensitivity of Tissues and Organs Based on Clinical Radiotherapy

Level of Radiosensitivity (rad)*	Tissue or Organ	Effects	Cell Type
High: 200 to 1000	Lymphoid Bone marrow Gonads	Atrophy Hypoplasia Atrophy	Lymphocytes Spermatogonia Erythroblasts Intestinal crypt cells
Intermediate: 1000 to 7000	Skin Gastrointestinal tract Cornea Growing bone Kidney Liver Thyroid	Erythema Ulcer Cataract Growth arrest Nephrosclerosis Ascites Atrophy	Endothelial cells Osteoblasts Spermatids Fibroblasts
Low: >7000	Muscle Brain Spinal cord	Fibrosis Necrosis Transection	Muscle cells Nerve cells Chondrocytes

*The minimum dose delivered at the rate of approximately 200 rad per day that will produce a response.

postulated by Bergonie and Tribondeau, stated in essence that the radiosensitivity of a tissue is dependent on the number of undifferentiated cells in the tissue, the degree of mitotic activity of the tissue, and the length of time that cells of the tissue remain in active proliferation (Table 20-3). Although exceptions exist, the preceding is true in most tissues. The primary target of ionizing radiation is the DNA molecule, and the human cell is most radiosensitive during mitosis. Thus each group of cells or tissues may respond directly, relative to its radiosensitivity, depending on the aforementioned factors.

Because tissue cells comprise primarily water, most of the ionizing interactions occur with water molecules (Figure 20-18). These events are called *indirect effects* and result in the formation of free radicals such as OH^-, H^+, and HO_2. These highly reactive free radicals may recombine resulting in no biological effect whatsoever, or they may combine with other atoms and molecules to produce biochemical changes that may be deleterious to the cell. The possibility also exists that the radiation may interact

with an organic molecule or atom, which may result in the inactivation of the cell; this reaction is called the *direct effect*. Since ionizing radiation is nonspecific in that it will interact with normal cells as readily as with tumor cells, cellular damage will occur in both normal and abnormal tissue. The deleterious effects, however, will be greater in the abnormal cells because a greater percentage of the abnormal cells are undergoing mitosis; they also tend to be more poorly differentiated. In addition, normal cells have a greater capability of repairing sublethal damage than do tumor cells. Because of these reasons, greater cell damage will occur to abnormal than to normal cells for any given increment of dose.

Radiation therapy may be performed using equipment that generates a radiation beam, which is administered externally (Figure 20-19) or through the internal placement of radioactive substances in body tissues or cavities (Figure 20-20). In the latter instance, the radioactive material is usually enclosed in metal tubes or needles and is removed from the body after the

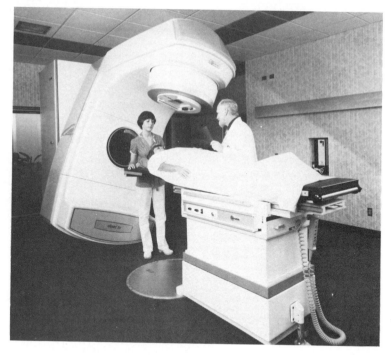

Figure 20-19 Radiation therapy unit. (Courtesy of Varian Radiation Division, Palo Alto, California.)

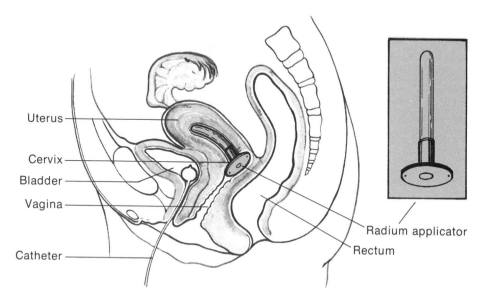

Uterus

Cervix

Bladder

Vagina

Catheter

Radium applicator

Rectum

Figure 20-20 Radium implantation in treatment of cervical cancer. The radium applicator (inset) is placed in the uterus, and the energy rays from the radioactive radium are allowed to come in direct con-tact with the cancer site. Note the catheter inserted into and decompressing the bladder so that it will not come in contact with the uterus.

gene unit of heredity located in the chromosome and made mostly of DNA

glands secreting organs

glaucoma disease characterized by abnormally high pressure within the eye, resulting in blindness

glucagon hormone that aids in the breakdown of glycogen in the liver

glucocorticoids steroid hormones that stimulate production of glucose from noncarbohydrate sources

gluconeogenesis formation of glucose from noncarbohydrate sources

glycocalyx carbohydrate-rich outer covering on the surface of cells

glycogenesis formation of glycogen from simple sugars

glycogenolysis breakdown of glycogen into simple sugars

glycoprotein carbohydrate–protein compound; a conjugated protein

glycosuria presence of glucose in urine

goiter enlargement of the thyroid

granulocyte cell with granules in the cytoplasm

growth hormone (GH)

gustation sense of taste

gyrus smooth surface of an organ

half-life time (specific for each radioactive substance) required for radioactive material to decay to half its initial activity

helical spiral

hematocrit formed element content of the blood

hematopoietic producing blood cells

hemodialysis removal of wastes from blood through a semipermeable membrane

hemodynamics forces connected with circulation of blood

hemoglobin an iron–protein compound that carries gases in the blood

hemolysis disintegration of red blood cells, which results in the appearance of hemoglobin in the surrounding fluid

hemophilia sex-linked, hereditary disease characterized by prolonged coagulation time and abnormal bleeding

hemorrhage bleeding through vessel walls

hemostasis checking flow of blood through any part of the body

hermaphroditism condition of having both male and female sex organs

hernia weakened opening in the abdominal wall

heterolysis destruction of cells of one species by lysins of a different species

histamine vasodilating substance in many cells

histocompatibility tolerance of host tissue to donor or foreign tissue, such as occurs in transplants

homeostasis consistency and uniformity of the internal body environment, which maintains normal body functions

hormone chemical substance produced in one organ, which, when carried to another organ by the circulation, stimulates the latter organ to functional activity

hyaline glossy membrane formed in the newborn lung

hydrocortisone steroid hormone that helps the body resist stress

hydronephrosis accumulation of urine in the kidney due to an obstruction

hydrophilic affinity for water

hydrophobic tending to repel water

hydrostatic pressure pressure created by fluid content

hyperactive increased activity or overactive

hypercapnea high carbon dioxide content of the air or blood

hyperemia swelling due to increased blood supply

hyperglycemia excess of sugar in the blood

hyperkalemia elevated potassium concentration in the blood

hypermenorrhea prolonged menstruation

hyperopia farsightedness

hyperplasia increased size of an organ or tissue

hyperpnea overbreathing due to abnormally rapid respiratory movements

hypertension elevated blood pressure

hypertonic property of higher osmotic pressure than some other solutions

hypertrophy increase in size of a tissue or organ

hypoactive diminished activity or underactive

hypogastric positioned below the stomach region

hypoglycemia deficiency of sugar in the blood

hypophysis pituitary gland

hypotonic property of lower osmotic pressure than some other solutions

hypoxia insufficient oxygen in the body tissues

immunity properties of the host that protect it from foreign agents

immunodeficiency disease disease due to failure of some immune function

immunoglobulin antibody against a particular antigen; protective immunity

immunosuppression   use of drugs to weaken immune response

incontinence   inability to control the passage of urine or feces

infarction   death of tissue due to loss of blood supply

infectious   capable of producing disease in a susceptible host

inflammation   series of reactions in tissues produced by microorganisms or other irritants and marked by redness of the affected area

insertion   attachment of a muscle to the more movable bone

inspiration   active mechanism creating a vacuum in the lungs

insulin   hormone that regulates carbohydrate metabolism

integument   covering, especially the skin

interoceptors   receptors within organs concerned with maintenance of the internal environment

interstitial   between cells

interstitial cell-stimulating hormone (ICSH)   pituitary hormone that stimulates androgen production in the testes

intoxication   pathological state produced by a drug, serum, alcohol, or any toxic substance

intrapulmonary   space within the alveolar sacs

intrathoracic   space in the thoracic cavity between the pleura

intussusception   infolding of one segment of the intestine within another segment

inversion   turning a body part toward the body midline

involuntary   performed without free will

ion   charged particle

ionization   production of ions from neutral atoms or compounds

ischemia   lack of blood in an area of the body

isometric   contraction of a muscle without shortening its length

isotonic   condition of equal osmotic pressure between two different solutions

isotope   element that has the same atomic number as another but a different atomic weight

jaundice   yellowness of skin and eyes

joint   point of connection between two or more bones

keratin   tough fibrous protein produced by keratinocytes

kinesthetic   referring to the ability to sense movement

kyphosis   increased curvature of the thoracic spine, giving a hunchback appearance

lactation   secretion of milk by the mammary glands

lacteal   one-celled vessel of the lymphatic system

lactogenic hormone (LTH)   pituitary hormone that stimulates milk production

lacuna   small hollow, depression, or pit

lamina   thin, flat layer in a portion of tissue, consisting of layers of cells; also a flat plate; e.g., the laminae of vertebrae

leukemia   disease of the blood-forming tissues marked by increase in the number of leukocytes (leukocytosis)

leukocyte   white blood cell

leukocytosis   increase in the number of leukocytes caused by the host body's response to an injury or infection

leukopenia   decrease in the number of leukocytes

leukorrhea   vaginal discharge other than blood

ligament   band of fibrous tissue, connecting bones and strengthening joints

lipid   fat, oil, or their derivatives

lordosis   forward curvature of the lumbar spine

lumbar   refers to the lower back

luteinizing hormone (LH)   pituitary hormone that stimulates formation of the corpus luteum in the ovary

lymph   a fluid found in the tissue spaces that contains most of the components of blood

lymph nodes   oval structures located along a lymphatic vessel that filter foreign matter and produce lymphocytes

lymphangiogram   injection of an opaque dye into a vein for x-ray purposes

lymphatic   pertaining to lymph nodes

lymphokines   soluble substances produced by lymphocytes that can affect other cells

lymphoma   proliferation of lymphatic tissue

lysis   rupture of a cell

macrocytic   large cell

malignant   referring to disorders that tend to worsen and cause death

malleolus   hammer-shaped protuberance

mastectomy   removal of breast tissue

mastication   act of chewing food

matrix   intercellular substance of a tissue

matter   substance

mediastinum   wall dividing the thoracic cavity

medullary   centrally located soft tissue

meiosis   special method of cell division, occurring during the development of sex cells (ova and sperm) in which the number of chromosomes is reduced

melanin   dark pigment found in skin, hair, and retina

melanocyte stimulating hormone (MSH)

melanocytes   pigment cell of the skin that produces melanin

menarche   time of life when the menstrual cycle begins

meninges   three membranes that envelop the brain and the spinal cord

menopause   period of life when menstruation normally ceases; change of life

menorrhagia   excessive menstrual flow

menorrhalgia   painful menstruation

menstruation   monthly event characterized by a bloody discharge from the uterus

mesenchyme   embryonic connective tissue

metabolism   physical and chemical processes by which living organisms produce the necessary energy to maintain life

metric   relating to the meter as a basis of measurement

metrorrhagia   irregular bleeding from the uterus

microcytic   small cell

micturition   urination

mineralocorticoids   steroids that control salt metabolism

mitosis   form of nuclear division characterized by complex chromosome movements and exact chromosome duplication

mixture   two or more substances that are not chemically combined

molecule   smallest unit of a particular substance that exists in nature

monochromatic   having only one-color vision

monosomy   when one of a pair of homologous chromosomes is missing

mosaic   inlaid network of pattern of small pieces

motor end plate   axonic terminals of motor neurons

motor unit   that which produces movement

mucin   substance secreted by mucous membranes that contains mucopolysaccharides

mucoprotein   compound composed of proteins and mucopolysaccharides

murmur   abnormal sound indicating a pathological condition of the heart valve

muscle   fibrous bands connected to bone that produce movement by contracting

myalgia   muscle pain or aching

myelin   fatty protective sheath around a nerve

myelocytic   produced in the bone marrow

myeloma   tumor found in the bone marrow

myocardium   part of the heart wall made up of muscle

myofibril   contractile fibers within a muscle fiber

myopathy   any disease of the muscles

myopia   nearsightedness

myosin   muscle protein found on the A-bands

myositis   inflammation of a muscle

necrosis   tissue death, usually in a localized area

neoplasm   new growth; a tumor

nephron   functional unit of the kidney

nerve fiber   extension of the nerve body

neuralgia   severe pain along the course of a nerve

neuritis   inflammation along a nerve

neuroglia   supporting cells to the nervous system

neurohypophysis   posterior pituitary gland

neuron   basic functional unit of the nervous system

neurotransmitters   chemical substance able to transmit an impulse between two structures

neutron   particle found in the nucleus of an atom that does not carry a charge; neutral

norepinephrine   hormone that stimulates the sympathetic nervous system

nucleic acid   one of a class of molecules composed of joined nucleotide complexes; the principal types are deoxyribonucleic acid (DNA) and ribonucleic acid (RNA)

nutrition   utilization of food for growth

nystagmus   involuntary side-to-side movements of the eyes

obesity   excess fat

occlusion   closing of an opening or passage

oncology   study and treatment of tumors

oncotic pressure   osmotic pressure exerted by colloids

oocyte   immature ovum or egg cell

oogenesis   process of formation of ova or egg cells

ophthalmoscope   instrument used to visualize the retina

opsonization   combination of antibody and antigen that makes them susceptible to phagocytosis

organelle   tiny specific particle of living material present in most cells and serving a specific function in the cell

orgasm   culmination or climax of sexual intercourse

origin   attachment of a muscle to the less movable bone

osmosis   passage of molecules of a pure solvent, such as water, from a solution of lesser concentration to one of greater concentration

osseous   bony or bonelike

ossification   process of forming bone or the conversion of fibrous tissue, or cartilage, into bone

osteoblast   young bone-forming cell

osteoclast   cell that absorbs bone tissue

osteocyte   mature bone cell

osteogenic   derived from bone

otoliths   calcium carbonate masses of the inner ear

ovulation   expulsion of the ovum from a follicle in the ovary

ovum   female sex gamete

oxytocin   hormone from the posterior pituitary which stimulates smooth muscle contraction

palsy   loss or impairment of nerve or muscle function

pancreas   abdominal gland that secretes enzymes for digestion and hormones that regulate carbohydrate metabolism

papilla   any small projection or elevation

paralysis   loss of muscle function; inability to move

parathyroid glands   a set of small glands behind the thyroid that produces a hormone to regulate calcium level in the blood

particles   small portions of matter

particulate radiation   pertaining to having small particles that emit energy

parturition   birth

pathogenic   capable of producing disease

peduncle   group of nerves

pepsin   enzyme produced in stomach responsible for chemical breakdown of proteins

perfusion   passage of fluid through the vessels of an organ

pericardium   serous membrane that lines the sac enclosing the heart; also the reflection that attaches itself to the heart itself

perichondrium   fibrous membrane, covering cartilage

periosteum   fibrous membrane, covering bone tissue

peristalsis   rhythmical waves of smooth muscle contractions

peritoneum   large serous membrane that lines the abdominal cavity and is reflected over the organs within the abdominal cavity

permeability   the extent to which molecules of various kinds can pass through cellular membranes

permeable   membrane that allows passage of all particles

phagocytosis   process by which a cell engulfs and digests a particle or substance

phosphocreatine   source of energy found in muscle

photon   unit of energy of a light wave

photoreceptor   receptor sensitive to visible light

pigmentation   coloration by deposition of pigments

pineal gland   cone-shaped gland in the middle of the brain that produces melatonin, which inhibits secretions of male sex hormones

pinocytosis   process by which a cell engulfs and digests a droplet of liquid

pituitary gland   almond-shaped gland at the base of the brain that produces hormones that regulate many body functions

placebo   chemical substance given in place of medication

placenta   organ within the uterus through which the fetus derives its nourishment

plasmalemma   flexible cell membrane

platelet   a thrombocyte

pleura   membrane covering the lungs

pleurisy   inflammation of the pleura

plexus   network or tangle of interweaving nerves, veins, or lymphatic vessels

pneumonia   inflammation of the lungs

pneumothorax   air in the thorax

polycythemia   abnormally large number of red blood cells

polymerize   process of joining small compounds to form a compound of high molecular weight

polyuria   abnormally large quantity of urine

precipitation   conversion of a soluble substance to an insoluble substance

presbyopia   vision of older adults

pressoreceptors   receptors sensitive to mechanical stimuli

process   prominence or projection

progestin   hormone of the ovary that stimulates uterine (endometrial) development

prolactin   hormone of the pituitary that stimulates milk secretion

pronation   lying face down or moving the arm so that the palm of the hand is facing backward

proprioceptors   receptors that provide the body with information about its position in space

prostaglandin   hormonelike chemical with a variety of effects

protein   complex nitrogenous compound of high molecular weight

proton   particle found in the nucleus of an atom, carrying a positive charge

protoplasm   building material of all organisms

protraction   movement of the jaw forward

pruritis   itching

pseudostratified   a false layered effect

pulse   rhythmical contraction and relaxation of muscles along a vessel

pustule   small, pus-containing elevation on the skin